Mathematics Methods for the Elementary and Middle School

Mathematics Methods for the Elementary and Middle School

A Comprehensive Approach

Gary G. Bitter
Arizona State University

Mary M. Hatfield
Arizona State University

Nancy Tanner Edwards
Missouri Western State College

Allyn and Bacon

Boston, London, Sydney, Toronto

Copyright © 1989 by Allyn and Bacon
A Division of Simon and Schuster
160 Gould Street
Needham Heights, MA 02194

Series Editor: Sean W. Wakely
Developmental Editor: Elizabeth Brooks
Production Administrator: Annette Joseph
Editorial-Production Service: Lisa M. Mosca, Production Concepts
Text Designer: Denise Hoffman, Glenview Studios
Cover Administrator: Linda K. Dickinson
Cover Designer: Lynda Fishbourne
Manufacturing Buyer: Tamara McCracken

Logo and BASIC are registered trademarks of Apple Computer, Inc.
Math Explorer is a registered trademark of Texas Instruments, Inc.

Library of Congress Cataloging-in-Publication Data

Bitter, Gary G.
   Mathematics methods for the elementary and middle
school.

   Includes bibliographies.
   1. Mathematics—Study and teaching (Elementary)
2. Mathematics—Study and teaching (Secondary)
I. Hatfield, Mary M., 1943–   . II. Edwards,
Nancy Tanner, 1943–   . III. Title.
QA135.5.B5343   1989   510'.7'12   88-16835
ISBN 0-205-11775-9

Printed in the United States of America
10 9 8 7 6 5 4 3 2   93 92 91 90 89

To Kay, Larry, and Lyman,
and to our children and grandchildren,
who sacrificed many days of family activities
so this book could be written

Brief Contents

Contents

1

The Past, Present, and Future
of Mathematics Education

2

How Children Learn Mathematics
and Solve Problems

3

Geometry

4

Measurement

5

Number Readiness–Early Primary Mathematics

6

Numeration

7

The Whole Number System

8

Algorithms

9

Number Theory

10

Rational Numbers–Common Fractions

11
Decimals, Ratio, Proportion, Percent, and Rate

12
Graphing, Statistics, and Probability

List of Tables

List of Activities

Calculators

Computers

Estimation

Manipulatives

Math Across the Curriculum

Mental Math

Problem Solving

Preface

Mathematics Methods for the Elementary and Middle School provides pre-service and inservice kindergarten through grade eight teachers with ideas, techniques, and approaches to teach mathematics. The book emphasizes manipulatives, problem solving, diagnosis and remediation, estimation, mental math, math across the curriculum, calculators, and computers as an integral part of teaching mathematics. These topics are not addressed as separate issues. As is appropriate, they are treated as part of the developmental learning process.

Although the book can be used without the computer-related infusion ideas, the authors do not encourage this practice. In using this text as we developed it, we noted numerous benefits to students from hands-on computer experience. If the book is used with related field experiences, the actual computer use is extremely important. A data disk is provided for the BASIC and Logo programs and database and spreadsheet templates for readers to use without manually entering each of the programs.

The book models the approach of concrete-to-abstract developmental learning. Manipulatives are emphasized throughout the book. Base 10 blocks, geoboards, attribute blocks, Cuisenaire rods, and the balance are a few of the concrete materials emphasized. In addition, paper models of many of the manipulatives are provided in Appendix C for the readers to remove for their personal use.

Diagnosis and remediation techniques are included in each chapter where appropriate. Actual children's work is provided for preservice and inservice teachers to review and evaluate. Included in each discussion are different thought processing styles and methods of correcting common student errors. Teachers who work with special needs students will find these materials helpful with direct application for classroom use. The Instructor's Manual provides additional worksheets for further evaluation of common errors.

Exercises for preservice and inservice teachers are provided at three levels in each chapter. The levels are related to the cognitive domain of Bloom's Taxonomy:

A. **Memorization and Comprehension Exercises**
   low-level thought activities
B. **Application and Analysis Exercises**
   middle-level thought activities
C. **Synthesis and Evaluation Exercises**
   high-level thought activities

Many exercises are provided at each level. The instructor can select those most appropriate on the basis of availability of field experiences, computers, software, calculators, manipulatives, and similar factors. This option makes it possible to match student work to unique situations.

Scope and Sequence charts (based on several current elementary mathematics textbook series) are provided in each chapter to help the user of this book relate topics in the elementary mathematics curriculum to grade levels. They are integrated into the development of the math concepts as appropriate. Activities are provided throughout the book for the following categories:

Problem Solving—emphasizes strategies and higher-order thinking

Manipulatives—uses concrete experiences for developing concepts

Calculator—emphasizes problem solving and de-emphasizes tedious computations

Mental Math—builds methods of doing computation mentally

Computer, *BASIC programming*—uses exploration to learn math concepts

Computer, *Logo programming*—uses exploration to learn math concepts

Computer, *Spreadsheets, Databases, and Graphing*—uses exploration in problem solving

Estimation—emphasizes techniques to determine reasonable answers

Math Across the Curriculum—encourages interdisciplinary approaches to learning mathematics

The following pictures will be used to alert the reader to the emphasis of each category when it appears throughout the text.

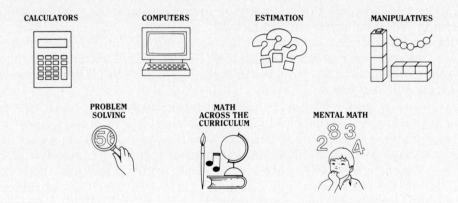

This book is a resource book for the teaching of mathematics in the elementary and middle school. Problem solving, cooperative learning, and manipulatives are emphasized at all levels. Different learning styles are emphasized, and concrete-to-abstract materials are a prerequisite for each development.

● The book has twelve chapters. Chapter 1 includes the past, present, and future of elementary mathematics. The philosophy is explored for emphasizing problem solving, manipulatives, the computer, the calculator, estimation, and mental math. Introductory computer experiences are provided in Logo and BASIC to ensure an anxiety-free introduction to the computer.

● Chapter 2 discusses how children learn mathematics and solve problems. Learning theories are briefly reviewed in relation to effective teaching models and lesson plan development. Problem-solving strategies are discussed and illustrated. Characteristics of good problem solvers are outlined and related to lesson planning. The chapter emphasizes the importance of good lesson planning in effective teaching.

• Chapter 3 discusses geometry. Its placement in the beginning of the book is to encourage teachers to use geometric ideas in their teaching. Concrete experiences with objects are the key to developing informal, plane, and solid geometry. Hands-on exploration is the theme of the chapter and is emphasized throughout the remainder of the book.

• Chapter 4, on measurement, follows geometry and emphasizes the metric and customary systems. The topics of linear, area, volume, capacity, temperature, mass, time, and money are carefully developed, including their related units of measurement. Activities are provided for each topic. Many of the activities encourage exploration to develop frames of reference for various units of measurement.

• Chapter 5 discusses number readiness and the beginning of the numerical aspects of mathematics. Number readiness is the initial introduction of numbers, including one-to-one correspondence, counting, and the general concept of number.

• Chapter 6 covers numeration and the development of counting systems. The base 10 numeration system is discussed in the context of the more general idea of numeration as it applies to any base system. Base 10 blocks and chip trading activities are developed as examples of proportional and nonproportional materials to use with children.

• Chapter 7 discusses the whole number system in relation to the basic facts of addition, subtraction, multiplication, and division. Concrete materials and mental arithmetic are emphasized in the development process.

• Chapter 8 discusses the algorithms for addition, subtraction, multiplication, and division using base 10 blocks followed by chip trading activities. The calculator and computer are an integral part of the development.

• Chapter 9 develops number theory, various divisibility rules, and numerical explorations. The middle school curriculum will find the topics extremely useful in developing number "sense." The Scope and Sequence chart identifies where each topic is emphasized in the mathematics curriculum.

• Chapter 10 covers rational numbers, specifically common fractions, using concrete materials. The operations are explored with several models. The approach emphasizes understanding of the algorithm rather than memorized procedures.

• Chapter 11 discusses decimals, ratio, proportion, percent, and rate. The base 10 blocks are used to develop an understanding of decimal fractions. Money is an integral part of the development. The calculator, estimation, and mental math are stressed in relation to ratio, proportion, percent, and rate.

• Chapter 12 explores graphing, statistics, and probability. Types of graphs and their interpretation are discussed and illustrated with examples. Computer activities are suggested for the graphing presentations. Statistics and probability theory are highlighted. Real-world examples are provided and activities to motivate students are presented. Math across the curriculum is emphasized throughout the chapter.

Throughout the book, preservice and inservice teachers will be referred to as *the teacher*. The book has been class tested in college courses and teachers have successfully used the activities with children. Using the computer as a tool offers many learning experiences for teachers as well as elementary and middle school children. The authors hope this careful, realistic presentation makes mathematics meaningful to elementary and middle school teachers and their students.

We wish to thank the following for their invaluable comments on the text: Jerry P. Becker, Southern Illinois University; Susan Brown, University of Houston; Hiram Johnston, Georgia State University. We wish to acknowledge our past students whose actual work appears as illustrations in the book. We also wish to thank Sue Canavan, Sean Wakely, Beth Brooks, Annette Joseph, and especially Violet J. Tanner for her careful reading of the initial manuscript. Special thanks to Lisa Mosca for making this book a reality.

Our special gratitude goes to the professors and students of Arizona State University and Missouri Western State College who participated in the two-year pilot study and gave valuable suggestions for improvement.

How to Use This Book

This book is to be used with classes that emphasize the teaching of elementary and middle school mathematics. Field experiences in relation to learning from the book are encouraged. Generally, the book is organized into six sections per chapter—introduction, teaching strategies, diagnosis and remediation, summary, exercises, and bibliography.

The *introductions* present a brief review of the concepts that have been covered in prerequisite college content courses. This is especially true in Chapters 5 through 12 where teachers need to see the mathematical understandings to which children will be exposed from an "adult" viewpoint. The mathematical content is presented in a more symbolic, formal structure. Definitions of mathematical terms have been checked with James and James (1976), *Mathematics Dictionary*, 4th edition, and with West, Griesbch, Taylor, and Taylor (1982), *The Prentice-Hall Encyclopedia of Mathematics*, as appropriate.

The *teaching strategies* present developmental activities that can be used with elementary and middle school students. These emphasize a variety of techniques in concrete/pictorial experiences, problem solving, cooperative learning, and computer/calculator technologies.

Because some readers may have a major in mathematics while others may have a more limited background, the authors suggest that the book be used in the following manner:

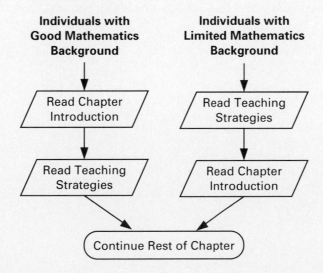

The *diagnosis and remediation* sections present techniques for correcting common error patterns and other frequently occurring difficulties of elementary and middle school students. Sample student work is included and analyzed for error patterns, a practice which we encourage in all teachers.

The *exercises* are written on three levels of difficulty. There are many exercises so that course instructors can choose which assignments are most appropriate for their course objectives. It is not the intent of the authors that everyone complete every exercise. The exercises are to encourage understanding of the content covered in the text as well as to prepare a preservice teacher for classroom instruction and to strengthen the inservice teacher's mathematics presentations. Many of the exercises can be used in future teaching, so teachers are encouraged to save them.

The *bibliographies* present quick references for teachers interested in further information on a topic. Readers who are asked to search professional journals for assignments will find the bibliography a good starting point.

The computer programs are to be integrated into the curriculum as tools to understand and teach mathematics. The BASIC and Logo programs are available on disk to be copied by each user, thus avoiding the tedious task of typing in the program. Readers need to be encouraged to explore with different responses and record results, to look for a pattern or discover a formula. Please note that a Logo disk is needed to run Logo and a spreadsheet and database program is needed to do the spreadsheets and databases. The assumption of the computer activities is that they will have direct transfer to the children in the classroom. Readers are not expected to do their own programming; however, individuals who have had more extensive work with computers may try extending computer activities or writing a new program to go along with a topic. The computer and calculator activities can be done as explorations in an individual setting or in cooperative learning (small, noncompetitive) groups where work is shared among students to draw conclusions.

An index is included for quick reference to key definitions. When in doubt, look up the word in the index. Paper models and patterns for several of the manipulatives are included in Appendix C and may be copied for individual practice or whole classroom use with children. A software evaluation form is included for software review assignments. Finally, the textbook evaluation criteria used by school districts is available for elementary textbook evaluations. Observation forms are provided for field experiences. These materials are referenced as needed in the text.

Mathematics Methods
for the Elementary and
Middle School

1

The Past, Present, and Future
of Mathematics Education

Key Question: *What insights must teachers have to survive the present and prepare for the future in mathematics education?*

Activities Index

Introduction

The content and instructional methods by which you were taught mathematics may not be appropriate for children living most of their lives in the twenty-first century. As mathematics educators, you must help prepare children for an ever-changing world by being aware of the expanding field of knowledge in mathematics. This text has three main goals: to build on the strengths of past mathematics education, to model the successful techniques of the present, and to prepare for changes in the future. Mathematics education has always been and will continue to be an expanding field of knowledge. As human beings grow in understanding and in awareness of new possibilities, mathematics education changes. It is not the stagnant body of knowledge that many people assume. Some basic mathematical principles, such as the commutative and associative properties, have remained the same for generations while others, such as fact strategies, are emerging as new insights. The ways to teach both the old and the new principles will change as more effective methods and new technologies develop. For example, the teacher of the early 1900s would be astonished to see today's students learning mathematics with a computer, creating their own mathematical equations, and checking to see if their answers are correct—all within a few seconds.

This chapter presents an overview of the pertinent course topics and the philosophy of the text. The ideas briefly presented here will be studied in greater detail in subsequent chapters.

Past

Reviewing past influences on mathematics by Thorndike, Pestalozzi, Grube, Dewey, Brownell, Spitzer, Polya, and others is beyond the scope of this book. Instead, the book begins with the change in mathematics brought on by the launching of Sputnik by the Soviet Union in 1957. Our country scrambled to keep up with the Russians by accelerating the pace of curriculum change in science and mathematics. What followed was the *New Math* revolution of the early 1960s that was promoted as a means of helping children understand the *why* behind computational processes. Emphasis was on the underlying structure of mathematics with new content being added on such topics as set theory, number systems, and number properties. In addition to the New Math, a strategy known as *Discovery Learning* was introduced as the necessary balance to understanding computations, as it was designed to provide the how to facet of learning mathematics.

Attention soon focused on the vocabulary and expressions used in teaching mathematics, as they were often ambiguous. Changes within the educational terminology, delineating more clearly the differences between curriculum, instruction, and teaching, were begun with MacDonald and expanded under Piaget. The developmental level of the child became a factor in the content of the curriculum.

During the past two decades, Piaget's theory dealing with the intellectual development of a child prevailed over most educational research. Many inferences were drawn from Piaget's theory regarding teaching and instruction: the classroom structure, organization, and management; the nature and sequence of curricular content; and the learning of mathematics. Much of the revision work done in mathematics textbooks in the 1970s reflected Piaget's effect upon the scope of children's education (Driscoll, 1982).

Present

In today's educational system, the Piagetian theory is being challenged by an information-processing theory, which suggests that understanding and comprehension are the mainsprings upon which cognitive science is based. Research in the learning of mathematics is related more to the intake of information, the working memory, and the interaction between working and stored information in the long-term memory. An information-processing system is often described as relating to computer or microcomputer systems.

Research (Capper, 1984) suggests that young children invent or construct much of their own mathematical knowledge. Although their thinking is immature, they enter school with some well-developed ways of using mathematics. Some educators feel that using a child's own conceptions of mathematics in programs of teaching and instruction will reduce a child's tendency to develop anxiety toward mathematics later. Very often it is the view that educators hold about the importance of mathematics in the classroom that affects the anxiety which students develop toward it. Learning associated with problem solving is strengthened, according to research, by the student's heightened awareness of metacognitive elements of the problem-solving process. *Metacognition is one's knowledge of how one's own cognitive processes work.* Evidence implies that metacognitive skills may be the reason for the difference between expert and novice problem solvers (Driscoll, 1982).

Teaching and learning mathematics is a lifelong process. Technological change outdates old skills and increases the importance of others. Slide rules, log and trig tables, and mechanical calculators have given way to programmable calculators and sophisticated electronic computers.

Technology is also changing the teaching profession, moving the teacher from chalk and blackboard to manipulatives, interactive video, and computer-assisted instruction. Research into education and psychology provides new insights into the ways in which children learn mathematics. Mathematics is required for all, with some students receiving the basic math instruction and some students using the full spectrum of instructional aids—computers, videotape and disc, and calculators. The full impact of this policy and its possible educational imbalance may not be known for years to come (National Science Board [NSB], 1983).

The Training of Math Teachers

The position of the National Council of Teachers of Mathematics (NCTM) (1981) on the training of math teachers was as follows:

> Prospective teachers of mathematics at any level should know and understand mathematics substantially beyond that which they may be expected to teach. They should be able to relate that mathematics to the world of their pupils, to the natural sciences, and to the social sciences. Teachers should be able to provide illustrations of the role of mathematics in our culture and should have sufficient understanding of its nature and philosophy to interpret the various strands of the curriculum beyond a superficial level. They should also have a knowledge of the historical development of mathematics in the different cultures of the world so that they can illustrate its cross-cultural nature using examples that will appeal to students from the diverse cultures represented in our society.

Future

The technological revolution is causing educators to reconsider their previous goals in determining future needs. Societies themselves are in a state of flux. New demands are being made on the populace. There are new expectations and new employment patterns. Mathematics is affected by both the new technology and the subsequent rapid sociological change it produces. Pedagogically, the future looks promising. It is becoming apparent that education must be a lifelong pursuit. No longer will education be completed when a student graduates from high school or college.

Although teachers will continue to play a central role in education, their role will change. Technology will affect the way teachers present their material in the classroom. How great an impact technology has on learning will depend on whether teachers prepare themselves to use technology to help students learn.

To assure that all students have the best education possible, educators must anticipate and guide the educational change. Educators must learn how to use the microcomputer effectively to achieve the highest learning potential for all students. They must set educational and technological goals which will influence designers of educational software and developers of hardware. Teachers should prepare for future developments as they go into preservice and inservice training. As students gain more experience on

their home computers, their parents will exert more pressure for computer-assisted instruction in the classroom. Students themselves will begin to demand more and better instruction as their world calls for greater knowledge and understanding.

As the twenty-first century approaches, educators must consider how to proceed with the teaching and learning of mathematics. Should mathematics instruction proceed with categories of arithmetic, algebra, geometry, or should it change to concentrate on dichotomies such as finite-infinite, continuous-discrete, and exact-approximate dichotomies? There is no doubt that the structure of the mathematics curriculum should be renovated. The renovation should include curricula to develop proof and deductive reasoning, which are highly characteristic of logical thinkers. Mathematics curricula should be designed to build an accepted body of mathematical knowledge which can serve as a basis for further learning. However, the future aim of teaching mathematics is to increase the body of mathematics knowledge so students will be able to apply it. The ability to apply mathematics will be vital in the 1990s and into the next century (National Council of Teachers of Mathematics [NCTM], 1984).

Curricular reform in school mathematics K–12 is the intent of three major national projects. The first is NCTM's document, "Curriculum and Evaluation Standards for School Mathematics " (Commission on Standards for School Mathematics, 1987), which will have far-reaching effects on shaping the mathematics curriculum. The standards are a set of benchmarks by which schools might measure their mathematics curriculum. The work focuses on major strands of the K–12 curriculum and their sequential development. Grade levels have been grouped as K–4, 5–8, and 9–12. The standards address specific needs at these grade levels. The recommendations are directed at the ways schools can change the manner in which mathematics instruction occurs as well as the emphasis different contents receive.

The NCTM standards create a vision of what it means to do mathematics in terms of what students need to do in learning mathematics and what teachers should do in teaching mathematics. The standards call for instruction to be based on problem situations whereby students gain mathematical power by investigating, exploring, making conjectures, verifying answers, and communicating results. This constructive view of learning means a redefined role for the teacher, from a dispenser of knowledge to a facilitator of learning. An assumption about instruction is that all students should experience the full range of topics addressed in the standards. The focus is on what it means to be mathematically literate in a world that relies on calculators and computers to carry out mathematical procedures and in a world where mathematics is rapidly changing and extensively being applied in many fields. The goals for all students across grades K–12 are: to become a mathematical problem solver, to learn to communicate mathematically, to learn to reason mathematically, to learn to value mathematics, and to become confident in one's own ability.

Another project is the Curriculum Frameworks of the Mathematical Sciences Education Board (MSEB) (Ralston, 1988) of the National Research Council. This task force examined the forces shaping the mathematics curriculum of the future and the forces that must be addressed in order to transform the present curriculum into the curriculum needed for the year 2000.

The third major effort is the Project 2061 (Rutherford, 1987) mathematics work of the American Association for the Advancement of Science. This project looked at what eighteen-year-olds might need to know for the year 2061

in order to function effectively in society. Although the year 2061 was chosen (the year Halley's comet returns), the major focus of this project is aimed at the year 2020.

Along with these plans for curriculum reform are concerns about the mathematical preparation of elementary and secondary school teachers. Recommendations by the Holmes Group and the Carnegie Forum on Education and the Economy have caused certification programs of many institutions to shift from four- to five-year programs of teacher education. The Mathematical Association of America's Committee on the Mathematical Education of Teachers (COMET) (Mathematical Association of America, 1987) prepared a set of guidelines for all previously certified teachers. These guidelines indicate direction for continued program development rather than recommend details for particular future programs.

It is becoming apparent that education is failing to provide the intellectual tools which will be needed in the twenty-first century. Our rapidly changing world is outpacing the many who are neglecting mathematics for nonmathematical subjects (NSB, 1983).

In the loud cry and major upheaval of the back to basics movement, little time was spent wondering whether the basics of the twenty-first century will differ from the basics of the nineteenth century. The twenty-first-century education must include communication and higher-order problem-solving skills in its basic makeup, along with scientific and technological literacy. No longer do readin', writin' and 'rithmetic suffice. Tomorrow's citizen must have highly developed thinking tools in order to function effectively. The National Commission on Excellence in Education (1983) proposed sweeping and drastic change in the breadth of student participation, in the methods and quality of teaching, and in the preparation and motivation of children. They also proposed that the content of courses be changed dramatically. The commission suggested that the change be initiated through a strong and lasting commitment to quality mathematics education, with mathematics training being started earlier and on a broader range. The National Science Board (1983) contends that every child can develop an understanding of mathematics, science, and technology if the subjects are appropriate and skillfully introduced.

Learning Mathematics for the Present and the Future

Mathematics can be taught to different groups of students through different curricula, which differ in style and content, as well as in ways of studying. Teachers who can teach mathematics by targeting a particular student's learning style are able to expand that student's future prospects. Educators need to eliminate the assumption that their students have no mathematical knowledge before they enter the classroom and bring the out-of-class mathematics into the classroom. Educators should identify and reinforce out-of-school mathematical experiences and knowledge. In a strictly utilitarian sense, pupils filter through their senses all kinds of mathematical activities. As an example, architecture acquaints students with two-dimensional representations of three-dimensional objects. There are many other applications where mathematics are used in everyday life.

Teachers need to ask questions and give assignments that require the student to reflect on and report their mathematical knowledge and behaviors. Garofalo (1987) suggested examples of questions and discussions that teachers could use to make learning mathematics more than a memorized activity.

1. What errors do you make? Why? What can you do about them?
2. Think about solving problems in mathematics. What type do you perform better? What type are the hardest for you to solve?
3. How can you get better at solving problems?
4. The teacher should also have students realize these aspects of problem solving:
   a. There are many different approaches to use in solving problems.
   b. Not all problems can be solved in a single math period; some take a longer time.
   c. There can be non-routine problems where an operation (add, subtract, multiply, divide) is not used.

People can know things on a variety of levels. For example, people can know Euclid's *Elements* as a statement. They can know that Euclid's *Elements* contains thirteen books and has a total of 465 propositions. They can know how to figure out the deductive theory of geometry as developed in the work. They can have a knowledge of how to utilize their knowledge—how to put all geometry to use in their working life (Howson, Nebres, and Wilson, 1985).

Communication of knowledge through the use of television, as in school lessons or current events on the evening news, is a way of life that people have come to understand. Media knowledge is often reduced to either correct or incorrect information presented through visual and verbal stimuli, as in computer games and drill and practice exercises. However, students must respond to school knowledge, thereby making it a less passive activity and a more active part of their learning. We have become a passive society. We watch television, but don't understand how it works; drive cars, but don't know how they run; own videos, computers, and computer software, but don't understand how they operate. While computers and television and videos are marvelous tools for instruction, they must be used to augment that instruction, not replace it. The integration of passive and active learning can be most effectively accomplished by the teacher when the students' experiences are brought into the classroom and used to illustrate the learning experience. Unfortunately, heretofore, the instruction of mathematics has not been designed toward the utility of mathematics, but instead has sought to validate it. As students increase their educational level, they decrease their study of mathematics unless they are entering fields in which a mathematics background is essential. Mathematics, then, is often received only by an elite mathematical group who are more able to cope with a variety of approaches (NCTM, 1984).

As we explore dimensions of future trends, these topics will be discussed: technology (calculators and computers), problem solving, math across the curriculum, mental math, and estimation. The chapter concludes with the discussion of two major issues in mathematics: math anxiety and women in mathematics. One definite trend of the future will be the increased use of technology in mathematics instruction. To be effective and valuable, it must be integrated into the mathematics program in a meaningful manner.

Trends in Technology

There is tremendous potential in computer and calculator technology. Calculators have become universal devices and computers are on their way. Microcomputers are excellent interactive devices for communication. A microprocessor, coupled with television via satellite, cable, or closed circuit communication, will expand the field of knowledge through telecommunications. The use of videodiscs and computer graphics can relieve the classroom of its monotony for students. With the use of computers, calculators, and the interactive devices used with them, students can be motivated as well as learn at their own pace. Two technological devices will be discussed—calculators and microcomputers.

Calculators

The impact of technology has caught up with schools and pressured them to use calculators and computers in the classroom. Evidence shows that calculators can improve the average student's basic skills both in working exercises and in problem solving. Many teachers do not allow the use of calculators in doing math but research supports calculator use throughout the math curriculum (Hembree, 1986). Most reports on the future of mathematics encourage the de-emphasis on the teaching of computation and the emphasis on higher-ordered thinking skills. In other words, they encourage the use of the calculator for doing the tedious computation as well as encourage higher-ordered thinking. The learner needs to know the basic facts and estimation skills to determine the reasonableness of a calculator or computer-produced result.

Microcomputers

The escalation of the use of microcomputers in schools in the United States is rapidly elevating microcomputers to the status of universal devices. Projections indicate that by the late 1990s, every classroom in the United States should have enough microcomputers to allow each student daily time on the computer. Therefore, a problem arises in having teachers sufficiently trained on the use of the computer as a tool in their teaching. If the educational system is to fully take advantage of technology as a classroom teaching tool, all teachers must be trained to use technology as a tool for learning. The NCTM (September, 1987) position on the training of teachers of mathematics was as follows:

> All preservice and in-service teachers of mathematics should be educated on the use of computers in the teaching of mathematics and in examining curricula for technology-related modifications. Teachers should be prepared to design computer-integrated classroom and laboratory lessons that promote interaction among the students, the computers, and the teacher. Mathematics teachers should be able to select and use electronic courseware for a variety of activities such as simulation, generation and analysis of data, problem solving, graphical analysis, and practice.
>
> Mathematics teachers should be able to appropriately use a variety of computer tools such as programming languages and spreadsheets in the mathematics classroom. For example, teachers should be able to identify topics for which expressing an algorithm as a computer program will deepen student insight, and they should be able to develop or modify programs to fit the needs of classes or individuals. Keeping pace with advances in technology will enable mathematics teachers to use the most efficient and effective tools available.

Students can use the computer in many formats. The most popular format is utilizing commercial software. *Software refers to the program, language, or instructions used by the computer* (Biehler & Snowman, 1986). Drill and practice software can be used to memorize material such as the basic facts, formulas, and basic mathematics concepts. Teachers can use the computer for diagnostic and competency testing, in which the program branches to questions according to the student's answer. The student can use the computer as a tutor, learning new subjects through questions and answers in an interactive dialogue with the computer. The student can use the computer as a tool, using applications such as word processing, spreadsheets, databases, and graphing. Finally, the student can learn to program the computer in languages such as Logo and BASIC to solve problems.

Computer study provides a method by which students can learn at their own individualized pace. Computers promise much as a tool of learning and as a provider of an environment in which learning can occur. There are several components involved in computer-assisted instruction (CAI). There are cybernetic environments (microworlds) which may combine elements according to given rules. There are also educational games which are used to develop skills in mathematics comprehension and problem solving. Real-world phenomena can be explored through the use of a microcomputer-based instrumentation system. Databases can be used to store information to be made available for student problem solving. Computer tools include graph-plotting routines, word processing, spreadsheet programs, and general purpose problem solvers. Logo, a special purpose computer language, allows intellectual development learning environments to stimulate the child's mentality. The student may discover properties of the real world through the use of computer simulations' flexible universes. In mathematics, computer-assisted discovery learning provides an active and self-directed learning environment. Although computers assist teachers as educational tools, even at their highest potential, computers cannot replace teachers.

The new technology has brought new ways and pertinent new material into the classroom to teach conventional mathematics. Although software is constantly improving, there is still much to be done to create better computer programs for educational training. This should be done in conjunction with, and parallel to, new developments in curriculum. For computer technology to impact education, attention and development must be consistent. Even in its infancy, computer-assisted instruction has been shown through various studies to improve initial results and retention. Students who studied using computer-assisted instruction were proven to learn faster than their non-CAI studying peers (Kulik, 1985).

Educators should be looking toward more integration of the microcomputers into the classroom as ratios of computers to pupils become higher. Through the use of blackboard-sized high resolution displays, an instructor can build the use of the microcomputer into regular class lessons. Mathematics and the spirit in which mathematics is undertaken can be af-

fected by software that is tied to the achievement of well-enunciated and easily described aims (NCTM, 1984). Three aspects of using the computer as a tool to teach mathematics will be discussed: computer-assisted instruction, applications, and programming.

Computer-Assisted Instruction: *Computer-assisted instruction (CAI) is defined as the presentation of subject matter to be learned by the student through the use of the computer.* There are several types of computer-assisted instruction software programs. These programs are usually categorized as drill and practice programs, tutorial, simulation, and problem solving. Each has strengths for involving the students' understanding of mathematics, and students should have experiences with all types of programs.

Drill and Practice Programs: Drill and practice programs generate problems on various skills and record the student's results. In most instances no instruction is provided. The computer randomly generates the problems and the student responds. As a student's success ratio reaches a certain plateau the computer generates problems of greater difficulty. There are more drill and practice math programs than any other type. Research indicates that these programs can produce successful results (Kulik, 1985). However, many educators criticize these programs as boring and expensive when compared with paper and pencil drill and practice, and therefore, perhaps not a very good use of technology.

Tutorial Programs: Good tutorial computer-assisted instruction takes an average of 100 hours to develop for every one hour of instruction. Therefore these types of programs are not available in large numbers. A tutorial program is considered to be a teaching program. In other words, the program carefully teaches a mathematics concept. The intent is that no other support would be required to understand the particular topic. The future will see most topics of the mathematics curriculum available as tutorials on the microcomputer or interactive video formats.

Simulation Programs: Many computer activities are based on the concept of simulation, using the computer to create realistic situations and circumstances to which students respond and discover the consequences of their choices. For example, a popular program simulates the experience of owning and operating a small business. Elementary students begin the activity with a certain amount of money in the bank. This money must be used for things such as supplies and advertising. The computer is programmed to present various conditions such as days when sales are high and days when business is poor. The students use data provided by the computer to make decisions such as how much of the product to buy and what price to charge. The objective is, of course, to make a profit. This program can be used in the study of social studies and economics which supports math across the curriculum.

Problem-Solving Programs: The goal of many mathematics teachers is to find software which enhances problem solving and problem-solving strategies. Higher-ordered thinking skills are also desired. These programs use ideas which require various strategies to solve the problem. The computer can generate too much information, too little information, require estimation, and show pattern recognition. Ideally these programs could involve individual or group learning. The future holds great promise for these types of programs in the teaching of mathematics.

Application Programs: Word processing, databases, spreadsheets, and graphics are the most popular application programs available on the microcomputer. The world of work utilizes each. Most writing and correspondence involves word processing. Businesses use spreadsheets and databases to keep track of customers and financial information as well as to predict trends.

Spreadsheets are software designed to investigate and develop thinking skills and organize and calculate data based on predetermined formulas. Students at all levels of mathematics can use spreadsheets. For example, young children can use them to budget their allowances. Older students may use them to create records for a hypothetical business in order to learn accounting procedures. A series of spreadsheets and databases will be shown where appropriate throughout this text so that teachers can see these ideas in action.

Engineers and architects utilize graphics programs to explore, design, and build various products. Graphics programs have been used to teach the principles of geometry with notable success (Olivier and Russell, 1986). Students can use graphics software to draw graphs and charts as well as manipulate geometric shapes on the computer screen. Images in computer graphics are appealing because students can see the data pictorially. In fact, graphics may provide the most powerful use of technology (microcomputers) in helping students understand mathematical concepts and processes as well as develop spatial visualization.

Programming the Computer: BASIC and Logo are the two most popular computer programming languages available on the microcomputer in the elementary and middle school. *Programming* is the ability to give the computer directions to solve a problem or create a design.

Software Evaluation

Many times teachers are required to evaluate microcomputer software. This is a difficult task, requiring a large amount of time to ensure that the program meets the classroom objectives. An example of a simple software evaluation form is found in Appendix A and will be used to evaluate mathematics software. The evaluation involves an analysis of the accuracy of the content as well as the pedagogical clarity of the program. In addition it includes the evaluation of how well the program takes advantage of the microcomputer capabilities such as color, sound, branching, graphics, self-pacing, and immediate feedback. The evaluation of each is subjective and often varies with the evaluator's familiarity and understanding of the microcomputer.

BASIC is a popular programming language. BASIC stands for Beginners All Purpose Symbolic Instruction Code. Most microcomputers accept BASIC and no additional purchases are required. In some cases BASIC is permanently programmed into the computer. The language was developed in the early 1960s as a conversational type of programming language to match English. The intent was to make programming more accessible for people using computers, including elementary school students. BASIC uses few commands, but an exact syntax with proper notation is required. Simple BASIC programs are provided throughout the book. Either type in the program or ask your instructor for the disk with the programs already on it.

In BASIC the computer can be used as a calculator using + for addition, − for subtraction, * for multiplication, and / for division. Always press the return key when finished with a line of commands. Begin by turning on a computer with a system disk. Try the following activity:

Activity

COMPUTERS

Beginning BASIC Commands

Directions:

1. Review the previous paragraph for steps in starting BASIC programs.

2. Type the following commands when the blinking cursor appears:

```
PRINT 3 + 2 [RETURN]        The answer should appear
PRINT 3 - 2 [RETURN]        immediately.
PRINT 6 * 2 [RETURN]
PRINT 6 / 2 [RETURN]
PRINT 12 * 3 / 6 [RETURN]
```

Another way to use the BASIC language is to write simple programs. Remember to press the return key after each line. Try this one exactly as printed here, remembering to leave spaces exactly as in the following example:

Activity

COMPUTERS

Simple Programming

```
NEW
10 PRINT "MY MATH"
20 PRINT "CLASS IS GREAT."
30 END
RUN
```

The microcomputer will print out . . .

```
MY MATH
CLASS IS GREAT.
```

Logo is another popular programming language developed in the 1970s. It has been used with students of all ages including the primary grades. It contains two modes: turtle graphics and programming. *Turtle graphics allows you to move a turtle to make designs. Programming is telling the computer how to solve a program with Logo commands.* Logo activities will be used whenever appropriate throughout the book. Either type the listed program into the microcomputer or get a data disk from your instructor which already has the program on it.

You will need to have a Logo software program for your microcomputer to write Logo programs; this is called a master disk. Your instructor's disk must be the same brand of Logo as the master disk and the microcomputer must also be compatible. Refer to Appendix B for the list of Apple Logo commands and control keys. If you are using a different version of Logo be sure to get this information from the manual provided by the manufacturer.

You can now begin using Logo; only the master disk is needed.

Special Instructions

To begin using Logo follow these steps:

1. Insert the Logo disk into the disk drive.
2. Turn on the Apple microcomputer—the power switch is on the back left of the micro.
3. Turn on the monitor.
4. Press the RETURN key. Ignore the message regarding inserting your own disk. Wait until the screen is blank except for the message.

   ```
   WELCOME TO LOGO
   ?
   ```

   A flashing box appears in the upper left corner next to the question mark. It is called the cursor.
5. If you make typing errors, or typos, just retype the line. If you discover an error before pressing RETURN, back up the cursor using the left arrow (find it on the right side of the keyboard) and type the line from the point of error.
6. Type in the word SHOWTURTLE (or ST for short) and press RETURN. You are now ready to create some turtle graphics. What in the world are turtle graphics? I'm glad you asked.

Turtle Graphics

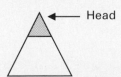
Head

In the center of the screen is a turtle (you may have thought it was just a triangle). The turtle moves about the screen going forward and backward. The command FORWARD (or FD for short), followed by a space and a number, tells exactly how many steps you want the turtle to take in the direction in which it is pointing. Now it is pointing straight up. Type in this command

(and always remember to press the RETURN key after you are finished typing a line):

```
FD 50   or   FORWARD 50
```

The turtle just marched 50 turtle steps forward. Make it go 25 more steps forward. Try having it go just one step and watch carefully as you press RETURN to see if you can determine any movement. One turtle step is very small.

The turtle can go backward, too. FORWARD and BACK are the two most important commands in Logo. The command BACK (or BK for short) causes the turtle to move in the opposite direction. Enter this line:

```
BK 155   or   BACK 155
```

Notice that the turtle went backward past the point where it began by 79 steps. What happens if you instruct it to go back 200 steps? Were you right? It moves so quickly you can hardly tell that it is moving down the screen and coming back on the top—since it ran out of space at the bottom of the screen. This is called wrapping around the screen.

Try to get the turtle into the center of the screen again. Just backtrack the instructions you have already given. An easier way to do this is to type the word CLEARSCREEN (CS for short), which is quicker than backtracking. It clears the screen and places the turtle in the center of it.

This clever turtle can also make turns to the LEFT (LT for short) and RIGHT (RT for short). Whatever direction the turtle is pointing, it will turn the number of degrees you tell it and in the direction you indicate (to the left or right). Enter these lines and observe the turtle as you press RETURN each time:

```
CS
RT 90
LT 90
RT 180
LT 360
```

The last line causes the turtle to turn completely around. The movement is so fast that it can hardly be perceived.

Combining these four commands: FORWARD, BACK, RIGHT, and LEFT, the Logo user can design an unlimited number of graphics. If you wanted to draw a square, here are some lines that you might begin with:

```
CS
FD 100
RT 90
FD 100
RT 90
```

Finish this program by adding lines that instruct the turtle to complete the square. Once you have mastered the square, experiment with other shapes. (Don't forget to use the CS command each time you begin a new design.)

Check out this command:

```
CS
REPEAT 4[FD 50 RT 90]
```

The square brackets in this line are special Logo characters made by holding down the SHIFT key and pressing N for the left bracket and M for the right bracket. Some microcomputers have special bracket keys. Remember you cannot substitute parentheses for brackets.

If you want to draw a square, the above program is one way to do it. It saves a lot of time compared to entering a series of FORWARD and RIGHT moves. The REPEAT 4 command tells the number of times you want the

(continued)

turtle to do whatever is inside the brackets. First the turtle goes forward 50 steps and then it turns 90 degrees; the second time around the turtle follows the instructions inside the brackets again, going another 50 steps forward and turning 90 degrees. It does the same two more times and stops because it only was supposed to do it four times. What happens if you change 4 to 8 and also change the numbers within the brackets? Enter this:

```
CS
REPEAT 8[FD 25 RT 45]
```

You can have the turtle appear when it is not on the screen by the command SHOWTURTLE (ST is the abbreviated form). There is also the opposite command called HIDETURTLE (HT for short). See what happens when you enter this command and press RETURN. The turtle is still in the same place, but you just can't see it. How can you prove this last statement? Simply type a FORWARD or BACKWARD instruction, indicate a number of steps, and watch what happens. Now make the turtle reappear.

Another thing you can do with the turtle is to tell it when to draw and when not to draw. PENUP and PENDOWN are turtle commands that stay in effect until you give the next pen command. When you first load Logo the pen is down and ready to draw. When the pen is up the turtle will move, but no drawing will occur. Here is another Logo activity for beginners of all ages.

Make a copy of the maze (Fig. 1.1) with transparency paper or wax paper, using a grease pencil or indelible marker. Attach it to the microcomputer monitor. Now guide the turtle through the maze using the appropriate Logo commands. If you need help refer to the commands found in Appendix B.

Computer-Managed Instruction: *Computer-managed instruction (CMI) is defined as programs used for administrative functions in educational settings.* Among the instructional management systems is the computer-assisted management of elementary mathematics instruction. The increasing sophistication of personal computers has accelerated their computing power to the capability of the former mainframe computer. Software, like computer-managed instruction (Bishop, 1982), can help instructors develop and sustain a continuous, diagnostic record of their students' mathematical skills. By reviewing the computer-generated data, teachers can determine a student's strengths and weaknesses by the objectives he or she has mastered.

Computer-managed software is available to maintain databases on each student. This includes the student's name, age, sex, teacher, standardized test score history, current mathematics unit, date of last access, unit of last access, amount of time used in last access, last unit successfully completed, total testing time used, birth date, and grade level. The computer recognizes student names when a student signs on and administers the appropriate unit test which includes questions on the objectives. The status of individual objectives and units is maintained, and a completed unit status is sequentially incremented onto the student's record. However, the teacher does have the option to override this sequential increment and set the student's assignment to any desired unit.

By having immediate access to this information, teachers can improve the quality of the time they spend with their students and enhance their achievements. By reducing the time necessary to construct, administer, and score diagnostic measures, and record the results, the quantity of time spent with students also increases.

Trends in Problem Solving

Research findings (Bitter, 1987; Capper, 1984) indicate that mathematics instruction should be used to enrich, deepen and widen students'

Start ▶

Finish ▼

Figure 1.1

problem-solving abilities. Researchers questioned how information should be organized in memory for the greatest beneficial use in problem solving. Findings of researchers (Capper, 1984) determined that teachers should know how to glean the most relevant elements to use in their mathematics classrooms from the myriad of reading materials on problem solving. They found that students' previous experiences, cognitive development, and impressions of mathematics affected their success in problem solving.

Results from the fourth mathematics assessment of The National Assessment of Educational Progress (NAEP) (Kouba et al., 1988) indicate students did best on problems that involved familiar settings. Around 90 percent of the third-grade students correctly solved one-step, whole-number addition word problems, and about 70 percent correctly solved one-step, whole-number subtraction word problems. Students in seventh grade performed well, with 90 percent to 95 percent responding correctly. Students performed poorly on two-step word problems. About 30 percent of the third-grade students correctly solved two-step addition/subtraction word problems, while 77 percent of the seventh-grade students correctly solved the problem. This performance level drops off greatly for seventh-graders once large numbers are included or settings are unfamiliar. Many of the errors occurred because students solved these problems using only one operation rather than two-steps as the problem required.

The fourth assessment also included items to assess the processes or procedures central to students' ability to solve problems such as logical reasoning, description or identification of word problems, and uses of problem-solving strategies. Performance on items involving logical reasoning indicates that many students (at least two-thirds of the third-grade students and over one-half of the seventh-grade students) have difficulty with logic items. When items involve several pieces of information or lack sufficient information to solve the problems, students performed poorly. Many third-grade and seventh-grade students indicated insufficient information to solve the problem although sufficient information was available in the problem. Students' responses to the mathematical methods items indicate that students do not appear to use problem-solving strategies such as determining the reasonableness of an answer or using an appropriate mathematical model.

Good problem solving encompasses four phases, according to Polya (1957). He defined the first phase as understanding the problem. Without understanding the meaning, students will not be able to find a correct solution. Once students understand the problem, they devise a plan. Polya suggested that the third phase is carrying out the devised plan. Good problem solvers then look back at the solution to verify its correctness.

Different classes of problems require different abilities. In younger children, problem solving and computational abilities have been found to be related, but in college students research (Driscoll, 1982) finds little relation between problem solving and computational abilities. Good problem solvers sift the key ideas out and use the relevant information to solve the problems. Although poor problem solvers may know the information, they do not capture the pertinent ideas. Of good and poor problem solvers, good problem solvers prove to be the more active. Successful problem solvers take a step-by-step approach to a problem. They are careful, showing concern and quick retracking when the ideas became confusing. They recheck, review, and reread to be sure errors have not occurred (Driscoll, 1982).

The difference between good and poor problem solvers is in the perception each has of the important elements of the problem. While good problem solvers distinguish between relevant and irrelevant information in a problem, poor problem solvers find this task difficult to do. Good problem solvers assess the structure of a problem rapidly, while poor problem solvers stumble. Good problem solvers recognize a common thread running through a wide range of problems and use that knowledge to aid them in solving the problem. Good problem solvers recognize a mathematical structure they have done or have seen done before. It appears that good problem solvers use a range of strategies in solving problems. Closely related to successful problem solving is thinking through problems, otherwise known as goal-oriented planning. More planning, more checking, and more reviewing of the problem are major attributes of the successful problem solver as suggested by researchers. Educators should help students develop an understanding of a variety of problem-solving resources which can be accessed when a problem needs to be resolved. Since problem solving cannot be taught per se, students should be provided with a wide variety of problem solving experiences and resources.

It has been found (Capper, 1984) that a teacher's attitude toward problem solving relates positively to the problem-solving ability of the class; therefore, to improve the problem-solving-abilities of a class, a teacher must be alert to the values he or she communicates to students. This can be done by modeling some aspect of problem solving to each individual class through posing problems. Also, a teacher can actively use strategies to push through to a solution. Another modeling method is to pose problems that spring from the problem just solved. Teachers can teach some aspect of problem solving such as drawing a diagram or making a sketch. This strategy is useful as a means of visualizing the problem in an uncluttered way. It also enables students to estimate solutions. Diagrams have often led to problem-solving solutions, and they help to approximate and verify, as well as give assistance in organizing the information.

Researchers call these activities tool skills as they are the foundation of successful problem solving. *Heuristics is the general suggestion or strategy that helps problem solvers in their approach to an understanding of a problem.* It also helps organize resources to use in solving problems. It involves goal-oriented planning, trial and error, and memory searches for similar problems. It also involves a search for patterns. With heuristics, the efficient problem solvers often work backward from a known objective to find a solution. They look back to construct a new, yet related problem. Teachers need to instruct students both how and when to use heuristics. Studies show that students who are taught heuristics regularly and in a variety of contexts organize their manner of approach to a problem (Driscoll, 1982).

There are many benefits to students being introduced to problem-solving strategies early and consistently. Two advantages of flexibility of thought are to see more than one correct answer and to see different ways of arriving at the same answer. Problem solving is an effective means of making mathematics more relevant to students. To students, formulas used to calculate an answer may not appear to have much to do with their daily lives. However, the same formula can be integrated into problems that are interesting to students, and students develop a sense that mathematical principles are valuable in solving problems that arise in the real world.

Students can absorb some aspects of problem solving from classroom problems and interaction with their teachers and peers. Although it is an unconscious process, goal-oriented planning is an element of problem solving that needs this type of nurturing. The student needs to be aware of the mental processes involved. Verbalizing their problem-solving practice in an encouraging classroom environment will help students develop this awareness (Driscoll, 1982).

It is recommended that teachers use a structured, open-minded approach when teaching problem solving. Their use of problem-solving tool skills and heuristics, and their examples of behaviors and attitudes of good problem solving will improve the students' abilities to think about mathematical problems. If teachers value the problem-solving process, and clearly show in their classroom that they value it, their students will think about the process involved in solving mathematical problems and gain the needed tool skills.

Problem solving in mathematics is one of the areas where students can benefit from cooperative learning (Burns, 1984). Placing students in cooperative groups is an effective way to stimulate cognitive development, increase self-esteem, and raise achievement. When students work together cooperatively to solve problems, their dialogue helps solidify their thinking and promotes positive attitudes and confidence. As you work the many problem-solving explorations in this text, you are encouraged to share the strategies you develop with others and gain new insights from their strategies.

Trends in Estimation and Mental Math

In general, most reports recommend the de-emphasis of long, written computation. More emphasis should be placed on estimation where exact answers are not required and on mental math where quick computation is done mentally. Research indicates that although there is considerable importance placed on estimation in elementary and intermediate mathematics curricula, there has been little research or development of estimation skills (Siegel, Goldsmith, & Madson, 1982). Part of the problem seems to be linked to the degree of difficulty in testing such skills, because the time needed to do so must be carefully controlled. Otherwise, the entire computation may be completed, rather than just the estimation segment (Reys, Rybolt, Bestgen, & Wyatt, 1982).

Studies have found that while there are a number of workable estimation strategies, misconceptions frequently inhibit the ability to estimate (Hildreth, 1983; Levine, 1981; and Rubinstein, 1983). More research needs to be done in the choice of strategies, i.e., how does someone determine the use of one particular strategy over another? In relation to teaching estimation, there needs to be a clearer definition of how it is best taught (Schoen, 1986). The text emphasizes estimation and mental computation throughout as meaningful and appropriate.

Trends in Math Across the Curriculum

Math across the curriculum is a phrase which refers to the integration of mathematical concepts in other subject matter areas of the elementary and middle school curriculum. Students learn more effectively when they can see how concepts are applied in other situations. An understanding of metric liters can help students find the most economical soft drink buys when planning a consumer report for social studies. An understanding of fractions can help the beginning music student see that each measure of music must add up to a whole. The examples are limitless, and the reader is encouraged to find other applications as various topics are presented in the text.

Issue: Math Anxiety

Math anxiety is a term given to the fear of failure when exposed to mathematics concepts and processes. Both men and women can suffer from math anxiety and avoidance of mathematics may result in some cases. Kelley and Tomhave (1985) found that both men and women elementary education majors scored higher than other groups when tested for math anxiety. Educators fear that math anxious teachers may produce math anxious students in elementary and middle school, effectively adding to the lack of interest in higher math courses in high school and college. More study is needed on the reasons for math anxiety and math avoidance. Some reasons

given have been an over-emphasis on one correct answer, use of ambiguous vocabulary, intolerance for unusual ways to solve problems, and pressure from timed tests. To reduce math anxiety, this text stresses many ways to reach solutions through different thought-processing strategies, estimation skills, problem-solving techniques, as well as two study approaches to the text (see Figure P.1 in Preface).

Issue: Women in Mathematics

Women earned 42 percent of the bachelor's degrees in mathematics in 1980 which was up 14 percent from the early 1950s. In the same time period, women earned 14 percent of the doctorates in mathematics—up 9 percent since the 1950s (Becker, 1987). Studies (Cooper, 1987) indicate that more women teachers may be taking advantage of relicensing opportunities to teach mathematics and science than male teachers, but surveys (Kemper & Mangieri, 1987) show that high school students about to make career choices still perceive female instructors as English teachers and male instructors as mathematics teachers. Research (Becker, 1987) has shown that successful programs encouraging women to be actively involved with mathematics start in the middle school to junior high years when decisions about careers begin to take shape. Groups have been organized for a diversity of age levels with a variety of mathematical interests as focal points, but the most successful groups have adopted a casual format, stressing cooperative grouping where concepts can be learned in a nonthreatening, social setting—a characteristic which research (Brody & Fox, 1980) shows is important to females as contrasted to their male counterparts. Middle school teachers need to be aware of such findings when planning math lessons for female students. Throughout the text, cooperative learning activities will be presented in the hope that they will promote a nonthreatening environment for the study of mathematics.

Summary

Mathematics of the past emphasized New Math as being encouraged by the Russian launching of Sputnik I. Back to basics was emphasized in the 1970s due to a decline in test scores and general illiteracy. The trends of the future include technology (calculators and computers), problem solving, math across the curriculum, mental math, and estimation. Trends as well as issues of the future include math anxiety, women in mathematics, and the role of the calculator and computer in the curriculum. The computer can provide tutorial, drill and practice, simulation, and problem-solving instruction. Programming in Logo and BASIC as well as the application of graphics, databases and spreadsheets can be used in the teaching of mathematics. As technology improves, the meaning of basics will need to be reconsidered and redefined to match the mathematics survival skills of the future. *The inherent question which needs to be considered throughout this book is: Are these the necessary mathematics skills required to be an effective citizen in the future?*

Exercises

Directions: Read all questions *before* answering any one exercise. Frequently the last question in one category leads to the first question in the next category.

A. Memorization and Comprehension Exercises
Low-Level Thought Activities

1. List 5 areas this text will emphasize. Hint: The pictures found in the Preface should help.
2. Describe how your mathematics training was different from your parents and grandparents.
3. Describe your mathematics training. What did you like? What did you dislike?
4. Pick a futurist and outline that person's prediction for education in the next century.
5. List 10 ways a computer can be useful in mathematics.
6. Identify several commercial software products for each of the following software types:
   a. drill and practice
   b. tutorial
   c. problem solving
   d. simulations
   e. spreadsheets
   f. graphics
   g. databases
7. Find a book featuring the history of mathematics and list the influences on the history of mathematics by the following:
   a. Pestalozzi
   b. Brownell
   c. Grube
   d. Spitzer
   e. Thorndike
   f. Dewey

B. Application and Analysis Exercises
Middle-Level Thought Activities

1. Search professional journals to see the changes in mathematics education over the years. Find the earliest journal in your library with articles about the teaching of mathematics. Compare it with journals from:
   a. the late 1960s through the early 1970s
   b. the mid to late 1970s
   c. the present

   What patterns or trends do you see? What do they tell you about mathematics education? How can they help you predict what might happen in the future?

   Follow the form below to report on the pertinent findings:

   Journal Reviewed: _____

   Publication Date: _____ Grade Level: _____

   Subject Area: _____

   Author(s): _____

   _____

   Major Findings:

   Study or Teaching Procedure Outlined:

   Reviewed by: _____

   Some journals of note are listed below to help you get started, but there are many more. Search your library to discover what is available. Other chapters will have similar assignments so learning the journals now will pay off later.

*Journal for Research in Mathematics
   Education*
Arithmetic Teacher
Mathematics Teacher
School Science and Mathematics

2. Review at least two articles on the teaching of calculators and two articles on the teaching of computers in elementary and middle school classrooms. Computer applications for classroom use are found in the journals in the following list. This is by no means an exhaustive list. Search the periodical section of your library for more.
The Computing Teacher
Electronic Learning
Teaching and Computers
Classroom Computer Learning
*The Journal of Computers in
   Mathematics and Science Teaching*
Computers in the Schools

3. Using the format in B1, find an article which discusses classroom use for each of the following:
   a. drill and practice software
   b. tutorial software
   c. simulation software
   d. problem-solving software
   e. graphics software
   f. spreadsheets
   g. databases
   h. Logo
   i. BASIC

4. After doing the above review of the literature, analyze and write a report on the changes in mathematics education since the early 1900s.

C. **Synthesis and Evaluation Exercises**
   High-Level Thought Activities

   1. Create a position paper on the pros and cons of using calculators on mathematics tests. State your recommendation after evaluating each side of the issue.

2. Speculate what you perceive the elementary or middle school classroom will be like in the beginning of the next century.

3. Write a letter to your students' parents explaining why you will be using a calculator and/or computer in your math class all year.

4. Using the software evaluation form in Appendix A, evaluate a simulation, a problem-solving, a tutorial, and a drill and practice piece of software.

5. Write a response to the following statement:
   We do not have to teach computation any longer since calculators can perform the computations faster and accurately.

6. Write a response to the following question:
   What steps will you take in your teaching to reduce math anxiety?

7. Write a response to the following question:
   How will you encourage girls to become actively involved in mathematics?

8. Write a position paper on curricular reform in mathematics education (K–12) using the following national reports as a basis of your paper:
   NCTM—Curriculum and Evaluation
      Standards for School Mathematics
   MSEB—Curriculum Framework of the
      Mathematical Sciences Education Board
   AAAS—Project 2061

9. Reflect on the goals for students in the NCTM Standards document. Write a descriptive paragraph on each goal as to its importance in the students' future and how classroom instruction can help achieve those goals.

10. What insights must teachers have to survive the present and prepare for the future in mathematics education?

Bibliography

Abe, Koichi. "The Role of the History of Mathematics Teaching in Teacher Training." In *Fifth International Congress on Mathematical Education: Using Research in the Professional Life of Mathematical Teachers*, edited by Thomas A. Romberg (Osaka Kyoiku Univ., Japan) May 1985.

Baron, Norman. "Computers in Schools: A Plea for a Thoughtful and Educated Approach." *Arithmetic Teacher* 34 (April 1987): 40–41.

Becker, Joanne Rossi. "Sex Equity Intervention Programs Which Work." *School Science and Mathematics* 87 (March 1987): 223–232.

Biehler, Robert F., and Jack Snowman. *Psychology Applied to Teaching.* Boston: Houghton Mifflin, 1986.

Bishop, Thomas D. "Microcomputer Assisted Management of Elementary Mathematics Instruction." *The Journal of Computers in Mathematics and Science Teaching* (Summer 1982): 14–16.

Bitter, Gary G. "Back to Basics with Math for the Future." *Electronic Education* 6 (January 1987): 22–23, 26.

———. "Educational Technology and the Future of Mathematics." *School Science and Mathematics* 87 (October 1987): 454–465.

Brody, Linda, and Lynn H. Fox. "An Accelerative Intervention Program for Mathematically Gifted Girls." In *Women and the Mathematical Mystique,* edited by Linda Brody, Lynn H. Fox, and Diane Tobin, 164–178. Baltimore: The Johns Hopkins Univ. Press, 1980.

Brownell, William A. "AT Classic: The Revolution in Arithmetic." *Arithmetic Teacher* 34 (October 1986): 38–42.

Burns, Marilyn. *The Math Solution: Teaching for Mastery Through Problem Solving.* Sausalito, Calif.: Marilyn Burns Associates, 1984.

Capper, Joanne. *Research into Practice Digest,* Vol. I, No. 1a and Vol.II, No. 1b. *Thinking Skills Series: Mathematical Problem Solving: Research Review and Instructional Practice,* Washington, D.C.: Center for Research into Practice, 1984.

Carpenter, Thomas P. "Research on the Role of Structures in Thinking." *Arithmetic Teacher* 32 (February 1985): 58–60.

Chance, William. " . . . the best of educations": Reforming America's Public Schools in the 1980s. N.p.: The John D. and Catherine T. MacArthur Foundation, 1986.

Commission on Standards for School Mathematics of the National Council of Teachers of Mathematics. *Curriculum and Evaluation Standards for School Mathematics.* Working draft, Reston, Va.: The Council, 1987.

Cooper, Bruce S. "Retooling Teachers: The New York Experience." *Phi Delta Kappan* 68 (April 1987): 606–609.

Cope, Charles L. "Math Anxiety and Math Avoidance in College Freshmen." *Focus on Learning Problems in Mathematics* 10 (Winter 1988): 1–15.

Desforges, Charles, and Anne Cockburn. *Understanding The Mathematics Teacher.* Barcombe, Lewis, E. Sussex, England: The Falmer Press, 1987.

Driscoll, Mark. *Research Within Reach: Secondary School Mathematics. A Research-Guided Response to the Concerns of Educators.* St. Louis, Mo.: Research and Development Interpretation Service CEMREL, Inc.; Washington, D.C.: National Institute of Education, 1982.

Garofalo, Joseph. "Metacognition and School Mathematics." *Arithmetic Teacher* 34 (May 1987): 22–23.

Gliner, Gail S. "The Relationship Between Mathematics Anxiety and Achievement Variables." *School Science and Mathematics* 87 (February 1987): 81–87.

Hembree, Ray, "Research Gives Calculators a Green Light." *Arithmetic Teacher* 34 (September 1986): 18–21.

Hildreth, David J. "The Use of Strategies in Estimation Measurements." *Arithmetic Teacher* 30 (January 1983): 50–56.

Horvath, Patricia J. "A Look at the Second International Mathematics Study Results in the U.S.A. and Japan." *Mathematics Teacher* 80 (May 1987): 359–368.

Howson, A. G., B. F. Nebres, and B. J. Wilson. "School Mathematics in the 1990s." 1985.

James, Glenn, and Robert James, eds. *Mathematics Dictionary,* 4th ed. New York: Van Nostrand Reinhold, 1976.

Johnson, David W., and Roger T. Johnson. *Cooperative Learning.* Minneapolis: Interaction Book Company, 1984.

Kelley, William P., and William K. Tomhave. "A Study of Math Anxiety/Math Avoidance in Preservice Elementary Teachers." *Arithmetic Teacher* 32 (January 1985): 51–53.

Kemper, Richard E., and John N. Mangieri. "America's Future Teaching Force: Predictions and Recommendations." *Phi Delta Kappan* 68 (January 1987): 393–395.

Kouba, Vicky L., Catherine A. Brown, Thomas P. Carpenter, Mary M. Lindquist, Edward A. Silver, and Jane O. Swafford. "Results of the Fourth NAEP Assessment of Mathematics: Number, Operations, and Word Problems." *Arithmetic Teacher* 35 (April 1988): 14–19.

Kulik, James A., Chen-Lin C. Kulik, and Robert L. Bangert-Drowns. "Effectiveness of Computer-Based Education in Elementary schools." *Computers In Human Behavior,* Vol. 1, 59–74. Elmsford, NY: Pergamon Press, Inc., 1985.

Levine, Deborah R. "Computational Estimation Ability and the Use of Estimation Strategies Among College Students." Ph.D. diss., New York Univ., 1980. *Dissertation Abstracts International* 41 (1981): 5013A.

Lumb, David. *Teaching Mathematics 5 to 11.* New York: Nichols Publishing Co., 1987.

Martinez, Michael E., and Nancy A. Mead. *Computer Competence: The First National Assessment.* Educational Testing Service, April 1988.

Mathematical Association of America. *Guidelines for the Continuing Mathematical Education of Teachers.* Washington, D.C., 1987.

McKnight, Curtis C., F. Joe Crosswhite, John A. Dossey, Edward Kifer, Jane O. Swafford, Kenneth J. Travers, and Thomas J. Cooney. *The Underachieving Curriculum: Assessing U.S. School Mathematics from an International Perspective.* Champaign, Ill.: Stipes Publishing Company, 1987.

Mohyla, J., ed. *The Role of Technology: ICME 5.* South Australia College of Advanced Education, Adelaide, South Australia, 1987.

National Commission on Excellence in Education. *A Nation at Risk.* Washington, D.C., 1983.

National Council of Teachers of Mathematics. *An Agenda for Action: Recommendation for School Mathematics of the 1980s.* Reston, Va., 1980.

————. "Guidelines for the Preparation of Teachers of Mathematics." Reston, Va., 1981.

————. "The Impact of Computing Technology on School Mathematics: Report of an NCTM Conference." Reston, Va., March 1984.

————. "The Use of Computers in the Learning and Teaching of Mathematics." Reston, Va., September 1987.

National Science Board Commission on Precollege Education in Mathematics, Science, and Technology. *Educating Americans for the 21st Century: A Report to the American People and the National Science Board.* Washington, D.C., 1983.

Olivier, Terry A., and Rebecca Gaye Russell. "Using Low-Resolution Graphics to Develop Problem Solving Skills." *The Computing Teacher* 13 (May 1986) 50–51.

Polya, George. *How to Solve It: A New Aspect of Mathematical Method.* 2d ed. Princeton, N.J.: Princeton Univ. Press, 1957.

Preston, Michael, ed. *Mathematics in Primary Education.* Barcome, Lewis, E. Sussex, England: The Falmer Press, 1987.

Ralston, Anthony, et al. *A Framework for Revision of the K–12 Mathematics Curriculum.* Task Force Report submitted to the Mathematical Sciences Education Board, National Research Council, January 1988.

Reed, Sally, and Craig Sautter. "Visions of the 1990s." *Electronic Learning* 6 (May/June 1987): 18–23.

Reinboldt, Werner C. *Future Directions in Computational Mathematics, Algorithms, and Scientific Software.* Society for Industrial and Applied Mathematics, 1985.

Resnick, Lauren B. *Education and Learning to Think.* National Academy Press, 1987.

Reys, Robert E., James F. Rybolt, Barbara J. Bestgen, and J. Wendell Wyatt. "Processes Used by Good Computational Estimators." *Journal for Research in Mathematics Education* 13 (May 1982): 183–201.

Romberg, Thomas. *School Mathematics: Options for the 1990s.* Chairman's report. Washington, D.C.: Office of Educational Research and Improvement, U.S. Department of Education, 1984.

Rubinstein, Rheta N. P. "Mathematical Variables Related to Computational Estimation." Ph. D. diss., Univ. of Montana, 1978. *Dissertation Abstracts International* 44 (1983): 695A.

Rutherford, James, et al. *What Science Is Most Worth Knowing?* Draft Report of Phase I, Project 2061; American Association for the Advancement of Science, December 1987.

Schoen, Harold, ed. *Estimation and Mental Computation.* Reston, Va.: National Council of Teachers of Mathematics, 1986.

Shumway, Richard J. "Why Logo?" *Arithmetic Teacher* 32 (May 1985): 18–19.

Siegel, Alexander W., Lynn T. Goldsmith, and Camilla R. Madson. "Skill in Estimation Problems of Extent and Numerosity." *Journal for Research in Mathematics Education* 13 (May 1982): 211–232.

Stevenson, Harold W., et al., "Mathematics Achievement of Chinese, Japanese, and American Children." *Science* 231 (14 February 1986): 693–699.

West, Beverly H., Ellen N. Griesbach, Jerry D. Taylor, and Louise T. Taylor. *The Prentice-Hall Encyclopedia of Mathematics.* Englewood Cliffs, N.J.: Prentice-Hall, 1982.

Wirszup, Izaak, and Robert Streit, eds. *Developments in School Mathematics Education Around the World.* Reston, Va.: National Council of Teachers of Mathematics, 1987.

2

How Children Learn Mathematics and Solve Problems

Key Question: *How do children learn mathematics and what are the techniques effective teachers use to help children solve problems?*

Activities Index

Introduction

Children learn some knowledge about mathematical concepts from informal, unstructured experiences in their environment, while other mathematical knowledge is received in a formal, structured educational setting. The objective of this chapter is to help the teacher understand how children learn mathematics and solve problems to plan competent mathematics instruction. A brief introduction of several learning theories will be presented. All the theories have practical application for the study of elementary and middle school mathematics and can be used in the structured setting of the classroom.

There are effective ways to learn mathematics and teachers can plan lessons to take advantage of them. This chapter explores effective learning through classroom instruction and the part that mastery learning plays in mathematics achievement. Diagnostic interview techniques will be discussed as they relate to planning remedial and advanced lessons to meet the individual learning needs of students.

The movement toward excellence in education has shown the need to teach innovative problem-solving strategies to students. Elementary and middle school textbooks are increasing their emphasis on thinking skills to solve problems. The important role of problem solving in mathematics will be discussed and problem-solving strategies will be outlined with examples given for classroom use.

Theories of Learning Applied to Mathematics

This chapter will show various theories as they co-exist in the development of students' understandings of mathematics. Table 2.1 shows the key points of three present schools of thought and some of the learning theories within each school. Key points of each theory have been included also. Some readers may find the table beneficial as a means for comparison. However, each theory and its mathematical implications are explained in more detail in the following paragraphs.

Cognitive Theories

The work of Jean Piaget and Jerome Bruner have had a great impact on the study of how children learn mathematics. Both believe that children must be allowed to experiment physically with the things around them if they are to learn. In mathematics these hands-on materials are often called *physical manipulatives* and may be defined as any three-dimensional, concrete learning material which models or represents the structure of a mathematical concept and/or the relationship among concepts. Every chapter contains many examples of physical manipulatives that teachers can use in mathematics instruction and problem solving.

However, Piaget and Bruner are not the only ones to discover the essential need for physical manipulation in the study of mathematics. Abundant non-Piagetian research points to the same phenomenon (United States Department of Education, 1986; Tunis, 1986). Numerous studies support the belief that children learn mathematics better if concrete manipulatives are used first in the presentation of new mathematical concepts and principles.

The first cognitive theorist we will discuss is Jean Piaget, a Swiss philosopher-epistemologist. He conducted extensive research over a sixty-year period (1920s to 1980) on the development of children's cognitive abilities. Piaget and his staff devised ingenious learning tasks, some of which tested children's understanding of mathematics. These tasks and their results will be presented throughout the book where appropriate. Special attention to Piagetian research as it relates to number readiness will be found in Chapter 5. The focus in this chapter will be Piaget's theory only as it applies to the general learning of mathematics.

Table 2.1 The Three Schools of Thought

Cognitive School	Behaviorist School	Information Processing School
Piaget:	**Behavior Modification:**	**Thought Processing:**
Stages of Development	Reinforcement Theory	Simultaneous Synthesis
Sensorimotor	Immediate Feedback	Successive Synthesis
Preoperations	Programmed Learning	
Concrete Operations		**Learning Styles:**
Formal Operations	**Bandura:**	Visual
	Social Learning	Auditory
Bruner:	Modeling	Tactile
Modes of Reality	Imitating	
Enactive		
Iconic		
Symbolic		

Piaget's theory is age and stage related, which means that people go through definite stages at certain age periods in their lives. Each stage must be completed before a person can attain the next stage. Interestingly, Piaget found that the development of mathematics proceeds through the same stages in a collective sense (taking many generations) that each person goes through the same stages individually in a relatively short amount of time. People must act on their environment at each stage until bits of knowledge form schemes (interlocking the bits of knowledge together to perform meaningful actions) which, in turn, form changed mental structures. These changed mental structures thrust a person into the next developmental stage. In a literal sense, the person no longer sees the world the same way. Notice that through the approximate age of twelve, every stage requires action on the environment. In mathematics, this means that elementary-age children must learn the operations of addition, subtraction, multiplication, and division through the use of hands-on materials that allow them to physically examine what is happening when an arithmetic operation is performed.

Name of Stage	Age	Mental Structures Permit:
Sensorimotor	Birth–2 yrs	Actions on objects
Preoperational	2 yrs–6 yrs	Actions on reality
Concrete Operational	6 yrs–12 yrs	Actions on operations
Formal Operational	12 yrs +	Operations on operations

When students enter the formal operational stage, Piaget's supporters believe that mental transformations can take place without the need of concrete materials. Students are then capable of understanding abstract concepts that cannot be easily proved in the physical world. An example would be the existence of irrational numbers having no repeating decimal pattern that can be keyed back to a basic building unit in our numeration system. Some studies have shown that not all people may attain formal operational thought, and still others may attain if far beyond the age of twelve. For this reason, some eighth graders have trouble understanding irrational numbers and concrete models are needed.

Another cognitive theorist is Jerome Bruner, a Harvard psychologist, who studied how people select, retain, and transform knowledge.

Bruner (1966, 1983) believes that learning is an active process which permits people to go beyond the information given to them to create new possibilities on their own. His theory is not age or stage related. According to Bruner, there are three modes of representing reality that occur in the same order but interact throughout a person's life.

Mode	Definition
Enactive	Action on reality in concrete ways without the need for imagery, inference, or words
Iconic	Pictorial need to represent reality; internal imagery that stands for a concept
Symbolic	Abstract, arbitrary systems of thought

When new or additional knowledge is to be learned, people must progress through one mode to the next for learning to occur. Mathematics educators frequently refer to the three modes as *concrete, pictorial*, and *symbolic*. Every elementary mathematics textbook starts with the pictorial and progresses to the symbolic. However, every series stresses the need for the teacher to present each new concept with concrete materials first.

Mathematics educators have expanded on the initial three-phase model of Bruner. They desire the middle level to encompass more than just the pictorial or iconic representations of Bruner. Baratta-Lorton emphasized the importance of the child's active explorations with mathematics and encouraged delaying the symbolic level. There is no pressure to ever abandon the concrete or pictorial in this model (Labinowicz, 1980).

Mary Baratta-Lorton Model

Intuitive Concept Level
Logical relations among objects;
manipulatives;
free to explore; free from calculation

Connecting Level
Familiar activities now done with math symbols;
higher level of abstraction

Symbolic Level
May still use concrete objects or pictorial representations of the concrete but emphasis is on symbols and what can be learned from them

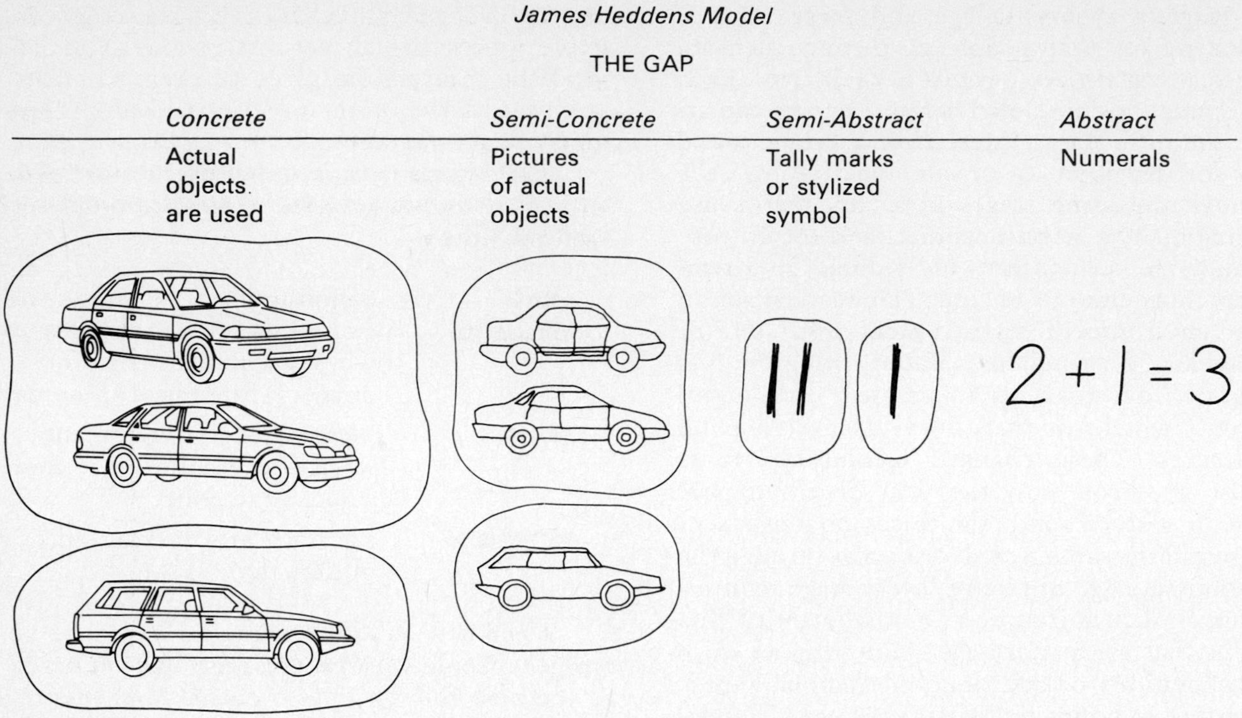

James Heddens Model

THE GAP

Concrete	Semi-Concrete	Semi-Abstract	Abstract
Actual objects are used	Pictures of actual objects	Tally marks or stylized symbol	Numerals

$$|| \ | \qquad 2 + 1 = 3$$

Figure 2.1

The other cognitive theorists whose writings apply to mathematics are Heddens and Schultz. The Heddens Model (1986) emphasizes that fine-line changes are occurring in the middle transitional positions. Figure 2.1 shows the model a child might use to discover the meaning of $2 + 1 = 3$.

Schultz (1986) points out that there is a difference between a student actively using objects in the concrete, pictorial, symbolic phases and watching others do the manipulating. All of the above models presuppose that students are doing the active hands-on exploration by themselves. This has implications for computer software as well. The best software may be programs where students can directly affect the image on the screen. Logo is an excellent computer language because of its interactive ability. This text uses the Logo programming language throughout to show how it can be applied to the study of mathematics.

Behaviorist Theories

Behaviorist theories stress the stimulus-response approach to learning. Each bit of information (stimulus) can be linked to a desired response if the correct contingency is presented to students. Research on behavior modification shows that the best contingency is *positive reinforcement* or an increase in a desired response when a positive stimulus is added to the learning environment. While there are many researchers in the field of behavior modification, one of the best known if B. F. Skinner (1968). If the desired response to $3 + 4 =$ is to be learned, a positive stimulus such as adding points toward a higher math grade (or any other reinforcement viewed as desirable by the student) must accompany the problem if and only if the correct response of 7 is given. Behavior modification also stresses immediate feedback so the correct stimulus-response will be associated together and false responses like $3 + 4 = 8$ will not become a part of what is learned. Research on effective teaching has shown that a system of Every Pupil Response (EPR) results in more learning (Hunter, 1985). Every student shows each answer immediately to the teacher. The teacher can catch misconceptions and/or sloppy errors quickly before the student associates an incorrect response with the question asked by the teacher. Figure 2.2 shows a variety of teaching material that can be used to see EPR results quickly during a lesson.

Immediate feedback is an important component of *programmed learning*, and is defined as a series of small steps to which a student is asked to respond with immediate feedback on the answer's correctness. While programmed learning has been presented on simply constructed learning machines for many years, computers have made programmed learning accessible to more and more school children in a more efficient manner.

This text includes many examples of how the computer programming languages of BASIC and Logo can be used to aid mathematics instruction. The programs are on a computer disk provided by your instructor. If you do not have a disk, each program can be typed on the computer as printed in each activity. The number of spaces left between typed letters and punctuation are important in both programming languages. Type the examples exactly as they are printed. The following computer activity allows the reader to experience immediate feedback.

Individual Chalkboards
or Plasticized Sheets

Yes – No Cards
or = and ≠ Cards

Top corner shows the
Desired Response
When the Student
Holds It Up

Multiple Choice Cards

Figure 2.2

Activity

COMPUTERS

Computer Exploration in BASIC

To Start a BASIC Program from a Disk

1. Press the CAPS LOCK key down so that the typing will be in all capital letters.

2. Place the disk BASIC side up in the disk drive slot and turn on the computer.

3. When the red light goes out, type the words: LOAD and the name of each program. The program is named ADD so the words

        LOAD ADD

   would be typed and then press the return key.

4. The red light will go on as the program is being loaded from the disk into the computer.

5. When the red light goes off, type RUN and press return to see what the program does.

6. When you have finished, call up another program by following the steps 1–5. If you wish to end the computer work, remove the disk first and then turn off the computer.

7. Refer back to these instructions as you work through the text.

Directions for the Program Named ADD

1. The purpose of the following program is to see the use of immediate feedback in computer learning. The program will allow students to add basic number combinations given randomly six at a time.

2. Follow the preceding instructions to start the program. Then go to step 3 for further explanation of how to use the program. (If you have no disk, type the following program:)

```
10 PRINT "ANSWER EACH ADDITION"
20 PRINT "PROBLEM."
30 PRINT
40 LET I = 1
50 LET A = INT(9 * RND (1)) + 1
60 LET B = INT(9 * RND (1)) + 1
70 LET C = A + B
80 PRINT A;" + ";B;" = ?"
90 INPUT Z
100 IF Z = C THEN 130
110 PRINT "THAT IS NOT CORRECT."
120 GOTO 140
130 PRINT "GREAT JOB!"
140 PRINT
150 LET I = I + 1
160 IF I > 6 THEN 190
170 PRINT "HERE'S ANOTHER ONE."
180 GOTO 50
190 PRINT "THIS IS THE END."
200 END
```

3. Type RUN and try the program. Answer some incorrectly to see what happens. Line 160 tells the computer to stop the program after six problems have been given. If a teacher wants to give more problems to students six can be changed to any desired number.

4. This program will be used again in Chapter 7 when the basic facts are studied.

Now the same program will be seen as written in the Logo programming language so that the reader can compare and contrast the two styles of programming.

Activity

COMPUTERS

Computer Exploration in Logo

To Start a Logo Program from the Disk

1. Follow steps 1–3 from page 10 to start Logo. Type ERALL and press return to erase all current Logo commands before starting a new program. This prevents the statement, OUT OF SPACE, from appearing on the monitor.

2. This program is saved under the name ADD on the Logo disk you have received from the instructor. Type:

```
LOAD "ADD
```

Only one space between the words and only *one* double quotation mark (") at the beginning of the name is required. If you do anything else, you will get an error message and you will have to start over again. Follow the directions in step 3.

(If you have no disk, type both TO ADD and TO EXAMPLES as printed here. Each time you type END the computer will respond with a phrase "_____ DEFINED". When this happens, Logo is telling you that the procedure you just typed now has meaning and is ready to be used.)

```
TO ADD
TEXTSCREEN
MAKE "N "1
PR [ANSWER EACH ADDITION]
PR [PROBLEM.]
PR []
EXAMPLES
END

TO EXAMPLES
MAKE "A "1 + RANDOM 10
MAKE "B "1 + RANDOM 10
TYPE :A TYPE "+ TYPE :B TYPE "=
MAKE "C :A + :B
MAKE "X FIRST READLIST
```

(continued)

```
IF :X = :C [PR [GREAT JOB !]]
IF :X < :C [PR [NOT CORRECT.]]
IF :X > :C [PR [NOT CORRECT.]]
MAKE "N :N + 1
IF :N > 6 [PR [THIS IS THE END.]]
IF :N > 6 [STOP]
PR [HERE'S ANOTHER ONE.]
PR []
EXAMPLES
END
```

3. Type ADD and the program will give you six problems to answer. Again, answer some incorrectly to see what the program does. Look at the procedure TO EXAMPLES as printed here. If you want to give a student more than six problems, what part of the procedure should you change?

4. This program will be used again during the study of basic facts in Chapter 7.

There are far more elaborate programs than the two you have just seen. Some programs can branch to probable causes for an error and tutor students to help them to understand their mistakes. However, even these relatively simple programs show the value of immediate feedback for students. By saving these two programs on an initialized disk of your own, you will have two programmed learning activities ready for your own elementary classroom. The directions for saving programs may be found in Appendix B for Logo. For BASIC, insert your disk into the drive and type SAVE ADD and the program will be saved under that name on the BASIC disk.

Another researcher in the area of behavior modification is Albert Bandura. Bandura's *Social Learning Theory* (1977) stresses observational learning where a response is linked with a stimulus after a person sees the consequences of another person's responses. The person responds by imitating the model if the consequences are viewed as desirable. Much is found in current educational literature about *cooperative learning*, which is based on the idea that students can learn from each other, coordinating their efforts to complete learning tasks (Slavin, 1983, 1985). Both parties will imitate each other's successful responses and model new concepts learned from the other, especially if there is a cooperative group and individual incentive (reinforcement) structure.

Information Processing Theories

Information processing is the study of how humans encode, store, and retrieve information. The theory uses the computer as its model. Designers created computers based on what was known about how humans acquire information. Computers, in turn, can help researchers gain insight into the more minute ways in which humans encode, store, and retrieve information (Kirby, 1984). Information processing is the newest of the three schools of thought and it is premature to designate definitive leaders. Two aspects of information processing are thought processing and learning styles.

Thought processing is defined as the strategies used to organize and classify new information or skills to obtain order out of a confusing series of stimulus events. Two of these strategies have a direct relationship to how children perceive mathematics. *Simultaneous thought processing* requires stimulus material to be presented all at once (simultaneously), seeing the whole before its parts. A person begins to look for patterns and relationships to break down the whole into its respective parts to arrive at appropriate solutions. *Successive thought processing* requires stimulus material to be presented from one component part to the next (successively), leading from detail to detail until the whole is seen. A person begins to look for patterns and relationships between details, building the respective parts into the whole to arrive at correct solutions.

At the beginning of this chapter Table 2.1 presented the three schools of thought on learning theories. Readers who prefer simultaneous thought processing may find themselves referring to the table as they read these paragraphs to see how all of the material fits together. Other readers may have profited little from it and skipped immediately to the paragraphs that outlined details of each theory. Such readers would be using successive thought processing. Material has been presented using both thought processing modes in this text. Perhaps you will be able to analyze your own processing preferences as you work in the text.

Some people may adopt processing styles depending on their perception of how difficult the material is from task to task. Figures 2.3, 2.4, 2.5, 2.6, 2.7, and 2.8 show typical mathematical problems seen in elementary and middle school books and worksheets. In each set, both problems are teaching the same concept. The ones on the left of each set require students to process the information by:

1. counting the objects one by one until the entire set is calculated, or
2. constructing the information one part at a time to get the answers.

This is successive synthesis—the part-to-whole processing strategy.

The problems on the right require students to process the information by:

1. receiving the entire set of answers all at once, deciding which sets do not belong, and partitioning them away from the whole, or
2. seeing the entire relationship pattern that one part has to another.

Shade the part to show the fraction.

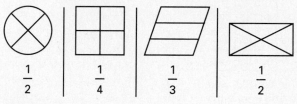

Mark the fraction for the shaded part.

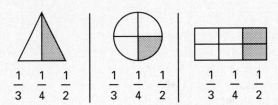

Figure 2.3

Find the answer to each picture.

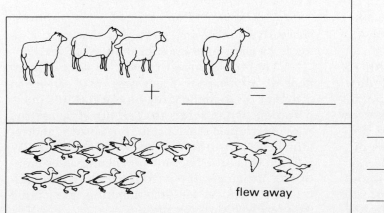

Finish all the number combinations using all of the picture below.

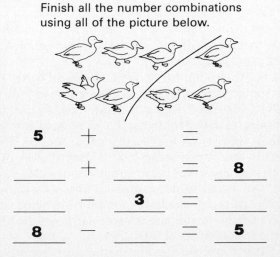

Figure 2.4

(+)	(−)
9 + 3 = ___	12 − 8 = ___
7 + 1 = ___	9 − 4 = ___
2 + 3 = ___	10 − 2 = ___
10 + 4 = ___	7 − 3 = ___
8 + 4 = ___	8 − 5 = ___
6 + 6 = ___	10 − 4 = ___

(12)	(4)
(9 + 3)	12 − 8
10 − 2	3 + 1
6 + 6	10 + 4
7 + 3	2 + 3
8 + 4	9 − 4

Answer each number combination in the + and − circle.

Find all the combinations that equal the circled number.

Figure 2.5

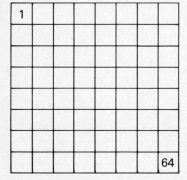

1							
							64

1	2	3	4	5	6	7	8
9	10	11	12	13	14	15	16
17	18	19	20	21	22	23	24
25	26	27	28	29	30	31	32
33	34	35	36	37	38	39	40
41	42	43	44	45	46	47	48
49	50	51	52	53	54	55	56
57	38	59	60	61	62	63	64

Write the numerals from 1 to 64 in the blocks going across. What patterns do you see with eights and fours?

Look at the filled-in chart above. What patterns do you see with fours ? With eights ? What addition combinations can you see easily?

Figure 2.6

This is simultaneous synthesis—the whole-to-part processing strategy.

Some students are very confused with all the material seen at once as shown on the right side of Figures 2.3 to 2.8. They may know the answers to the problems but they may not be able to show what they really know because processing the information in this format is so difficult. Other students are equally confused when they are asked to determine an answer without seeing all the possible responses from which to deductively reason the answer as shown on the left side of Figures 2.3 to 2.8.

The diagnosis and remediation section of each chapter will present ways to teach difficult mathematics concepts using simultaneous and successive processing strategies that may help students learn mathematics more effectively.

The term learning styles is another component of information processing that can mean different things depending on the author's interpretation. Dunn refers to learning style as "the way individuals concentrate on, absorb, and retain new and different information or skills" (1985). This broad definition encompasses emotional, sociological, environmental, psychological, and physical factors, including perceptual data received through the senses.

It is beyond the scope of this text to include all of the above factors. Therefore, we have chosen to comment only on that which we feel has a salient relationship to the study of

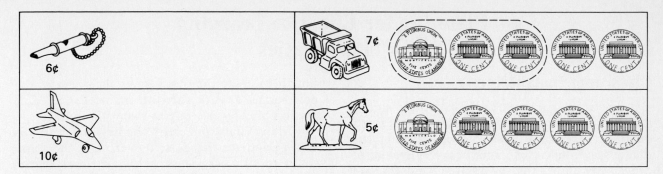

Draw the circle to show the coins you would need to buy each toy. Put a 1, 5, or 10 in the middle of the coin to show a penny, nickel, or dime.

Figure 2.7

Circle the correct amount of coins needed to buy each toy.

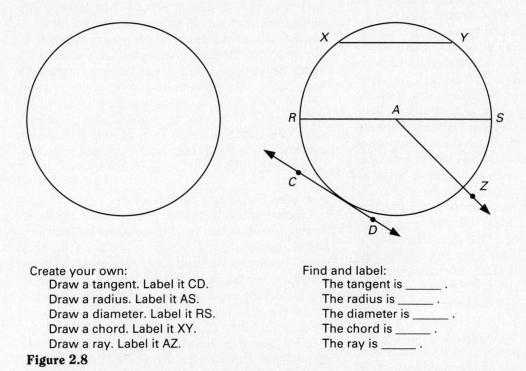

Create your own:
 Draw a tangent. Label it CD.
 Draw a radius. Label it AS.
 Draw a diameter. Label it RS.
 Draw a chord. Label it XY.
 Draw a ray. Label it AZ.

Figure 2.8

Find and label:
 The tangent is _____ .
 The radius is _____ .
 The diameter is _____ .
 The chord is _____ .
 The ray is _____ .

mathematics—the perceptual learning style. *Perceptual learning style refers to a person's preference for material presented through one or more of the five senses.* The visual, auditory, and tactile modalities are the ones most appropriate for the study of mathematical concepts.

Many researchers believe that education does not currently have the expertise or the empirical instruments to identify learners by per-

ceptual learning style (Glass, 1983). The authors of this text agree that identification of such learners and prescriptive teaching for them may be premature. Therefore, a variety of visual, auditory, and tactile techniques and processing strategies will be shared throughout the text in the hope that teachers will use a variety of learning aids to reach as many students as possible when teaching difficult mathematical concepts.

Planning for Effective Learning of Mathematics

Classroom Instruction

Classroom instruction can be structured so that students attain a greater understanding of mathematics. Research in lesson plan presentations, diagnostic interview techniques, and evaluation procedures add to the successful elementary and middle school experience in mathematics.

Studies (Good and Grouws, 1979; Good, Grouws, & Ebmeier, 1983; and Confrey, 1986) have found that teachers who plan five sequencial steps in lesson presentations and use a dialogue format known as elaborating techniques have students with higher mathematics achievement. Known as the Missouri Mathematics Program (MMP), the five-step format has proven more successful than traditional approaches. Research (Bracey, 1986) has also shown that cooperative learning produces higher math achievement in computation and that teachers preferred cooperative learning and MMP over other methods when compared in research settings. Glasser (1987) advocates classrooms where students have a sense of control over the environment and a sense of power to aid their own learning. This concept fits well with cooperative learning since everyone is contributing to the success of all members. The authors recommend that cooperative learning and MMP be used together to foster effective teaching of mathematics.

The five-step format (adapted from the MMP) can be used as a start to planning an effective mathematics lesson. One exercise at the end of each chapter will ask the reader to plan a model lesson in which the following five steps are included. Some states mandate actual time to be spent in math instruction. In such cases, the times suggested for each of the following steps should be increased or decreased proportionately to meet the state mandate.

- **Step 1: Review**
  The teacher and students review what was covered in the previous day's lesson for no more than ten minutes at the beginning of the lesson. This may include going over homework, but other methods of review are encouraged as well. Practice with mental computation and estimation are also possibilities to use here. Ideas for mental math and estimation are presented throughout the text and may be used in this part of the lesson. Mondays may be spent in a more general review of the past week's work. Every

fourth Monday may be a general review of the past month's work for skill maintenance. This session may last as long as twenty minutes.

Many educators claim the first five minutes of a period are the most powerful in the sense that attention is keenest at this time. Active participation should be included if possible, in which every child is expected to give a response.

- **Step 2: Development**
  The teacher presents a new idea or an extension of a previous mathematics concept. Students should know the objectives of the lesson and have what is commonly referred to as "an anticipatory set" about the goals of lesson. Explanations and discussion with active interaction between teacher and students are presented. Demonstrations with examples of a concrete, pictorial, and symbolic nature are included. Some educators recommend at least 50 percent of the class time be spent on development. If this is done, it is wise to combine development with controlled practice to make sure students are following the presentation of new material when longer time slots are involved.

- **Step 3: Controlled Practice**
  Students are asked to respond to a series of examples while the teacher watches carefully to see if any misconceptions occur. It is during this part of the lesson that the EPR materials are most beneficial for the teacher and students. Both development and controlled practice may be interwoven together, allowing a total time of twenty minutes. Guidelines for the day's cooperative learning activity can be given at this time. They should include specific details on responsibilities for group and individual rewards based on math achievement of the lesson material.

- **Step 4: Seatwork**
  Students work on their own or in cooperative learning groups to practice or extend the learning concept originally presented by the teacher in step 2.
  Approximately fifteen minutes of a forty-five minute period should be spent on step 4.

- **Step 5: Homework**
  The MMP format calls for assignments to be made at the end of each math class so students

learn to listen for them. The authors feel this plan should be followed only on days when reinforced practice is necessary. Homework should never be given unless the teacher is certain the students will practice using correct procedures. It takes much longer to unlearn an inappropriate or incorrect technique than to learn it correctly in the first place.

Homework assignments should include one or two review problems. The homework assignment is an excellent time for students to explore answers to problem-solving questions where more thought time may be needed for quality answers. Less time may be needed in the primary grades and more time may be needed in the middle school grades, but it is important that students not become fatigued with the amount of homework to the point that negative learning takes place.

Middle school teachers need to be aware that even twenty minutes of math homework coupled with homework from five or six other academic classes can add up to a large amount of work per night. This is an area where math across the curriculum ideas could be utilized, coordinating with what is being studied in the other academic areas. Coordinating homework schedules with colleagues may also be beneficial.

Good and Grouws also found that teacher observation of students' overt behavior while they worked was not sufficient to see if students truly understood a mathematical concept. If students were allowed to share their thoughts as they worked, mathematical achievement rose. Therefore, effective classrooms must include time for students to think aloud. This student interaction allows time for verifying answers, which works best in cooperative groups. Students who have trouble expressing their thoughts should be taught *elaborating techniques* (ways to guide their thought structures) to reach appropriate mathematical conclusions. An example of an elaborating technique may be helpful at this point. Some readers may have had no previous experience with the Logo computer language. A procedure like the one in the previous section could be frightening the first time it is tried. Here is an elaborating technique to help students understand some of the things Logo can do with numbers, using a short procedure. The student's words and the teacher's words should be read alternating back and forth as a dialogue.

Logo Program

```
TO FIND
MAKE "A "1
MAKE "B "2
PR :A + :B
END
```

Student's Elaborating Techniques

I can't tell what this program will do.

I see the word MAKE used more than any other word, but I don't know what the "A and "1 means.

I MAKE a picture with red paint.

Picture and red paint.

"1 tells what to do to A. So "2 must tell what to do to B, and A + B must be 1 + 2 which is 3. But I still don't know what PR means.

The program just answered 3. So PR must mean I get some kind of an answer

I was right.

With Teacher's Guidance

Key words are sometimes helpful in solving any mystery. What do you see?

Create a sentence with the word MAKE in it . . . just like you do in everyday language.

What were the important words after MAKE in your sentence?

So red paint told you what you would do to the picture. Now what do you think "A and "1 are doing in the Logo program? Use the same wording.

You could run the program and see if that helps you find out.

Yes, anytime you see PR, it means something is going to be done that you can see or it will be PRINTED so you can see the answer.

You were right.

Notice in the preceding dialogue, the teacher never told the answer but guided responses to: 1. promote transfer of learning from that which was familiar to that which was unknown, 2. direct problem solving on the student's own, and 3. guide the encoding process so that the student would associate PR with getting that which is seen on the screen. These three points are essential steps in developing the elaborating technique with students.

Classroom instruction must also encompass diagnostic interviews between learner and teacher. Mathematical achievement improves as teachers understand students' misconceptions about mathematics and move to remediate the problems before they become too complex. The diagnostic interview technique has proven helpful in assessing problem areas. *The diagnostic interview technique* is the teacher's observation of a student's thought process through questioning to learn the degrees of understanding and/or misconception of a mathematical concept or principle. Interviews may occur individually or in a group. It is recommended that teachers tape the interviews to improve their questioning technique over time. Audio or video tapes work well as long as they do not distract students from sharing their thoughts. The wait time between teacher question and student responses may require adjustment to meet the individual or group needs. There is no one ideal wait time, but it is widely known that most teachers wait only a second or two between responses. This is definitely not long enough.

Teachers should ask questions as students are working with a concept or principle. Teachers must remember that they are not correcting student misconceptions at this point; they are learning what the misconceptions are. Labinowicz (1985) suggests strategies to handle an interview with competence. Three of the strategies are:

1. Discouraging the "Parroting" Responses of Students
Students frequently give an explanation back to the teacher just as the teacher has worded it in a teaching session. To make certain that the student truly understands the concept, Labinowicz suggests that the teacher ask, "How would you explain this to a first grader who doesn't understand it?" (1985, p.29)

2. Asking Students to Justify Their Answers
Students may answer an example correctly but have no idea what the reasoning is behind it, and some students can reason correctly but still come up with erroneous answers. Sometimes they have trouble with both. Students should be asked to prove that their reason or their answer is the correct one.

3. Keeping a Student Elaborating on a Procedure or Reason
Teachers must remain non-committal and non-judgmental if they are to learn what students really think. Enough encouragement should be provided without implying that the student explanation is correct or incorrect.

Use words like   *"Keep going"*
                   "I'm listening"
                   "Tell me more"

rather than   "Okay"
                  "Good"
                  "All right"

Words like okay, good, and all right imply that the answers are correct. These responses should be avoided because a teacher does not want to imply that a wrong answer is correct when it will need to be retaught later, possibly confusing the student even more.

When a teacher moves from the role of observer to the role of teacher, it is helpful to use the pupil-teacher elaborating technique demonstrated earlier in the lesson plan presentation. It guides the student to correct responses without negating the work a student has done in the interview session.

Evaluation Procedures

Evaluation of student progress follows after lessons are taught and diagnostic interview techniques are used. *Evaluation procedures* are the methods of assessing students' cognitive understanding of mathematical concepts and principles. There are several ways to evaluate student progress. Those that are most commonly used in mathematics will be discussed in the following paragraphs.

Evaluation can be on different levels of the cognitive domain as identified in the *Taxonomy of Educational Objectives: Handbook I: Cognitive Domain* (Bloom, Engelhart, Furst, Hill, & Krathwohl, 1956). It is one thing to answer a memorized number fact, such as $3 + 4 = 7$, and quite another to evaluate the correctness of one's own created number combinations for the family of 7 without previous study of probable answers. The first example is on the lowest level of the cognitive domain, memorization, while the second example is on the highest level, evaluation. Bloom (1984) reports that over 90 percent of the evaluation activities in American schools are on the lowest level of the cognitive domain. With the movement toward excellence in education coming from the recommendations of four commission reports, namely the Carnegie, Holmes, National Education Association, and the National Governors Association (Keppel, 1986), failure to perform higher-ordered thinking skills has been taken to task. Activities, such as the problem solving discussed later in this chapter, are growing in importance for all educational levels, kindergarten through adulthood. Evaluation of objectives in the mid- to upper-ranges of the taxonomy are being stressed in more classrooms. Therefore, exercises at the end of each chapter in this text have been divided into three levels, emphasizing the middle and high levels of the cognitive domain.

Memorization and Comprehension Exercises
Low-Level Thought Activities

Application and Analysis Exercises
Middle-Level Thought Activities

Synthesis and Evaluation Exercises
High-Level Thought Activities

Further elaboration of the *Taxonomy of Educational Objectives* is left to the general methods courses in education. It is hoped that teachers will become familiar with how the taxonomy works in the evaluation of mathematics by performing activities in this book designed to assess the level of one's own thought development.

Another evaluation procedure that receives popular attention is mastery learning. *Mastery learning* refers to the approach which "gives pupils multiple opportunities to master goals at [their] own pace" (Biehler & Snowman, 1986, p. 579). The goals are written as observable instructional objectives, based on the levels of the taxonomy's cognitive domain. Specific terminal behaviors tell how well each objective is met. A criterion is set for each goal. If the student can perform the objective at the desired criterion level, the student is said to have mastered the goal. Local school systems and/or the state decide on which objectives will be placed at each grade level. Several states have threatened to withhold state funds from local public schools if a mastery learning plan is not followed, and still other states are demanding to see positive results from the money that has been spent on assessment procedures (Burnes & Lindner, 1985).

Mastery learning depends on a working knowledge of the scope and sequence in mathematics. *Scope* refers to the breadth or extent to which any one concept is studied. *Sequence* refers to the order of succession from one concept to another. In mathematics, instructional objectives are most often coordinated with the scope and sequence of topics generally agreed upon by writers of elementary textbooks. Appropriate objectives cannot be written without knowledge of the scope and sequence of topics. Each concept is traced from its inception in an early grade to its completion, which may be several years later in an upper grade. Every major elementary and middle school textbook series includes a scope and sequence chart for teachers.

Concepts may take many years to master fully. The first time a new concept is introduced, it is usually classified as a topic to which the student will have "exposure." Mastery is not expected at that point. Most concepts are allowed a minimum of a year's exposure time, although there is no one rule that can be applied.

A scope and sequence chart (based on several current elementary mathematics textbook series) will be presented in Chapters 3 through 12. It is hoped that teachers will familiarize themselves with the mathematics concepts to be taught at each grade level. Teachers need to know what concepts students have had in previous years if they are to meet the needs of students in the present grade. Beginning teachers often fail to explain concepts adequately because they assume that students have had more exposure to topics than they actually have had. Scope and sequence charts can also help teachers plan instruction for academically gifted students and remedial students. Both groups are likely to need concepts not included in the present grade level textbook. The charts tell the recommended order in which such topics can be taught. There are several simulation activities in this text that require the use of scope and sequence charts to aid in the planning of lessons and in diagnosing learning difficulties of elementary and middle school students.

Competency-based testing in mathematics is another evaluation procedure. Potential graduates are tested to determine if they possess the basic skills of mathematics needed to survive in society (Biehler & Snowman, 1986). These tests have also been called minimal competency tests because the examinations often emphasize low-level skills as the basic skills. While some states may continue to support competency-based testing, mastery learning, with its emphasis on multiple skill levels, is expected to gain more prominence as an evaluation procedure.

Traditional methods and interpretations of student evaluation are being challenged by the NCTM *Curriculum and Evaluation Standards for School Mathematics* (Commission on Standards for School Mathematics, 1987). It proposes that assessment is more than testing and should involve the processes of observing and conjecturing to determine what and how students think about mathematics. The NCTM curriculum standards convey a spirit of mathematics that emphasizes exploration and creation. Therefore, the standards are inseparable from including dialogues between teacher and student in order to assess thinking.

To help implement these evaluation standards, different assessment techniques are required. Testing instruments that ask for the identification of a single correct response will no longer be appropriate. Many assessment techniques are needed that include open interviews, teacher probing, observations of individual and group work, and observations of students communicating mathematics. Mathematics educators must search for valid, novel methods of assessing students' understanding of mathematical concepts. Resnick (in press) argues that it may be productive to consider mathematics not as a subject that is well-structured and tightly organized, but rather as a knowledge domain subject to interpretation and meaning construction.

Problem Solving

Problem solving is the oldest intellectual skill know to humanity. The ability to understand a problem, relate it to a similar problem or to past experiences, speculate about the possible solution, and carry through until the problem is solved is basic to human survival. Without the ability to solve problems, human beings would have become extinct. Even the caveman needed to solve problems dealing with food gathering and climate conditions.

Problem solving as an educational method has received a good deal of attention. Some people may remember solving mathematical problems dealing with how many hours two painters would require to paint a room if one worked twice as fast as the other. For many people, this is an unpleasant memory because they felt as though they were stumbling blindly through problems without much guidance from the teacher or the textbook. Sometimes teachers resist teaching problem-solving skills because of their own frustrating experiences with problem solving.

Teachers differ in their opinions of the best approach to teaching mathematics. Some may favor an algorithmic approach where all mathematical problems are reduced to a formula that can be solved by inserting and manipulating numbers. Other teachers believe that teaching mathematics through problem solving helps students develop mathematical skills and apply those skills to situations they encounter in their daily lives. Therefore, it is important to establish a definition of problem solving before proceeding.

The term problem solving is broad and refers to a complex of cognitive activities and skills. George Polya, a noted scholar in the area of problem solving theory, held that solving a problem is finding the unknown means to a distinctly conceived end (Polya, 1957). To solve a problem is to find a way where no way is known, to find a way out of a difficulty, to find a way around an obstacle, to attain a desired end that is not immediately attainable, by appropriate means.

The ingredients of problem solving are a problem with no immediately obvious solution and a problem solver who is capable of trying to find the solution by applying previously learned knowledge. Problem solving involves simple word problems, real-world applications, nonroutine problems and puzzles, and the creation and testing of mathematical conjectures leading to new discoveries (Branca, 1980). It is helpful to remember that one essential aspect of problem solving is basic logic. Consider that most story problems contain words like *and, not,* and *or,* and this can present a problem in the elementary classroom where logic is not usually taught. Yet, in order to solve problems, students must master logical means for finding a solution. Means for solving such problems will be presented later in this chapter.

There are three common methods of interpreting problem solving:

1. As a goal
2. As a process
3. As a skill

When problem solving is interpreted as a goal, it does not rely on particular procedures and methods, specific problems, or even mathematical content. Instead it becomes an end in itself. As a process, problem solving is seen as an opportunity to exercise certain methods, strategies, and heuristics. Heuristics involves the discovery of the solution to the present problem on one's own. Problem solving as a skill demands more attention to specific types of problems and methods of solution. The interpretation that the teacher brings to problem solving will determine the approach taken in the classroom.

Problems are classified generally into three types: open-ended, discovery and guided discovery. The open-ended question has a number of possible solutions, so the process of solving the problem becomes more important than the answer itself. Discovery questions usually have a terminal solution, but there are a variety of methods the student can use to reach the solution. Guided discovery questions, by far the most common type, include clues and even directions for solving the problem so that the student does not become overly frustrated and give up.

Whatever approach is taken, it is important to understand the process of solving problems. Polya (1957) identified four phases of the process:

1. The student understands the question and is motivated to answer it;
2. The student has learned facts and strategies that are useful in solving problems;
3. The student applies various strategies until the problem is solved;
4. The student checks the solution to see if it is correct. If not, another strategy must be tried.

Accepting Polya's phases as a working model, the teacher is presented with step-by-step instructions in teaching. The problem must be designed and posed in such a way that: a) the student is capable of deciphering clues and determining what information is being requested, and b) the student becomes involved in the problem and is interested in solving it. Many books and teaching aids on the market present more or less effective problem-solving exercises. A number of computer-assisted instruction programs rely on problem solving as a way of teaching organizational and decision-making skills. Other teachers can be an excellent source of advice about what types of problems are effective.

Teachers who want to create their own problem-solving exercises can help to ensure that they are effective by following several rules. Davis and McKillip (1980) recommend a set of standards to follow when writing problems or modifying problems that have been written by someone else.

Standards for Creating Problem-Solving Exercises

1. The level of difficulty should be appropriate to the age and skill level of students. Substitute smaller numbers for larger when working with younger students, for example.

2. The reading level should be appropriate to the age and reading level of students. It should be a simple matter to identify the essential information in a problem and restate it in language that students can understand. It may also be helpful to include key words in a vocabulary lesson to assure that students will recognize them when they appear in a problem. This also reinforces math across the curriculum when students use mathematics terminology as a part of the English lesson.

3. Subjects of interest to students should be used. Problems should relate to topics that are familiar to students. The social context of students can provide clues about the topics they are likely to find interesting. Another option is, if necessary, to reword problems so that they are up-to-date and relevant.

4. The given information and the goal of the problem should be identified carefully. What information does the student need to solve the problem? What information will the solution reveal? How much extraneous information, if any, should be included in the problem to heighten the challenge of solving it?

5. Acting out the problem or representing it visually may make it more compelling to students. How does the problem apply to everyday life?

Problem solving involves several aspects of learning and teaching—how to decode the problem (a reading skill) and how to translate the answer to a meaningful end (a writing skill). Many educators support the technique of having students write the numerical answer in a full sentence. This approach forces the students to reflect on their answer as they translate it. It may also provide an opportunity to consider the reasonableness of the answer. Asking students to verify and justify their answers in a written form often helps them clarify their thinking.

Strategies for Teaching Problem Solving

Strategies are methods by which a problem can be solved. The strategies used to solve problems are determined by two factors: 1. the skill and sophistication level of the student, and 2. the range of mathematical tools that the student has previously mastered. The degree to which the student is able to compare a problem to a familiar situation, problem, or experience tends to dictate which strategy the student will use. The more complex the problem to be solved, the more strategies may be required to solve the problem. Therefore, students need to learn as many strategies as possible to become effective problem solvers. Research (Suydam, 1982a) shows that students may become more proficient problem solvers by using various strategies. Eleven strategies are described here, although some educators have identified as many as sixteen (Suydam, 1982b). Sample problems are given for each strategy.

Estimation and Check: Estimation is the strategy of proposing an approximate answer to determine a range within which the solution might fall. The assumed answer is checked in relation to the solution. Estimation can be done on a daily basis in the classroom. Such a strategy is effective in two types of problem solving:

1. in problems where there is too little data to allow for elimination of unlikely answers, and
2. in problems that deal with very large unknown quantities.

Activity

COMPUTERS

ESTIMATION

Estimating the Size of the Logo Screen

Directions:

1. You will need only the Logo master disk to start working this problem.

2. It is very helpful to know how many turtle steps (number of movements) that the turtle can move in any direction. It will help you in planning several activities throughout the book. The turtle will wrap around and appear on the other side of the screen if you ask the turtle to move too far in any one direction.

3. Estimate how many turtle steps it will take to reach the edge of the screen in all four directions. (This is the case of not having enough information because you do not know how much each turtle step is worth on a metric scale.)

4. Check your estimate and see how many turtle steps you need. Place your answer on the monitor in Figure 2.9. You can keep it as a reference for later work.

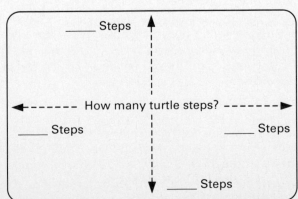

Figure 2.9

Looking for Patterns: Some problems are designed in such a way that the only way to solve them is to identify patterns in the data given in order to predict the data not given. Students need to practice examining given data to see whether it reveals a predictable pattern. Once the pattern is established, the student can calculate the unknown data in order to solve the problem. Putting data into table format will help show the pattern.

Example 1: The peg puzzle can be used to create patterns to make a generalization. In the activity, the object is to exchange the red and white pegs (Fig. 2.10) by moving in a prescribed order.

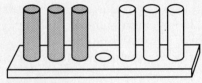

Figure 2.10

Directions:

1. Move a peg forward one space into an empty hole, jump one peg if it is a different color. Pegs can only move forward.
2. Play with a partner—one player moves the whole time while the other partner counts the moves. Then switch positions.
3. First start with one pair (a peg of each color), find the minimum number of moves to exchange pegs from one side to the other. Record in the following table. Try again with two pairs and find minimum number of moves. Look for a pattern in the table:

# of Pairs (x)	# of Moves (y)
1	3
2	8
3	15
4	?
5	?

Example 2: Students must be made aware that patterns are not always as projected from a few samplings. Figure 2.11 is an example of a time when the beginning pattern does not hold.

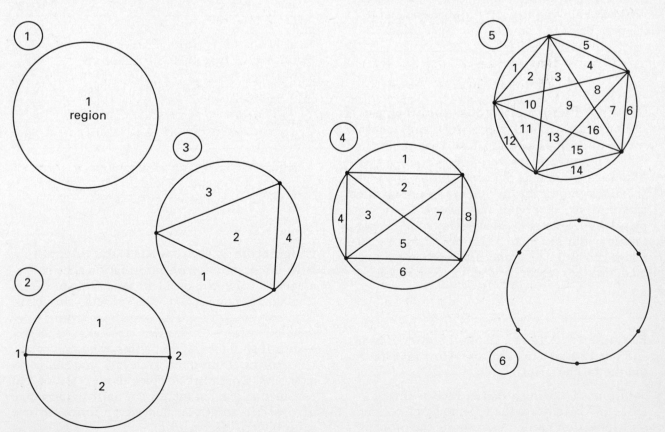

Figure 2.11

What is the maximum number of regions formed by connecting all the points in a circle?

# of Points	# of Regions
1	1
2	2
3	4
4	8
5	?
6	?

Find out what the next one would be.

Insufficient Information: Sometimes not enough information is supplied in the problem.

Example: How much will it cost to buy a 5 pound bag of dog food today if it cost $.20 less last week?

Solution: The insufficient information is what the actual cost of the dog food was last week. Without knowing it, $.20 cannot be added to the base price to find out this week's price.

Drawing Pictures, Graphs, and Tables: If students have difficulty grasping a problem that seems too complex or abstract, they might find it helpful to create a visual image to assist them. Aids such as drawing pictures, graphs, and tables provide a graphic means of displaying numerical data in a way that students can see. Graphs can help to demonstrate relationships among data that may not be apparent immediately. A visual image can also help students keep track of intermediate steps required to solve the problem.

Example 1:   A farmer planted 8 rows of beans and put 12 beans in each row. How many beans did the farmer plant?

Solution: Drawing a sketch such as this one will quickly indicate the repeated addition or multiplication needed for solving the problem.

Example 2:   Who won the bowling tournament? Who had the highest score?

Solution: Create a bowling average computer database (a computerized method of organizing and storing information that can be easily retrieved) to keep continuous, up-to-date records of bowling results each week. The information can be ordered using the database and the table is convenient for displaying the data to solve the problem(s). A sample database template (the outline of the records) follows:

		Bowling Average			
Name	Week	Game 1	Game 2	Game 3	Average
Bitter	1	155	185	175	172
Hatfield	1	182	175	180	179
Edwards	1	158	192	188	179
Smith	1	177	182	188	182
Bitter	2	182	189	188	186
Hatfield	2	182	178	168	176
Edwards	2	155	187	177	173
Smith	2	177	166	197	180

Elimination of Extraneous Data: Many problems are designed with information that is necessary to solve the problem as well as information that is not needed. The students' first task is to sort through the information given to determine what is necessary and what is extraneous. If students fail to do this, they may waste time trying to produce irrelevant and insignificant data. Operating on clues about relevant and extraneous data, students can narrow the range of possible solutions instead of trying to use meaningless information.

Example: Rambo is a 35-pound dog. He eats a 5-pound bag of dog food that costs $3.14 every week. How much does it cost to feed him for four weeks?

Extraneous information: Can the extra information be found?

Developing Formulas and Writing Equations: It is often useful to invent a formula into which one can plug numbers to arrive at an answer. Students learn that routine formulas often have real-world applications and that numbers stand for objects and concepts in a mathematical formula. It is helpful to give students exercises that require the translation of ideas into words and numbers.

Example: Translate "everyday for 6 weeks" into numbers.

Solution: $7 \times 6 = 42$

Modeling: Constructing a physical representation or model of the problem is another way of helping students to conceptualize the operations necessary to solve a problem. Students may even be more interested in a problem that they can manipulate manually, as evidenced by the popularity of puzzles in teaching arithmetic and geometry. Teachers who have access to microcomputers and graphics software can make good use of the modeling strategy. Students can be given problems that ask them to construct a variety of shapes. Computer graphics can be used to motivate students to take an active interest in the problem-solving process (Olivier and Russell, 1986).

Example: The Logo computer programming language will be used throughout the text. It allows students to do things such as:

> Create an octagon on the computer screen. Then divide it into eight equal triangles.

Solution: Students draw octagons and then experiment with drawing lines through the octagon to create triangles. This program appears in Chapter 3.

Working Backwards: Geometric proofs call for the strategy of working backwards. More common everyday problem solving calls for this skill as well.

Example 1: Prove that angle x is 60 degrees.

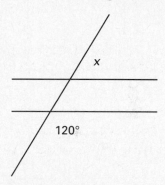

The answer is given, that x = 60 degrees. Now the student needs to work backwards to find out how to get the answer.

Example 2: The lumber truck left enough lumber to build three recreation rooms but only one-half of that is needed for even the biggest recreation room. This room will be only $\frac{2}{3}$ as large as the biggest recreation room. Will I have enough if I offer to give $\frac{5}{8}$ of it back?

Flowcharting: Borrowing the concept of flowcharting from the field of computers, teachers can assist students in visualizing the process of solving a problem by using a flowchart. A *flowchart* is a detail-by-detail outline of steps that must be taken and conditions that must be met before the solution is reached. It is not necessary to use the various symbols that programmers use in flowcharting. If presented like the example in Figure 2.11, the shape of the boxes can help young children decide where to place each decision. Children can be shown the process of problem solving by a chart with directional arrows, showing the way one step leads to another or requires testing for a certain condition.

Activity

Problem Solving

Example:

Draw a flowchart that shows all of the sequential steps required to buy an ice cream cone.

(Note: This kind of activity has also been featured in elementary reading texts, an area supporting math across the curriculum.)

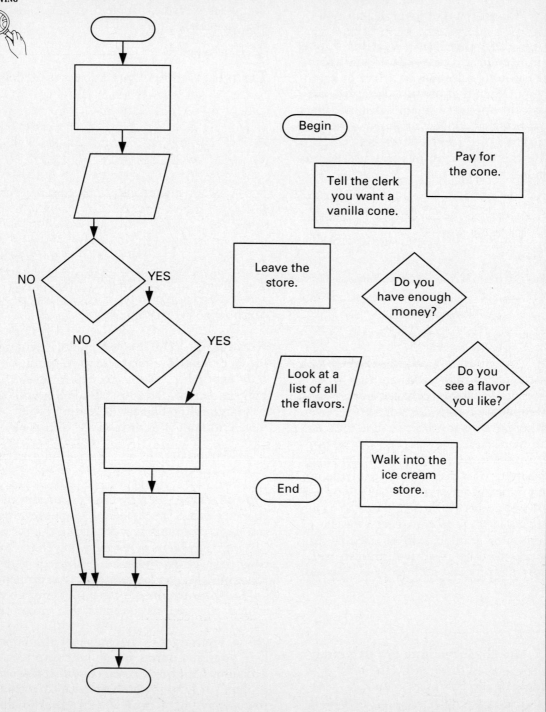

Acting Out the Problem: This is similar to the modeling strategy, although it differs in that it does not depend on physical objects or visual aids that students can manipulate. Students are more likely to view the problems as real-life situations if acting is encouraged and it will be easier for them to see the steps involved.

Example: If a salesperson sells $1.50 worth of groceries to one customer, $3.25 to another, and $2.75 worth of groceries to a third customer, how much has the salesperson sold altogether? What is the total dollar value if that same amount is sold every day for four days?

Solution: Students take the roles of the salesperson and the customers in the store. They may actually exchange play money to simulate earnings and expenditures, but the visual aids are not necessary for the students. They can see just by acting that the process is one of addition for the salesperson and one of subtraction for the customer.

Simplifying the Problem: Substitute smaller numbers that can be handled with quick estimation skills so students can test the reasonableness of their answers before doing the problem with the original numbers. Have students substitute basic facts or easy number combinations so they can get a sense of the appropriateness of the operation chosen.

Example: A person bought a car for 25 percent less than the original price of $3495. How much was paid?

Solution:

Think: 25 percent off $100 would be $75.

Think: 25 percent off $1000 would be $750.

Think: 25 percent off $3000 would be 3 × $750 = $2250

Each answer fits into the general pattern,

$$.25 \times \text{old price} = x$$
and
$$\text{old price} - x = \text{sale price}$$

Then the original numbers can be supplied. Another way to approach this problem would be to substitute 10 percent of $3000. Think about how that answer is obtained, then work up to 25 percent of $3000, and finally to the actual numbers.

Effective Teaching of Problem Solving

Effective teaching of problem solving involves techniques, other than the strategies discussed, to make students feel more comfortable with the problem-solving experience (Suydam, 1982). The key principles are underlined below and may be used as a checklist for teachers:

• *Choose problems carefully, paying special attention to interest and difficulty level.* Students are not motivated to solve problems that they find boring, irrelevant, too easy, or too difficult. Other teachers can be excellent advisors on appropriate problems to present. Collect students' work when they are asked to generate word problems and use these settings as examples.

• *Put students in small groups and allow them to work together on problem-solving exercises.* Some teachers have found that students are more successful at problem solving when they work in pairs or teams (Biglan and Kirkpatrick, 1986). Cooperative learning enables the exchange of ideas about which strategies to use and provides assistance in estimating and testing results. Each student needs to experience several ways of conceptualizing the problem and selecting strategies. Cooperative learning establishes an environment that supports problem solving. "A small group structure has the potential to maximize the active participation of each student and reduce individual isolation. When organized in small groups, more students have the opportunity to offer their ideas for reaction and receive immediate feedback. This provides a setting that values social interaction, a needed element of children's learning." (Burns, 1984, p.41)

• *Identify wanted, given, and needed information.* Before students can solve a problem, they must understand the three types of information that are involved in the problem itself:

a. Wanted information—the solution
b. Given information—presented in the problem
c. Needed information—information not presented in the problem but required to solve it

It may be useful for students to create a list or table showing what data applies to the three categories above. Include the likely source of the needed information since students cannot solve a problem if they do not have access to the information required.

• *Pose the problem in such a way that students clearly understand what is expected of them.* No matter how motivated students are, they will not be able to solve problems unless the problem and accompanying instructions are stated clearly and simply. Ask students to repeat directions to check for understanding. Foster an environment in which they feel comfortable asking questions when they do not understand. If necessary, introduce them to vocabulary words that are used in the problems. This also supports English as a subject for math across the curriculum.

• *Present a wide range of problems.* Students may become bored if they are expected to work similar problems repeatedly. They should practice a strategy or problem type only until it is mastered. At that point, it is time to introduce a new type of problem so that students continue to feel challenged.

• *Present problems often.* Make problem solving a frequent part of class instruction so that students do not see it as an isolated skill but as an ongoing, familiar, and necessary process.

• *Provide opportunities for students to structure and analyze problems.* Discuss the makeup of a problem as well as its components. Students will develop a language for discussing strategies and patterns for analyzing problems that would not be possible otherwise.

• *Provide opportunities for students to solve different problems with the same strategy.* This will provide the practice necessary for students to master a given strategy. Students will also develop an important sense that strategies are flexible and can be applied to a wide range of situations. Also encourage students to solve the same problem with different strategies. This makes them aware that they have choices about how to approach a problem.

• *Help students select an appropriate strategy for a particular problem.* At first, students have little experience on which to base judgments about which strategies are most effective with specific types of problems. Trial and error are frustrating ways to choose a strategy. If students are assisted in selecting the best strategy and in defining the qualities that make it effective, students gain experience that will aid them later when making judgments on their own.

• *Help students recognize problems that are related.* Students need to be trained to see structural relatedness of problems. Teachers need to create related problems by varying the data and condition of a single problem. Ask students to list key words and features of the problems to identify their similarities.

• *Allow students plenty of time to solve problems, discuss results, and reflect on the problem-solving process.* Students need to be able to discuss their methods and rationales as a way of organizing and processing their experience. As mentioned earlier in this chapter, research shows that elaborating techniques are rewarded by higher achievement in mathematics.

• *Demonstrate to students how they can estimate and test their answers.* This can save students the frustration of wasted time and effort in problems where the range of results is not immediately obvious. Students also need to be shown how they can work a problem backwards or use other methods of testing their answers.

• *Discuss how the problem might have been solved differently.* With many problems, a variety of strategies will result in correct solutions. Ask students what other methods they could have used after a problem is solved successfully. Help students see that various approaches to the problem are acceptable. The selection of strategies depends, in part, on learner style and preference. Various strategies that can be used with different learning styles will be mentioned where appropriate throughout this text.

Teachers can use the following checklists to periodically rate themselves and their students on the necessary skills to become good problem solvers and teachers of good problem solvers.

Good Problem Solvers
- understand concepts, terms
- note likenesses, differences—analogies
- identify critical elements
- note irrelevant details
- evaluate and select alternative solution routes
- estimate, approximate, and check for reasonableness
- switch methods readily
- generalize from few examples
- learn from mistakes
- have less anxiety, more confidence
- transfer learning to similar problems
- remember mathematical structure of problem—forget context, details (Meiring, 1980)

Teaching Strategies for Problem Solving
- present many, varied problems
- teach variety of problem-solving strategies, plus overall plan
- give opportunities to analyze problem situations
- encourage using a strategy to solve many problems, and many strategies to solve one problem
- have students determine question asked, necessary and unnecessary information, process: discuss why appropriate
- have students seek similarities across problems
- provide time for discussion, practice, reflection
- have students solve some problems in small groups
- have students estimate, analyze estimates, test reasonableness of answers
- have students use dramatization, manipulatives, models, pictures, diagrams, charts, tables, graphs as aids to solving problems
- teach students to select main idea, make inference, construct sequences

- help students to simplify a problem
- vary the working in problems of the same type
- consistently ask questions encouraging involvement and thinking
- encourage students to look back and reflect on solved problems (Meiring, 1980)

Metacognition has been suggested as an aid for problem solving. *Metacognition* is one's knowledge of how one's own cognitive processes work. Good problem solvers can learn from observing their own actions and mental processes as they work on problems. The following checklist is taken from Capper (1984).

A Good Problem Solver:
1. Reads and rereads the problem to ensure clarity in problem understanding.
2. Plans and schedules appropriate problem-solving activities.
3. Separates the important from the unimportant information in the problem.
4. Breaks the problem down into smaller units.
5. Connects information to what is already known.
6. Organizes the information contained in the problem.
7. Supplies missing or implied information.
8. Keeps track of time limits.
9. Monitors and evaluates progress. Recalculates solution or verifies recorded data or idea units within the problem.

Capper (1984) cites research to show that elementary and middle school students can improve their problem-solving skills if a teacher makes them routinely aware of their own thought processes in relation to the preceding chart.

Summary

No one learning theory or method of instruction will suit every situation all the time for every person. An effective teacher will use all theories wisely, coordinating the difficulty of the task with stages of cognitive development and a student's individual learning style and thought processing preference. While consideration of all the factors may look like an impossible task for any teacher, conscious practice of several approaches over time will help a teacher become secure and familiar with balancing many factors at once.

One factor is problem solving. The benefits of introducing students to problem-solving strategies early and consistently are many. Flexibility of thought to see more than one correct answer is one advantage. Problem solving is an effective means of making mathematics more relevant to students. Formulas into which numbers are plugged to calculate an answer may not appear to students to have much to do with their daily lives. However, if the same formula can be integrated into problems that are interesting to students, students develop a sense that mathematical principles are valuable in solving problems that arise in the real world.

Problem solving integrates all areas of the curriculum since it draws on reading, writing, social studies, economics, science, etc. when developing real-world examples. Math across the curriculum helps the marriage of words and numbers, both of which must be understood and processed before a problem can be solved. The subjects of problems can convey information about nearly any topic, and students can learn a variety of facts while they are applying strategies to solve problems. Perhaps most importantly, students acquire the basic problem-solving skills and methods that will serve them well throughout their lives.

Exercises

Directions: Read all questions *before* answering any one exercise. Frequently the last question in one category leads to the first question in the next category.

A. **Memorization and Comprehension Exercises**
   Low-Level Thought Activities
   1. The overwhelming research showing what works in education supports the idea that children learn mathematics better if _____ is/are presented first.
      a. Problem solving
      b. Computer-assisted instruction
      c. Concrete manipulations
      d. All of the above
      e. None of the above
   2. Which learning theorist developed his theory from studying the evolution of mathematics?
   3. Name the problem-solving strategy that could be best used for each of the following examples:
      a. A train picked up passengers at the following rate: 1 at the first stop, 3 at the second stop, 5 at the third stop, and so on. How many passengers got on the train at the sixth stop?
      b. Translate "Five hours at $10.00 per hour" into numbers.
      c. How many different ways can five squares be arranged so that if two squares touch, they border along a full side?
   4. State the definition of problem solving in your own words.
   5. List the stages of Piaget, Bruner, Baratta-Lorton, and Heddens. Compare and contrast the theories.
   6. Name the three schools of thought in learning theory discussed in this text and state major contributions of each one to mathematics education.
   7. Look at a unit of instruction from a current elementary or middle school math textbook. Classify how many pages require simultaneous thought processing or successive thought processing to complete the task successfully. Does the textbook favor any one processing strategy over another? If so, what implications does this have for students?
   8. List the four steps in problem solving presented by Polya.

B. **Application and Analysis Exercises**
   Middle-Level Thought Activities
   1. Using the material gathered in A7 above, change a page requiring one processing strategy to the other processing strategy. Sketch how the new page would look.
   2. Apply the use of an EPR to teach the concept, $5 + 2 = 2 + 5$. Show three different ways to make an EPR.
   3. Apply the three points needed for a successful elaborating technique to teach one of the problem-solving strategies found in this chapter.
   4. Pick one grade level and analyze an elementary or middle school textbook to see what problem-solving strategies are used the most in that grade level. Find three word problems that could be solved by more than one strategy. Show the solutions using different strategies. Label each strategy as you use it.
   5. Analyze the information presented in this chapter, making a list of the important components to include in preparing mathematics instruction for elementary and middle school students.
   6. Using the evaluation form found in Appendix A, select commercial software for any three topics from this chapter. In the

selection, include drill and practice, simulation, tutorial, and problem-solving software. For each software program selected, describe how you would integrate it into a lesson plan.

7. To reinforce math across the curriculum:
   a. List five mathematical questions/problems that come from popular storybooks for children. Show how they are related to the topics of this chapter.
   b. List five ways to use the topics of this chapter when teaching lessons in reading, science, social studies, health, music, art, physical education, or language arts (writing, English grammar, poetry, etc.).

C. Synthesis and Evaluation Exercises
High-Level Thought Activities

1. After analyzing all the information in B5 above, create (synthesize) a model lesson plan for a mathematics concept of your choice. Evaluate how all the components can work together. Include as many of the following components in the plan as possible:
   a. The five-step lesson plan format
   b. Elaborating techniques for the concept
   c. Cooperative learning
   d. Every Pupil Response (EPR)
   e. An idea for math across the curriculum
   f. Concrete, pictorial, symbolic material in the correct sequence
   g. Find or make worksheets that include both simultaneous and successive thought processing strategies
   h. Plan activities for the auditory, visual, and tactile learner to include in the development or controlled practice part of the lesson plan
   i. Diagnostic Interview Strategies to help the students who may have difficulty with the concept
   j. List interactive computer software that could be used with the concept. Reviews of software programs may be found in:
      i. Teaching material catalogs
      ii. Reference section of the library
      iii. Professional journals and magazines (there are many available) some include:
         Teaching and Computers
         Classroom Computer Learning
         The Journal of Computers in Mathematics and Science Teaching
         Computers in the Schools
         Electronic Learning
         The Computing Teacher
   k. Plan at least one problem-solving activity that could be incorporated in the lesson, stressing one of the eleven strategies
   l. List three mastery learning objectives on which students will be evaluated at the end of the lesson. (i.e., What do you want the students to be able to do after you have taught the lesson that they could not do before?) List them as observable behaviors.

2. Evaluate a textbook series using the form in Appendix E. See what current trends in problem solving, computers, calculators, etc., are included in the books.

3. Using the class presentation form in Appendix E, present a lesson on problem solving, stressing two of the strategies discussed in the chapter.

4. Using the class observation form in Appendix E, observe a teacher working with an elementary or middle school class and evaluate the ingredients that made the lesson an effective one.

5. Using the teacher interview form in Appendix E, interview a teacher on the successful ways he/she has found to teach mathematics.
   (Notice that higher-level thinking skills are frequently multiple step procedures requiring a high degree of coordination of what has been learned previously.)

6. How do children learn mathematics and what are the techniques effective teachers use to help children solve problems?

Bibliography

Bandura, Albert. *Social Learning Theory*. Englewood Cliffs, N.J.: Prentice-Hall, 1977.

Biehler, Robert F., and Jack Snowman. *Psychology Applied to Teaching*. Boston: Houghton Mifflin, 1986.

Biglan, Barbara, and Susan Kirkpatrick. "Using the Computer to Enhance Problem-solving Skills." In *NECC 1986 Proceedings.* San Diego: National Educational Computing Conference, 1986.

Bitter, Gary G. "Microcomputer-Based Math Fitness" *Technological Horizons in Education* 15 (November, 1987): 106–109.

Bloom, Benjamin S. "The Two Sigma Problem: The Search for Methods of Group Instruction as Effective as One-to-One Tutoring." *Educational Researcher* 13 (1984): 4–16.

Bloom, Benjamin S., Max D. Engelhart, Edward J. Furst, Walker H. Hill, and Davis R. Krathwohl. *Taxonomy of Educational Objectives: Handbook I: Cognitive Domain.* New York: McKay, 1956.

Bracey, Gerald W. "Yay, Team! Different Teaching Strategies and Mathematics Achievement." *Phi Delta Kappan* 67 (January 1986): 392.

Branca, Nicholas A. "Problem Solving as a Goal, Process, and Basic Skill." In *Problem Solving in School Mathematics,* 1980 Yearbook of the National Council of Teachers of Mathematics, 3–7. Reston, Va.: National Council of Teachers of Mathematics, 1980.

Bruner, Jerome S. *Toward a Theory of Instruction.* New York: Norton, 1966.

———. *In Search of Mind: Essays in Autobiography.* New York: Harper & Row, 1983.

Burnes, Donald W., and Barbara J. Lindner. "Why the States Must Move Quickly to Assess Excellence." *Educational Leadership* 43 (October 1985): 18–20.

Burns, Marilyn. *The Math Solution: Teaching for Mastery through Problem Solving.* Sausalito, Calif.: Marilyn Burns Education Associates, 1984.

Capper, Joanne. *Research into Practice Digest* Vol. I, No.1a, and Vol II, No.1b. *Thinking Skills Series: Mathematical Problem Solving: Research Review and Instructional Implications.* Washington, D.C.: Center for Research into Practice, 1984.

Castaneda, Alberta D., Glenadine Gibb, and Sharon A. McDermit. "Young Children and Mathematical Problem Solving." *School Science and Mathematics* 82 (January 1982): 22–28.

Chance, Paul. "Master of Mastery." *Psychology Today* 21 (April 1987): 43–46.

Commission on Standards for School Mathematics of the National Council of Teachers of Mathematics. *Curriculum and Evaluation Standards for School Mathematics.* Working draft. Reston, Va.: the Council, 1987.

Confrey, Jere. "A Critique of Teacher Effectiveness Research in Mathematic Education." *Journal for Research in Mathematics Education* 17 (November 1986): 347–360.

Davis, Edward J., and William D. McKillip. "Improving Story-Problem Solving in Elementary School Mathematics." In *Problem Solving in School Mathematics,* 1980 Yearbook of the National Council of Teachers of Mathematics, 80–91. Reston, Va.: National Council of Teachers of Mathematics, 1980.

Engelhardt, Jon M. "Diagnostic/Prescriptive Mathematics as Content for Teacher Education." *Focus on Learning Problems in Mathematics* 10 (Winter 1988): 47–53.

Fennell, Francis (Skip). "Early Childhood and Mathematics Concerns Regarding Teacher Preparation." *School Science and Mathematics* 82 (January 1982): 11–21.

Glass, Gene V. "Effectiveness of Special Education." *Policy Studies Review* 2 (January 1983): 67–74.

Glasser, William. *Control Theory in the Classroom.* New York: Harper & Row, 1986.

Good, Thomas L., and Douglas A. Grouws. "The Missouri Mathematics Effectiveness Project: Experimental Study in Fourth-Grade Classrooms." *Journal of Educational Psychology* 71 (June 1979): 335–362.

Good, Thomas L., Douglas A. Grouws, and Howard Ebmeier. *Active Mathematics Teaching.* New York: Longman, 1983.

Gough, Pauline B. "The Key to Improving Schools: An Interview with William Glasser." *Phi Delta Kappan* 68 (May 1987): 656–662.

Heddens, James W. "Bridging the Gap between the Concrete and the Abstract." *Arithmetic Teacher* 33 (February 1986): 14–17.

Hunter, Madeline. Keynote address of the opening session conducted at the 40th Annual Conference of the Association for Supervision and Curriculum Development, Chicago, March 23, 1985.

Johnson, Marietta B., and Grace M. Burton. "Developing Number Concepts Through the Seasons." *School Science and Mathematics* 86 (November 1986): 559–566.

Keppel, Francis. "A Field Guide to the Land of Teachers." *Phi Delta Kappan* 68 (September 1986): 18–23.

Kirby, John R., ed. *Cognitive Strategies and Educational Performance.* New York: Academic Press, 1984.

Kouba, Vicky L., Catherine A. Brown, Thomas P. Carpenter, Mary M. Lindquist, Edward A. Silver, and Jane O. Swafford. "Results of the Fourth NAEP Assessment of Mathematics: Number, Operations, and Word Problems." *Arithmetic Teacher* 35 (April 1988): 14–19.

Krulik, Stephen, and Robert E. Reys, eds. *Problem Solving in School Mathematics.* 1980 Yearbook of the National Council of Teachers of Mathematics. Reston, Va.: National Council of Teachers of Mathematics, 1980.

Kulm, Gerald. "Seeing Relationships between Problems." *Math Lab Matrix* (Fall 1978): 2–3.

Labinowicz, Ed. *The Piaget Primer: Thinking, Learning, Teaching.* Menlo Park, Calif.: Addison-Wesley, 1980.

———. *Learning from Children: New Beginnings for Teaching Numerical Thinking, a Piagetian Approach.* Menlo Park, Calif.: Addison-Wesley, 1985.

Martinez, Michael E., and Nancy A. Mead. *Computer Competence: The First National Assessment,* Educational Testing Service, April 1988.

McKnight, Curtis C., et al. *The Underachieving Curriculum: Assessing U.S. School Mathematics from an International Perspective.* Stipes Publishing Company, 1987.

Meiring, Stephen P. *Problem Solving . . . a Basic Mathematics Goal,* vol. 1. *Becoming a Better Problem Solver,* vol. 2. *A Resource for Problem Solving.* Columbus, Ohio: Ohio Department of Education, 1980.

Mitchell, Charles E., and Grace M. Burton. "Developing Spatial Ability in Young Children." *School Science and Mathematics* 84 (May 1984): 395–405.

Olivier, Terry A., and Rebecca Gaye Russell. "Using Low-Resolution Graphics to Develop Problem Solving Skills." *The Computer Teacher* 13 (May 1986): 50–51.

Polya, George. *How to Solve It: A New Aspect of Mathematical Method.* 2d ed. Princeton, N.J.: Princeton University Press, 1957.

Ralston, Anthony, et al. *A Framework for Revision of the K–12 Mathematics Curriculum,* Task Force Report submitted to the Mathematical Sciences Education Board, National Research Council, January 1988.

Reinboldt, Werner C. *Future Directions in Computational Mathematics, Algorithms, and Scientific Software.* Society for Industrial and Applied Mathematics, 1985.

Resnick, Lauren B., *Education and Learning to Think.* Washington, D.C.: National Academy Press, 1987.

——— (in press). "Teaching Mathematics as an Ill-structured Discipline." In *Research Agenda for Mathematics Education: Teaching and Assessing Mathematical Problem Solving,* edited by Randy Charles and Edward A. Silver. Reston, Va: National Council of Teachers of Mathematics. [copublished with Lawrence Erlbaum Associates, Hillsdale, N.J.]

Rutherford, James, et al. *What Science Is Most Worth Knowing?* Draft Report of Phase I, Project 2061; American Association for the Advancement of Science, December 1987.

Schultz, Karen A. "Representational Models from the Learners' Perspective." *Arithmetic Teacher* 33 (February 1986): 52–55.

Skinner, B. F. *The Technology of Teaching.* New York: Appleton-Century-Crofts, 1968.

Slavin, Robert E. *Cooperative Learning.* New York: Longman, 1983.

———. *Learning to Cooperate, Cooperating to Learn.* New York: Plenum Press, 1985.

Smith, Lyle R. "Verbal Clarifying Behaviors, Mathematics Students Participation, Attitudes." *School Science and Mathematics* 87 (January 1987): 40–49.

Stevenson, Harold W., et al. Mathematics Achievement of Chinese, Japanese, and American Children, *Science* 231 (14 February 1986) 693–699.

Suydam, Marilyn N. "Update on Research on Problem Solving: Implications for Classroom Teaching." *Arithmetic Teacher* 29 (February 1982): 56–60.

———. "The Problem of Problem Solving." *Problem Solving* 4 (September 1982): 1–2, 5.

Tunis, Harry B., ed. "Manipulatives." [Focus issue]. *Arithmetic Teacher* 33 (February 1986).

United States Department of Education. *What Works: Research about Teaching and Learning.* Washington, D.C.: United States Department of Education, 1986.

3

Geometry

Key Question: *What are the key concepts of geometry that are needed by elementary and middle school students, and how can a teacher develop these concepts in the classroom?*

Activities Index

Introduction

Geometry provides many concrete experiences that can be used in the teaching of mathematics. Introducing geometry early in this book will inspire teachers to use the many objects in the world around us to teach mathematics. When considering topics included in the study of mathematics, geometry may be overlooked or mentioned much later than other areas. Geometry has not always been part of the scope and sequence of an elementary mathematics program. Textbooks prior to the 1960s generally included only measures of area and volume. There is still no consensus on which concepts should be presented at the various grade levels. Without a standard, nationally accepted core curriculum for geometry, it is difficult to address the problem. There are some teachers who see mathematics as primarily "arithmetic" and would choose to skip geometry but would never skip a chapter on long division with larger numbers. According to the results from the Second International Mathematics Study (McKnight et al., 1987, p. vii), "Achievement in geometry for the U.S. was among the bottom 25 percent of all countries, reflecting to a large extent low teacher coverage of the subject matter."

Students' performance on the National Assessment of Educational Progress (NAEP) items from the geometry section indicate that if an item could be solved visually performance was better than if the problem required more abstract thinking (Kouba et al., 1988). When seventh-grade students were given pictures of line segments, over two-thirds could determine if the line segments would form a triangle. Without the pictures, less than 10 percent could answer the question. This performance difference may indicate a need for help in applying their informal geometric knowledge to formal situations. The NAEP results indicate that seventh-grade students did well on some items involving identification of figures, but their performance was not uniform (Kouba et al., 1988). When given a choice of figures, 90 percent correctly identified parallel lines, but only 33 percent identified perpendicular lines. Another item asked students to identify a quadrilateral that was not a parallelogram. The choices were pictures of a square, a rectangle, a parallelogram, and a trapezoid. Very few students chose the parallelogram or rectangle. Over one-half of the students chose the trapezoid, the correct response, but one-fourth chose a square. It might be said that, " . . . seventh-grade students do not have a fully developed concept of the inclusion relationships among quadrilaterals; that is, they do not understand that squares are rectangles, rectangles are parallelograms, and thus, squares are parallelograms" (Kouba et al., 1988).

There remain many unresolved issues in the problems related to school geometry. A discussion of these and suggestions for their resolution are given by Usiskin (1987). He presents findings related to our students' continued poor performance, lack of geometry knowledge, and dimensions of geometry related to curriculum problems. "Geometry is too important in the real world and in mathematics to be a frill at the elementary school level or a province of only half of all secondary school students" (Usiskin, 1987, p. 30).

Informal Geometry

Many of us may think of geometry as a study of axioms, postulates, proofs of theorems, constructions, and so on. What geometry are we talking about: motion geometry, solid geometry, plane geometry, Euclidean geometry, or another type? The important issue is not the name, but rather the type of experiences we intend children to have as part of the elementary school geometry curriculum. Experiences in geometry should allow for the intuitive investigation of concepts and relationships. These activities should provide rich backgrounds and solid foundations for the generalizations about geometric relationships that come during the middle school grades. The study of geometry should encourage children to explore a variety of geometric concepts. This results in a study of "informal geometry."

The study of geometry is important for many reasons. One of the most important reasons is to develop adequate spatial skills. During childhood, children respond to the three-dimensional world of shapes as they play, build, and explore with toys and other materials. These early geometry experiences are useful in developing spatial abilities. These abilities must be nurtured through geometric activities. Spatial skills include interpreting and making drawings, forming mental images, visualizing changes, and generalizing about perceptions in the environment. Research by Guay and McDaniel (1977) noted a high positive relationship between the level of achievement in mathematics and spatial abilities with students of elementary school age. Two articles that suggest ways to improve spatial abilities in children are mentioned in the bibliography (Young, 1982 and Mansfield, 1985). Young says that our knowledge of the world is influenced by how we interpret and organize visual stimuli. He believes that children need activities dealing with proximity, separation, order, and enclosure. Mansfield urges the teaching of projective geometry to improve spatial abilities. This, in turn, will promote the ability to reason, to predict, and to represent knowledge in appropriate ways. Developing spatial skills is also an important part of everyday life.

Another important benefit from geometry is that it provides opportunities to develop logical thinking and reasoning. As Burns states, "Appropriate geometry experiences are useful for developing reasoning processes which in turn support the problem-solving skills children need to understand arithmetic as well as geometric concepts." (1984, p. 79). Many practical experiences in our lives involve problem-solving situations that require a knowledge of geometric concepts, such as making frames, determining the amount of wallpaper, paint, grass, or fertilizer to buy, and other work situations.

Geometry provides a fun, welcome break from the routine of arithmetic. Activities related to geometry help to keep children interested in mathematics. Some children who are not academically capable and have a low interest in other routine, tiring aspects of mathematics may show sparkle and enthusiam during the unit on geometry. For these children and for the aforementioned reasons, it is an unfortunate mistake when a teacher treats geometry as an optional branch of mathematics and chooses to skip this topic. In order to teach informal geometry effectively, teachers must be familiar with geometry and its extensive glossary of terms.

Geometry Terminology Review

Terms in geometry, as with anything else, tend to be forgotten when not in use over a period of time. The review shown in Table 3.1 appears in table form to reinforce vocabulary and definitions that will be helpful to know when working with students in the elementary and middle school.

Table 3.1   Geometry Terminology Review

Name of Shape or Term	Description
Polygon	A plane (simple), closed figure with *n* number of points, *n* sides, and *n* interior angles. (In elementary geometry, it is usually required that the sides have no common point except their endpoints.) EXAMPLE: No common point except endpoints / Other common points
Regular Polygon	Sides are congruent and interior angles are congruent. Congruent / Non-congruent
Triangles	Polygons with 3 sides and 3 interior angles.
Kinds of triangles	
Equilateral	All 3 sides and its 3 interior angles are equal.
Isosceles	Only 2 sides and 2 interior angles are equal.
Scalene	No sides and no interior angles are equal.
Right	One right angle. (The side opposite the right angle is called the hypotenuse and the other 2 sides are called the legs of the right triangle.)
Obtuse	Has one interior angle greater than 90 degrees.
Acute	All interior angles less than 90 degrees.
Quadrilaterals	Polygons with 4 sides.
Kinds of quadrilaterals	
Square	Has 4 right angles and 4 equal sides.
Rectangle	Has 4 right angles with opposite sides congruent and parallel. (All squares are rectangles.)

(continued)

Name of Shape or Term	Description
Rhombus	Has 4 equal sides with opposite sides being parallel. (A square may be considered a special case of the rhombus by most mathematicians.)
Parallelogram	Opposite interior angles congruent and opposite sides are congruent and parallel.
Trapezoid	At least one pair of opposite sides are parallel and nonparallel sides may be congruent. (If congruent, it is called an isosceles trapezoid.)

The relationship of quadrilaterals may be seen in the following Venn diagram:

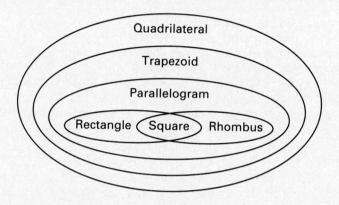

Other Polygons

Pentagon	Has 5 sides and 5 interior angles.
Hexagon	Has 6 sides and 6 interior angles.
Heptagon	Has 7 sides and 7 interior angles.
Octagon	Has 8 sides and 8 interior angles.
Nonagon	Has 9 sides and 9 interior angles.
Decagon	Has 10 sides and 10 interior angles.
Dodecagon	Has 12 sides and 12 interior angles.

Point	Non-dimensional location in space.
Path	Set of points passed through in going from one point to another. (curved, broken, straight)
Line	A straight path that is endless in both directions.
Line segment	A straight path.
Ray	Set of points contained in the infinite extension of a straight path in one direction beyond an endpoint.
Angle	Union of two rays with the same starting point. (The point is known as the vertex.)
Interior Angles	Angles made by adjacent sides of polygon and lying within the polygon.
Perpendicular lines	Lines that intersect at right angles.
Parallel lines	Lines that will never intersect.
Plane	A surface such that a straight line joining any two of its points lies entirely on the surface.

Scope and Sequence

Kindergarten
- Identify circular, square, triangular, and rectangular shapes.
- Identify cubes, cylinders, rectangular, prisms, and spheres.
- Identify objects that are the same length and determine which object is longer, taller, or shorter.
- Solve problems by recognizing and extending patterns.

First Grade
- Identify circles, squares, rectangles, and triangles.
- Identify congruent figures.
- Identify cubes, cylinders, cones, spheres, and rectangular prisms.
- Identify shapes of same size.

Second Grade
- Identify circles, squares, rectangles, triangles, points, cubes, cylinders, cones, spheres, and rectangular prisms.
- Identify congruent figures.
- Identify lines of symmetry.
- Find the perimeter or area with grid paper.

Third Grade
- Identify solid shapes.
- Identify line segments and polygons.
- Identify congruent line segments and figures.
- Identify lines of symmetry.
- Identify plane figures.
- Find the perimeter and area.

Fourth Grade
- Identify lines, line segments, parallel and intersecting lines, rays, and angles.
- Identify and name polygons.
- Identify congruent figures.
- Identify lines of symmetry.
- Find the perimeter and area of figures.
- Solve problems involving area and perimeter.
- Identify solid figures.

Fifth Grade
- Identify parallel, perpendicular, and intersecting lines.
- Identify plane and solid figures.
- Measure angles; classify angles and polygons.
- Identify similar and congruent.
- Identify lines, rays, angles, sides, and vertices.
- Find perimeter and area of figures.

Sixth Grade
- Measure angles with protractor.
- Identify parallel and perpendicular lines.
- Identify plane and solid figures.
- Classify lines, angles, triangles, and quadrilaterals.
- Identify similar figures.
- Identify congruent lines, angles, and figures; identify corresponding parts and lines of symmetry.
- Find area and perimeter.
- Use a compass for constructions.
- Identify translation and reflection images.

Seventh Grade
- Identify solid and plane figures.
- Identify corresponding parts of congruent figures.
- Measure and classify angles; classify polygons, quadrilaterals, and triangles.
- Identify properties of parallel and perpendicular lines.
- Determine the measure of corresponding parts of similar triangles.
- Identify and determine lines of symmetry.
- *Construction:* angle bisector, using protractor, compass, parallel and perpendicular lines.
- *Transformations:* reflections, translations, dilations.
- Find perimeter of polygons.
- Find area of rectangle, parallelogram, triangles, and trapezoid.
- Find surface area of prisms.

(continued)

Eighth Grade
- Identify solid and plane figures.
- Measure and classify angles; classify polygons, quadrilaterals, and triangles.
- Identify properties of parallel and perpendicular lines.
- Identify congruent triangles according to their corresponding parts. (SSS, SAS, ASA)
- Identify corresponding parts of similar figures.

- *Constructions:* lines, congruent triangles, polygons, bisectors.
- *Transformations:* reflections, translations, dilations.
- Identify lines of symmetry.
- *Right triangles:* tangent, ratios, hypotenuse, and legs.
- Find area, perimeter, and surface area.

Teaching Strategies

Activities for informal geometry are characterized by using manipulative materials, exploring concepts in both two- and three-dimensional space, and discovering relationships. Patterns are provided in Appendix C for attribute blocks (pieces), geoboard, metric ruler, protractor, and compass—all used with plane geometry. Patterns for the polyhedra used in the discussion on solid geometry are also included in Appendix C. The focus is to provide experiences that will produce meaningful definitions and properties of geometric ideas across time. For example, after numerous experiences with materials, the child will accumulate facts and discover the properties of a square. With a teacher's guidance, the child's thinking will gradually sharpen so that when a definition is needed, it will hold meaning for the child. As you read through the rest of this chapter, examine the activities and the process thinking underlying them. Keep in mind that children need many experiences from a variety of materials to ensure the development of geometric concepts.

Two Dutch educators, Pierre van Hiele and Dina van Hiele-Geldof, studied children's acquisition of geometric concepts and the development of geometric thought. These findings hold many ideas that should be considered when reading this chapter. The van Hieles concluded that children pass through five levels of reasoning in geometry in much the same way that Piaget said children must proceed through the stages in cognitive development (Chapter 2). The five levels are described by Burger and Shaughnessy (1986) as:

- *Level 0—Visualization:* The student reasons about basic geometric concepts, such as simple shapes, by means of visual considerations. A square is a square because it *looks* like one.

- *Level 1—Analysis:* The student reasons about geometric concepts by means of an informal analysis of the parts and attributes. A square is a square because it has four equal sides and four right angles.
- *Level 2—Abstraction:* The student logically orders the properties of concepts, forms abstract definitions, and can distinguish between the necessity and sufficiency of a set of properties in determining a concept. A square's definition is dependent on some properties that are related to other shapes. The student sees that a square can be both a rectangle and a parallelogram.
- *Level 3—Deduction:* The student reasons formally within the context of a mathematical system, complete with undefined terms, axioms, an underlying logical system, definitions, and theorems. This is the level needed to perform well in a high school geometry class.
- *Level 4—Rigor:* The student can compare systems based on different axioms and can study various geometries in the absence of concrete models.

The van Hieles (Crowley, 1987) proposed that progress through the five levels is more dependent on instruction than on age or maturation. They submitted five sequential phases of learning: inquiry, directed orientation, explication, free orientation, and integration. According to the van Hieles, instruction developed according to this sequence promotes the acquisition of a level. Examples of level-specific student behaviors can be found in Burger and Shaughnessy (1986). Research has shown that materials and teaching can be matched to the levels and promote growth (Burger, 1985, and Burger & Shaughnessy, 1986).

The first three levels (0–2) should be experiences in informal geometry during the elementary and middle school years. The van Hieles asserted that children should have a wide variety of exploratory experiences. Children should move through those levels with an understanding of geometry that will prepare them for the deductive study of geometry in high school and perhaps college. Many students experience great difficulty handling high school geometry and it may stem from the fact that inadequate experiences were provided in previous grades. The deductive study of geometry should be delayed until the student has developed the mental maturity required for this study. Unfortunately this is not often the situation, so the student takes the class, receives low grades, develops a negative attitude, and may not choose to proceed further in mathematics classes.

This chapter discusses two main aspects of geometry: the study of two-dimensional objects known as plane geometry, and the study of three-dimensional objects known as solid geometry. As you read this chapter, consider how you can relate two-dimensional shapes to three-dimensional shapes and what models you can use to teach the goals of geometry in an intuitive manner.

Plane Geometry

Plane geometry is the study of two-dimensional figures, their properties and relationships. Early activities could include classifying or sorting cut-out shapes or commercial materials such as pattern blocks or attribute blocks (also called attribute pieces). The child needs concrete models of the shapes to feel the shape and relate the name to its properties—number of edges, corners, and other attributes.

Attributes:	Values:
Size	Large, small
Shape	Circle, square, rectangle
Color	Red, yellow, green, blue
Thickness	Thin, thick

An important aspect of teaching geometry is to have students became familiar with the properties of shapes. A good activity is to have students find representatives of shapes in their environment. For example, the borders of the following could be selected: the file cabinet's sides are rectangles; the ceiling tiles are rectangles; the floor tiles are squares; the wastepaper basket rim is a circle; the flag is a rectangle; the traffic sign for yield is a triangle; the stop sign is an octagon. Give children a set of shapes to sort into some classifying scheme—shapes with three sides and the others; shapes with four corners and the others; shapes with no corners and the others. With attribute blocks, or cut-out shapes of different colors and sizes, many other classifying systems are possible. The following activity offers one way to reinforce properties of shapes as well as logical reasoning skills.

Activity

PROBLEM SOLVING

One and Two Difference Trains

Materials: Attribute blocks (or similar cut-outs see Appendix C)

Procedure: Place the attribute blocks in a pile on the floor or table. Start with one block. Have students take turns adding one block to the train so that each block is placed next to one that is different from it in just *one* way. Have the student verbalize the difference as each block is added. For example, the starting block is a green, small triangle. The first player adds a green, *large* triangle. The next player adds a *yellow* large triangle. The next player adds a yellow, large *square.* Play until all blocks are in the train. This may require some rearranging, which is allowed at the end of the round. Players can also see if they can get the end of the train to join to the beginning block. This may also require some rearranging.

Variation: Do the above activity and have each piece differ from the others by two attributes. For example, start with a thick, red, large square. The next player places a thick, *green, small* square in the train—two attributes are different (color, size). Thickness may be achieved by pasting attribute pieces on styrofoam, painted to match the color of each piece.

Making shapes on a geoboard offers one of the best opportunities to explore the properties of shapes. A geoboard is usually a 6- x 6-inch board or plastic square with 25 nails or pegs placed into a 5 x 5 array. A geoboard pattern is provided in Appendix C. The reader may wish to plasticize or laminate the pattern in Appendix C or use it as a model when making a wooden geoboard. (Other arrangements of nails or pegs are possible, such as circular, isometric or other dimensions of an array.) The most common arrangement is shown in Figure 3.1. Rubber bands can be stretched around the nails or plastic pegs to create shapes or designs. This is an invaluable teaching device for geometry as it allows children to create shapes, designs, and patterns to express ideas about geometry in a quick, accurate manner. Where it is impossible to ask a young child to make a square or octagon, the geoboard quickly provides such experiences.

Some teachers fear that rubber bands may become projectiles and have found that using the term "geobands" helps reduce the temptation. Whatever procedure you use, provide adequate opportunities for free exploration with the geoboard. Have older children copy their design or creation onto geometric dot paper (a replica of the 5 x 5 array using large dots for the nails), thereby keeping a record of their work. The following are some activities for using a geoboard with children.

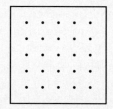

Geoboard in 5 X 5 Array
Figure 3.1

Activity

MANIPULATIVES

Using a Geoboard

Procedure:

Copy This Shape

Hold up a large card with a shape on it. Then ask children to copy the shape onto their geoboards. Discuss the name and properties of the shape. Then make a shape on a transparent geoboard using a projector and have the children copy the shape. Rotate the shape (Figure 3.2) and see if the children still call it by the same name. Many children have a fixed mindset about the way some shapes should look and if the geoboard is rotated, the shape changes in their mind.

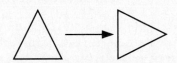

This triangle ⟶ if rotated
may not appear as the same
triangle to some children.
Figure 3.2

Make This Shape

Ask students to make a shape with four sides. Have them compare their shapes and discuss the names and possibilities. Ask children to make a shape with four equal sides, a figure with six sides, or a figure that touches six nails. Make the largest shape possible. Make the smallest shape possible. Take the smallest shape and lift the rubber band from one corner and stretch it around another nail. What shape is this? How many sides does it have? Make a figure with five corners. Make a figure with the same number of sides and corners.

The possibilities are endless for this activity. Allow children to show their shapes to the class and always discuss the names and properties. Many different shapes are possible so encourage sharing.

(Look at the Geometric Terminology Review, Table 3.1, and see which shapes can be created on a geoboard.)

Dividing Our Shapes

Have children make the largest square possible on their geoboards. Then have them take another rubber band and divide the shape into two equal (congruent) regions. What are the names of the regions formed? Use another rubber band to divide one of the regions again. Now what shape is formed? Have children show the different possibilities to other classmates.

How Am I Classifying?

Have students make a shape on their geoboards. Then decide a way to classify them into two groups. Call children to the front and ask them to put their geoboard figure into the appropriate set. Line the geoboards in the chalk tray. Have children determine how the teacher is classifying the shapes. For example, one set of geoboards might be figures with the geobands touching six nails and all other geoboards are in the other set. Another example is that one set of geoboards consists of figures with four sides and all others are in the other set.

Logo computer programs can help students see that the same shape can be rotated in many different ways around a central point. This is a good beginning activity to see what Logo can do as well.

Activity

COMPUTERS

PROBLEM SOLVING

Rotating Shapes with Logo

Directions:

1. This program is saved under the name SHAPES

2. Type LOAD   "SHAPES and call up each individual shape one at a time as described in step 3.

    (If you have no disk, type the following procedures:)

```
TO TRI
REPEAT 3 [FD 40 RT 120]
END

TO HEX
REPEAT 6 [FD 40 RT 60]
END

TO SQ
REPEAT 4 [FD 40 RT 90]
END

TO CIR
REPEAT 36 [FD 10 RT 10]
END
```

(continued)

3. Type:

```
TRI CS HEX CS SQ CS CIR
```

just to see what each will do.
Type CS to clear screen after each set of examples.

4. To see how Logo rotates shapes, type: RT 45 TRI
Now type: LT 120 TRI

5. Do the same thing with HEX, SQ, and CIR, moving the turtle a different number of degrees to the left and right to watch the figures rotate.

6. Beautiful designs can be created when shapes are rotated continuously, from a starting point, coming full circle 360 degrees to where the figure began.
Type: REPEAT 8 [RT 45 SQ]
(Notice that the number of degrees the square turns each time (45) has to match the number of turns (8) so that 45 x __ = 360 to get the figure around to its starting place.)

7. What would you type in the blank to bring the triangle design around to its starting place?

```
REPEAT___[RT 30 TRI]
```

Try it and see if you were correct.

8. Try designs like the one you created in step 7 using the HEX and CIR procedures.

9. If you have a color monitor, type: SWIRL and see what happens. The procedure will run on a monochrome monitor; it will just not be quite as spectacular. SWIRL is made up of the following commands. If you have no disk, type the following:

```
TO SWIRL
SETPC 2 REPEAT 12 [RT 30 TRI]
SETPC 3 REPEAT 8 [RT 45 HEX]
SETPC 4 REPEAT 12 [RT 30 TRI]
SETPC 5 REPEAT 8 [RT 45 HEX]
END
```

10. The Logo disk has two special procedures in its startup file called CIRCLER (for circle right) and CIRCLEL (for circle left). All you need to do is type: CIRCLER 30   and see what happens.

11. Now type: CIRCLEL 30   Now type: CIRCLEL 60
What do you think the 30 and the 60 measure, the radius or the diameter? Try additional circle commands using different numbers and see if you can answer this question.

Another way to have children experience the properties and attributes of shapes is by making shapes with yarn and soda straws. The straws can be cut into various lengths to form other shapes such as scalene triangles, parallelograms, and trapezoids. Cut straws in half if smaller shapes are easier for students to manage. Thread six or eight straws onto the yarn. Have the children take the number of straws needed for the shape; for example three straws will form a triangle, and push the other straws apart from them on the yarn. See Figure 3.3 for additional ideas about this device.

As relationships are explored, it might help to put the properties of the geometric figures into chart form. Write the name of the polygon, the number of sides, and the number of angles. Children in upper elementary grades classify tri-

There are two types of symmetry—line and rotational. The activities discussed earlier are examples of line symmetry. A figure with line symmetry has a line of symmetry, which can be determined and the exact same image appears on both sides of the line. To test for lines of symmetry, use a mirror, a Mira (a commercial device of red plexiglass), or have the figure traced and folded. To test for line(s) of symmetry using a mirror or Mira, place the mirror or Mira on the figure and move it until half the figure is reflected on the mirror or plexiglass. This portion should coincide with the portion of the figure behind the mirror. To test for symmetry using the folding method, have students fold the traced figure until the two halves match exactly. The tracing can also be flipped about the line of symmetry. Students can investigate familiar objects for symmetry (windows, wheels, human body) and will find many objects have more than one line of symmetry.

Symmetry can also be explored using geoboards and pattern blocks with older children. The figures or designs can be tested with mirrors and the figures can be reproduced with geometric dot paper or pattern block paper and folded to show lines of symmetry. Letters of the alphabet can be used for younger children.

Activity

Symmetry in Letters and Words

Directions:

1. Symmetry can be found in letters of the alphabet or with some words (i.e., wow). Use a small rectangular mirror and determine which students have names that begin with a capital letter with one line of symmetry or two lines of symmetry.

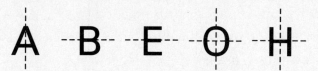

Lines of Symmetry for Letters

2. Ask the following questions:

   Whose initials are symmetrical? Which letters have rotational symmetry?
   Which shapes have an infinite number of lines of symmetry?

3. Have children look at things that have line symmetry—buildings, people, pictures, leaves, plants, and cars.

Renshaw (1986) suggests activities for having students examine the symmetrical properties of well-known trademarks or logos, such as Westinghouse Electric Corporation, General Electric Company, and others. For example, a tracing of the Chrysler Corporation trademark shows five lines of symmetry.

A figure that has rotational symmetry can be rotated or turned (less than a full turn) around a point until a matched figure results. An example of a figure with rotational symmetry is a square.

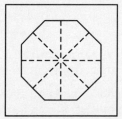

Testing for Lines of Symmetry

A quarter turn will produce a square, a half turn will produce a square, and a three-quarters turn will produce another matched square figure. The Chrysler Corporation trademark when tested for rotational symmetry has four turns that produces the matched figure. Examine patterns, decorations, or designs to find examples of rotational symmetry. Renshaw (1986) gives many suggestions about experiences for line, rotational, and point symmetry.

Angles of rotation can also be discovered with older children. Logo presents an excellent

Rotational Symmetry in Chrysler Symbol

way to work with these concepts in the classroom:

Activity

COMPUTERS

ESTIMATION

Angle Rotations with Logo

Directions:

1. This program is saved on the Logo disk under the name ANGLES.

2. Type: LOAD "ANGLES   and follow the directions in step 3.

   (If you have no disk, type in the following procedures:)

   ```
   TO ANGLE :D
   FD 40 BK 40 RT :D
   FD 40 BK 40
   END
   ```

   (The 0 in SETPC 0 is a zero, not the letter, O. It must be typed that way.)

   ```
   TO ROTATE :D :R
   HT REPEAT :R [ANGLE :D LT :D SETPC 0 ANGLE :D]
   END
   ```

3. Try making several angles. Place CS command between each example to clear the screen. Here are some angles to start with:

   ```
   ANGLE 90
   ANGLE 75
   ANGLE 120
   ANGLE 180
   ```
   ANGLE 180 — Notice that the line is extended down to show that 180 degrees is a straight line.

   ```
   ANGLE 360
   ```
   ANGLE 360 — Notice the line flickers but does not extend beyond original line. The turtle has made a complete trip around the figure to complete the 360 degrees but it moves so quickly that you only see the flicker. This is called *the total trip theorem*.

4. Try other angle measures: Check with a protractor to see if your command has measured the angles correctly.

5. The next procedure will show angles of specified degrees and rotate each angle so you can see it in different positions. The procedure is named ROTATE :D :R, where D stands for the number of degrees in the angle and R stands for the number of times you wish to rotate the angle. The colon (:) tells the turtle that the variable changes each time the procedure is run. Therefore, you can make many angles of different degrees and different rotations without the need to type new procedures each time.

A procedure is a specified set of commands that enables the turtle to carry out the commands in the same order each time the name of the procedure is typed.

Use a CS between each angle rotation. Here are some angles to try:

```
ROTATE 90 5
ROTATE 180 10
ROTATE 30 15
ROTATE 100 3
```

6. In a cooperative learning venture, use ANGLE and decide on the degrees you want. Hold down the control key and the L key. Now the directions are hidden. Let your partner guess the degrees. You can check with a protractor to verify the command. Hold down the control key and the S key and your answer will appear again. Do several times with each partner getting an equal number of chances to guess and create angles.

7. *For those who like a challenge:*
All the angles were "drawn" to the right. If you wanted to "draw" angles that would go to the left, what commands in the procedure named ANGLE :D would you have to change?

There are many activities that can be done with Logo to show line and rotational symmetry. The following activity uses a POLY (for polygon) procedure to create designs which can be tested for line symmetry (adapted from Craig, 1986).

Activity

COMPUTERS

Line Symmetry with Logo

Directions:

1. This program is saved under the name POLY

2. Type   LOAD "POLY   and follow the directions in step 3.
(If you have no disk, type the following procedure:)

```
TO POLY :SIDES :MULT
REPEAT :SIDES [FD 50 RT (:MULT * 360 / :SIDES)]
END
```

3. Type   POLY 5 1   (do *not* clear screen)
Then type   POLY 5 4
Does this shape have line symmetry? Remember, a figure is symmetrical if it can be folded along a line so that the halves match exactly. How many lines of symmetry? Type CS
Type this pair of procedures to see another design with line symmetry.

```
POLY 8 3     (do not clear screen)
POLY 8 5
```

Do another pair: POLY 7 3   POLY 7 4
Try some pairs of procedures on your own and determine if they are symmetrical designs. Use POLY to create a design with one line of symmetry, with two lines of symmetry, and with infinite lines of symmetry. The sum of the second number in each set must equal the first number in each set. The first number in each set must be the same.

The orientation of a figure may influence a child's interpretation and perception of that figure. If an isosceles triangle is usually drawn with its base horizontal, a rotation of the triangle (Fig. 3.5) may cause a change in perception about that figure causing the child not to recognize the isosceles triangle. Children may have such limited experiences with shapes that they may feel the triangle is not "right-side up."

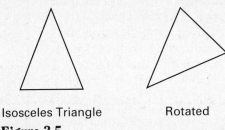

Isosceles Triangle   Rotated

Figure 3.5

Transformational geometry gives children some experience with changing their orientation and perspective about figures. This area of study deals with motion geometry and three basic motions: flips, slides, and turns. The motions follow certain rules and the figure that results may produce a transformation which is a mirror image of the original figure. Study Figure 3.6 to see the different motions. A flip creates a reflected image of the original shape and is found frequently in our environment. A slide motion is when the figure moves along a plane with no changes in the position of the points. A turn is a rotation around a point. Children can quickly acquire an intuitive notion of flips (reflection), slides (translation), and turns (rotation) by working with pattern cards and tessellations (Fig. 3.6).

These activities involve the ability to visualize images at both a perceptual and a representational level. While perceptual level is based on manipulation of objects and visual impressions, the representational level relies on mental manipulations, imagination, and thought. Performing a mental operation (e.g., Euclidean transformation) is a more difficult task and may be limited by a child's developmental level in that it requires formal operational thought (in a Piagetian sense). Kidder (1977) studied the child's ability to perform Euclidean transformations at a representational level; that is, the ability to visualize images under a motion and then to draw the images in correct final position. His studies suggest that the ability to learn at the perceptual level does not mean that the child can perform spatial tasks at the representational level. He recommends that the teacher foster spatial visualization by including perceptual-level transformational activities such as tracings, paper foldings, and visual recognition of motion images.

The National Council of Teachers of Mathematics (NCTM) curriculum standards (Commission on Standards for School Mathematics, 1987) call for the development of spatial sense that includes insights and intuitions about two- and three-dimensional shapes and their characteristics, the interrelationships of shapes, and the effects of changes made on shapes. Such experiences allow children to develop a more comprehensive understanding about shapes and their properties. Rich experiences with geometry will encourage children to see uses of geometry in their lives.

To extend their knowledge of plane figures, children need to understand many properties of figures: the number of sides, the length of the

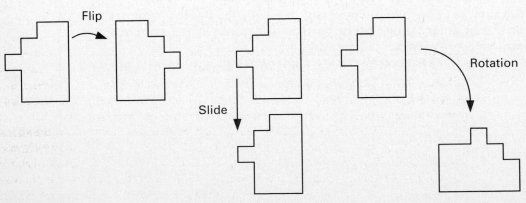

Flip

Slide

Rotation

Figure 3.6

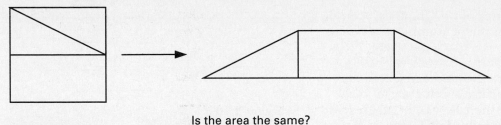

Is the area the same?

Figure 3.7

sides, the size of the angles, parallel and perpendicular lines contained in the figure, the number of angles. As these concepts are encountered in the elementary mathematics program, it is important that experiences be provided to allow for guided discovery of what these concepts mean. Paper folding or origami are excellent ways to develop the visualization, precise language, and careful thinking required to understand geometric principles. Inexpensive wax paper serves as an effective material for paper folding activities because it is clear and easy to see through and when creased, it leaves a visible line.

Logo activities, or turtle geometry as it is sometimes called, provide opportunities for children to study angle measurement of polygons in a problem-solving situation. In order for the turtle to draw a polygon, the student must develop a procedure. Decisions about the procedure are based on the number of sides in the figure. Procedures for drawing three-, four- and six-sided figures are fairly simple, almost intuitive in nature. When considering something such as a five-sided figure, problems arise since the angle the turtle turns is not evenly divisible. The "total-turtle-trip" theorem becomes important. This means that to make a closed figure, the turtle must turn a total of 360 degrees. Articles by Billstein and Lott (1986) and by Thomas and Thomas (1984) offer some ideas for exploring these concepts with children.

Area and Perimeter: *Area, the number of square units required to cover a surface*, and *perimeter, the distance around*, are two compo-

nents of the study of geometry. One aspect of a child's cognitive development that Piaget studied was the child's concepts of geometry. His tasks assessed conservation of length, area, and volume and were based on the child's ability to reason that the measure was unchanged by displacement. Many children aged 12–14 still had trouble being convinced that the area remains unchanged with different configurations such as in Figure 3.7 (Hart, 1984).

Results of the fourth NAEP mathematics assessment indicate a confusion between area and perimeter that continues through eleventh grade (Kouba et al., 1988). About one-third of the seventh-grade students computed area for perimeter, and vice versa. The students' performance revealed only a little more than 10 percent could find the area of a square given one side and the figure labeled as a square. Children also were not comfortable believing that perimeters could change but the areas were the same as in Figure 3.8.

Pentominoes provide an exercise in perception, logical reasoning, and help demonstrate some geometric principles of congruency. Pentominoes are made of five (pent-) congruent squares which must have every square touching at least one side of another square, Figure 3.9.

Area = 6 sq. units   Area = 6 sq. units

Figure 3.8

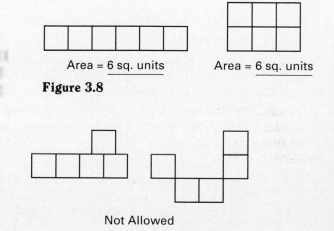

Pentominoes—Allowed   Not Allowed

Figure 3.9

Have children cut squares out of graph paper or use plastic tiles to form the pentominoes.

A common activity is to form the twelve different pentominoes that can be constructed. Individual squares such as mosaic tiles, which can be moved easily, work better with younger children. The shapes can then be cut from graph paper and kept as a record. Making the shapes out of paper allows children to test if the newly formed shape is unique. The shape, when flipped or rotated, may produce a figure that has already been found. If the two shapes are congruent, they are the same shape (Fig. 3.10).

If this task is too hard for younger children, it can be made easier by starting with three

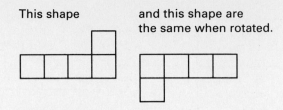

This shape   and this shape are the same when rotated.

Figure 3.10

squares or triominoes. There are only two of them. Next try the different arrangements of four squares or tetrominoes. Here are some additional activities (Cowan, 1977) using pentominoes.

Activity

PROBLEM
SOLVING

Pentominoes

Materials: One cut-out set of the 12 pentominoes [use graph paper (one-inch or centimeter)]

Procedure:

1. Do all the pentominoes have the same area?

2. Do all the pentominoes have the same perimeter? List the possibilities.

3. How do you know that you have found all possible arrangements?

4. How many pentominoes can be folded to make a cube without a lid? Is there something special about how those that can be folded look?

5. Using all 12 pentominoes, can you form a rectangle that is 6 units wide and 10 units long (a 6 x 10 rectangle)?

6. Using all 12 pentominoes, can you form a 5 x 12 rectangle?

7. Are there any other rectangular arrangements into which all pentominoes will fit? How do you know?

8. Use two different pentominoes. Can you make a 2 x 5 rectangle? Any others?

9. Use three different pentominoes. Can you make a 3 x 5 rectangle? Any others?

10. Continue with different numbers of pentominoes and determine what size of rectangles can be formed. Do you see a pattern? Can you make any predictions?

11. Using two pentominoes, find all the tessellations that can be made. Color graph paper to show the tessellations that are formed.

12. Using a checkerboard, try to place all the pentominoes on the board without overlapping the pieces. This can be played as a game with partners. Each player draws six pieces and in turn places a pentomino on the board until no more plays remain. The winner is either the one who places more pieces on the board or the one who places the last piece on the board.

Area and perimeter can also be explored with tangrams and geoboards. These teaching activities help develop an intuitive understanding about how area can remain the same regardless of various transformations. Students who rely heavily on perceptual cues gain understanding about conservation of area. The geoboard is an effective device to show area formulas for parallelograms and triangles. Rather than use the formula that requires the student to visualize a line segment as the height, the strategy with geoboards is to form a rectangle around the region, determine the area of the rectangle and from there determine the area of the other figure.

Activity

MANIPULATIVES

Discovering Area with Tangrams

Materials: Tangram sets

Procedure:

1. Use the two smaller triangles. Make a square. This square represents a square unit of measure. What is the area of the triangle?

2. Use the two smallest triangles. Make a parallelogram. What is the area of this shape? Compare the shapes and area.

3. Use the three smallest triangles. Make a square. What is the area? Use the same pieces and make a triangle, rectangle, trapezoid, and parallelogram. What is the area of each shape? If one edge of the square is assigned a value of one unit, compare their perimeters.

4. Use the five smallest pieces (all except the two large triangles) and form the same five shapes. Compare the area and perimeter of each shape.

5. Which shape do you feel has the greatest area? Why? Greatest perimeter? Why?

Activity

MANIPULATIVES

Exploring Area and Perimeter with Geoboards

Materials: Geoboards
Rubber bands (geobands)

Procedure:

1. On your geoboard, the distance between two adjacent nails is considered a unit of length •—• (as it appears on the geoboard). Stretch a rubber band around the first row of nails. What is the perimeter?

2. Make a figure with a perimeter of 4, 6, 7, 10. Be sure your figures have right angles for corners.

3. How many different ways can you construct figures with a perimeter of 8?

4. Make the smallest square possible. This is called one square unit of area. Make squares of all different sizes and areas.

(continued)

5. Make these figures on a geoboard: Perimeter 12, area 9; perimeter 8, area 4; perimeter 12, area 8; perimeter 10, area 4.

6. Make the unit square. Form a right triangle from the square. What is the area of the triangle? Construct a square on each side of the triangle. What is the area of each square? What is the relationship between these areas?

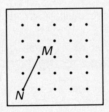

7. Construct a right triangle with an area of 3 square units. Build squares on each side. Is the relationship between the squares the same?

8. Consider line segment MN. Can you predict the area of the square which has this line segment as one side? Can you predict the area of the right triangle that has this line segment as one side?

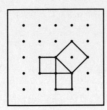

9. The Pythagorean theorem can be used to find the length of any line segment of the geoboard. First, find the right triangle that has that line segment as its hypotenuse (the side opposite the right angle). Determine the lengths of the other two sides and use the theorem that says:

$$a^2 + b^2 = c^2$$

10. Can you make squares with areas of 1, 2, 3, 4, 5, 6, 7, 8, 9, and 10?

11. How many square units are in the figures below?

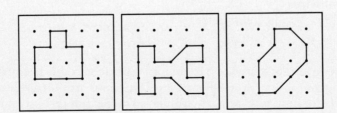

Activity

MANIPULATIVES

Additional Activities for Area with Geoboards

Materials: Geoboards
Rubber bands (geobands)
Geoboard dot paper (Appendix C, optional)

Procedure:

1. Put a band around the smallest square possible. This is called one square unit of measure. How many nails are touching the rubber band? How many nails are inside?

2. Stretch the band around 5 nails. What is the area of this shape? How many nails? How many nails are inside?

3. Stretch the band around 6 nails. What is the area? How many nails? How many nails are inside?

4. Continue in a similar manner including one more nail each time. Develop this data into a table:

Area (A)	Number of Nails Touching (T)	Number of Nails Inside (I)

5. What patterns do you notice? What predictions can you make?

6. Keep the number of nails inside as zero and increase the number of nails touching. What effect does this have on the area? How is the area increasing? What patterns can be found?

7. Keep the number of nails touching as four and vary the number of nails inside. What effect does this have on the area? How is the area increasing? What patterns can be seen?

8. If there is a relationship between these two variables, can a function be found to express that relationship?

9. Form a right triangle on the geoboard such as a base of 2 and a height of 3. Find the area by enclosing this figure with a rectangular region. The area of the triangle is half the area of the surrounding rectangle.

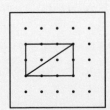

(continued)

10. Form an acute triangle with a base of 3 and a height of 2. Find the area by enclosing this figure with two rectangles. Inside each rectangle is a right triangle. Since the area of a right triangle is half the area of the surrounding rectangle, add the two areas together to find the total area of the acute triangle.

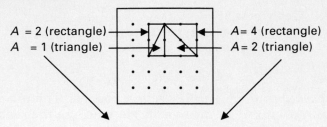

A = 2 (rectangle) A = 4 (rectangle)
A = 1 (triangle) A = 2 (triangle)

1 + 2 = 3 total area of acute triangle.

11. Form a parallelogram on the geoboard. Find the area by forming a rectangle around it and finding the area of the rectangle. This helps children see the relationship between the two figures and how the area formula for a parallelogram, $A = h \times b$, is computed.

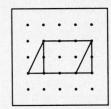

12. On the basis of data you collected and recorded in the table, write a formula to find the area of any region using only the number of nails touching (T) and the number of nails inside (I) as your data. This formula is known as Pic's theorem. It gives the area of a figure made with one rubber band on a geoboard given the number of nails inside and the number of nails touching on the outside.

Students will be able to check work done in the last geoboard activity by using the following Logo program. It also helps move a student from the concrete level to the pictorial level by doing the same activities as above with the Logo program.

Activity

COMPUTERS

Logo with the Geoboard

Directions:

1. This program is saved under the name GEOBOARD. Type ERALL and press return.

2. Type LOAD "GEOBOARD and follow the directions in step 3.

   (If you have no disk, type the following procedures:)

```
TO PIC
MAKE "RK READCHAR
```

```
IF :RK = "I [FD 10]
IF :RK = "M [FD 10]
IF :RK = "K [LT 90 FD 10 RT 90]
IF :RK = "L [RT 90 FD 10 LT 90]
IF :RK = "O [RT 45 FD 14.1421 LT 45]
IF :RK = "U [LT 45 FD 14.1421 RT 45]
IF :RK = ", [RT 135 FD 14.1421 LT 135]
IF :RK = "N [LT 135 FD 14.1421 RT 135]
IF :RK = "S [ST]
IF :RK = "E [STOP]
IF :RK = "C [CS GEOBOARD CLEARTEXT]
IF :RK = "F [FORMULA]
PIC
END

TO FORMULA
PR [HOW MANY NAILS TOUCHING?]
MAKE "T FIRST READLIST
MAKE "T :T / 2
PR [HOW MANY NAILS INSIDE?]
MAKE "I FIRST READLIST
MAKE "I :I - 1
MAKE "A :T + :I
PR :A
END

TO GEOBOARD
REPEAT 4 [FD 60 RT 90]
DOT [10 10] DOT [10 20] DOT [10 30]
DOT [10 40] DOT [10 50]
DOT [20 10] DOT [20 20] DOT [20 30]
DOT [20 40] DOT [20 50]
DOT [30 10] DOT [30 20] DOT [30 30]
DOT [30 40] DOT [30 50]
DOT [40 10] DOT [40 20] DOT [40 30]
DOT [40 40] DOT [40 50]
DOT [50 10] DOT [50 20] DOT [50 30]
DOT [50 40] DOT [50 50]
PU RT 45 FD 14.1421 LT 45 PD
END
```

3. Look at the computer keyboard and you will see the keys pictured in Figure 3.11 in the same relationship pattern. You can use these keys to move the turtle around the screen in the same relationship pattern shown with the arrows.

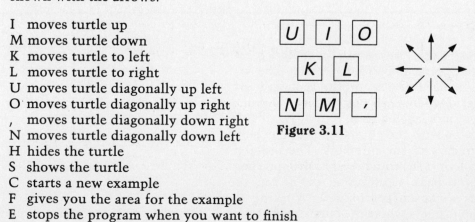

I moves turtle up
M moves turtle down
K moves turtle to left
L moves turtle to right
U moves turtle diagonally up left
O moves turtle diagonally up right
, moves turtle diagonally down right
N moves turtle diagonally down left
H hides the turtle
S shows the turtle
C starts a new example
F gives you the area for the example
E stops the program when you want to finish

Figure 3.11

(continued)

4. To start the program, type PIC [and press return key]. Do not wait for the blinking cursor. Press C (for starting the program on Pic's Theorem). Then press the return key. When the geoboard appears, use the preceding key explanation to move the turtle around the geoboard, making whatever figure you wish. If the turtle gets in the way of figuring the area, press H and the turtle will disappear.

5. After each example, press the F key to find if you guessed the correct area.

6. *For those who like a challenge:*
The formula is shown in the preceding procedure called FORMULA. If you study the commands, you will be able to see how the formula is figured.

Activity

COMPUTERS

Geoboard Areas Using a Spreadsheet

1. Develop a computer spreadsheet to compute the areas created on a geoboard utilizing Pic's Theorem. Follow the directions of the user's manual for the spreadsheet you are using (see Appendix E). The spreadsheet template could be as follows:

Geoboard Areas

INSIDE PEGS	OUTSIDE PEGS	AREA
0	4	1
0	6	2

2. The formula for computing the area uses Pic's theorem which is:

$$\text{Area} = \frac{\text{Pegs that touch}}{2} + \text{Inside Pegs} - 1$$

Enter this formula into the spreadsheet to compute the area.

3. Create various designs on the geoboard and enter the number of inside and the number of outside pegs in the spreadsheet template. The area will be shown in the area column. Check answers by counting the actual square units.

Geoboards can also be used to practice coordinate geometry and ordered pairs. The arrangement of nails or pegs on the geoboard can provide a grid where each place denotes a point and each point is labeled from 0 to 4 (Fig. 3.12). A strip of tape with the points labeled may help remind students of the numbers. The ordered pair notation of (1,3) would refer to over one peg and up three pegs. Connecting cubes can be placed over a chosen peg or nail. Working in pairs, students guess where the cube or cubes are located on the unseen geoboard in the same way

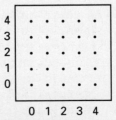
Figure 3.12

that the game Battleship is played. Another activity is to create a shape or design on the geoboard with a rubber band and, using ordered pair notation, have the child describe that shape to a partner who either makes the figure on a geo-board or on geometric dot paper. This encourages precise language and accuracy of naming the coordinates.

Activity

PROBLEM SOLVING

What Is the Figure?

Materials: Geoboards
Rubber bands (geobands)
Geoboard dot paper (Appendix C)

Procedure:

1. Connect these nails and record the figure formed:

   (1,1), (1,2), (1,3), (2,1), (2,4), (3,1), (3,2), (3,3)

2. Connect these nails and record the figure formed:

   (0,0), (1,0), (1,1), (2,1), (2,2), (3,2), (3,3), (4,3), (4,4)

3. Make a shape, record the coordinate pairs, and ask a friend to record the figure formed.

These activities only begin to illustrate the many experiences that can be provided for studying plane geometry in an informal way. Articles by Edwards (1977) and Young (1982) offer ideas and further investigations for discoveries in geometry.

Geometric Constructions: *Constructions are geometric drawings of angles and figures using a protractor, compass, and straightedge (a ruler without markings).* Constructions with a protractor or a compass are generally delayed until upper or middle school grades.

A *protractor is a geometric tool used to measure angles.* To understand measuring angles, children should have experiences with Logo where they see the relationship of moving the turtle by degrees or have action experiences of turning themselves or the hands of a clock. In this manner, students can see the effect of enlarging the size of the angle and compare how far they must move their arms or the hands of a clock to get certain angles. Without such preparation, they are unsure about what an angle is and they have a tendency to be misled about the length of the sides and the size of the angle. Often children say that angle MNO (Fig. 3.13) is greater than angle XYZ because they are comparing the length of the sides.

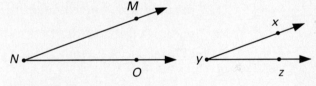

Figure 3.13

Students need to feel comfortable extending the length of the sides of an angle for a better measurement. Using a protractor (Figure 3.14) correctly means the child must know:

1. how to place the center of the protractor on the vertex of the angle,

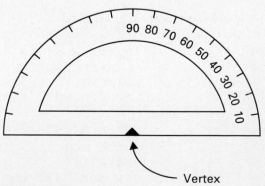

Figure 3.14

2. how to read the degrees of the protractor on the vertex of the angle, and
3. how to read the degrees of the angle along the edge.

This is particularly important with circular protractors and those with two sets of numbers in the half-circle formation.

Children should start with a protractor like the one in Appendix C where each degree is accounted for at its own unique position. Include measuring angles shown in various positions, not just angles with one side on a horizontal line. Angles of shapes should be measured to learn that the sum of the measures of the angles of a triangle is 180 degrees and the angles of a quadrilateral sum to 360 degrees.

The following Logo programs can help students check if they are using the protractor correctly. Students can work in cooperative learning groups. For example, one student decides the degree of the angle to be constructed and cuts appropriately sized pieces of wax paper (an estimation skill) while the second student manipulates with a protractor on wax paper. The third student measures the same angle size with the Logo protractor, and a fourth student holds the wax paper sample up to the computer display to compare the angles. Students can take turns doing each task.

Activity

MANIPULATIVES

COMPUTERS

The Simple Protractor Activities and Logo

Materials: Wax paper
Circle protractor (found in Appendix C)
Soft-lead pencil
Logo protractor program

Directions:

1. This program is saved under the name PROTRACTOR

2. Type   LOAD "PROTRACTOR   and follow the directions in step 3.

(If you have no disk, type the following procedures:)

```
TO A :D
PROTRACTOR
ANGLE
END

TO PROTRACTOR
HT SETPC 1
DOT [0 18] PU
SETPOS [0 18]
LT 90 FD 63 RT 90 PD
CIRCLER 63
PU LT 90 FD 18 RT 90 PD
CIRCLER 81
END

TO ANGLE
PU SETPOS [0 18] PD
SETPC 4 RT 90 FD 95 BK 95
LT :D FD 95 BK 95
REPEAT :D [RT 1 FD 20 BK 20]
PR [IF YOUR ANGLE MATCHES]
PR [THIS ONE, YOU USED THE]
PR [PROTRACTOR CORRECTLY.]
END

TO RESET
PU HOME SETPOS [0 18] PD
END
```

3. Using the circle protractor, measure an angle of 70 degrees and an angle of 300 on the wax paper. Start out simple, placing the vertex along the horizontal line. Label each angle on the wax paper.

4. Then check with the Logo program. Type the following:

   A 70   (It stands for an angle of 70 degrees.)
                Then hold the wax paper angle up to the screen.
   CS   (To clearscreen for next example)
   A 300   (It stands for an angle of 300 degrees.)
                Check to see how well your paper angle matches.

5. Try other angle sizes on your own. Notice that the angle being measured is highlighted so that students do not confuse it with the supplementary angle. This is especially important for angles greater than 180 degrees.

6. Now try measuring angles whose vertices are not along the horizontal line. The Logo program will rotate to the right or left by a certain number of degrees if you type:

   CS   (To clear screen)
   RT 60 A 45   (Rotates the protractor to the right 60 degrees
                   and makes an angle of 45 degrees)

   then try . . .

   CS
   LT 90 A 250   (What does this command do? Check with the
                   physical protractor.)

7. Make up other examples patterned after those in step 6.

The next activity helps students estimate the size of different angles using the preceding Logo protractor program. It can also be played using only the wax paper if no computer is available.

Activity

COMPUTERS

Guess the Size of the Angle with Logo

Directions:

1. Let a partner choose an angle size and type it on the computer using the directions in step 3 of the last activity. Your partner will cover up the last four lines on the screen so you cannot see the angle size.

2. Now estimate the size of the angle on the screen and type your choice. Here are two examples:

Partner 1 Types:	Partner 2 Types the Estimate:
A 80 RESET	A 60 (It will take a few seconds.)
CS	
A 190 RESET	A 210

(*Note:* RESET sets the program back to the beginning without erasing Partner 1's angle and allows Partner 2 to see how close his/her choice comes to Partner 1.)

(continued)

3. *For those who like a challenge:*
   Let Partner 1 rotate the protractor before drawing an angle. Then Partner 2 has a real challenge to estimate the angle size! Here's an example:

Partner 1 Types:	**Partner 2 Types the Estimate:**
RT 135 A 90 RESET	A 100

4. Make up more examples like those in step 3. Let partners switch roles frequently so both have a chance to practice estimation.

A compass is a measurement tool that is used to make circles or portions of circles. One uses a compass by placing the spike of the compass at the center point of the circle. The measure from the spike to the pencil is the radius of the circle; it is that distance which is measured on the compass. A compass may be difficult for students to manipulate because it takes a certain amount of coordination to swirl the compass around the radial point without slippage. A purchased compass (Fig. 3.15) usually has two measurement systems, one on either side of the arm.

The metric system (in centimeters) is on one side and the standard English measurement (in inches) is on the other side. The compass found in Appendix C is a simplified metric compass that is inexpensive for teachers to make and may be easier for some children to manipulate in beginning activities.

The following Logo programs help students check if they are using the compass correctly.

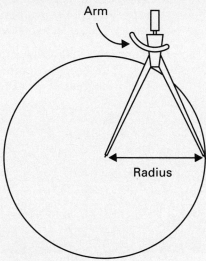

Figure 3.15

The same idea may be used for dividing tasks in cooperative learning groups, as was done in the preceding Logo protractor activities.

Activity

MANIPULATIVES

COMPUTERS

The Metric and Inch Compass with Logo

Materials: Wax paper
Soft-lead pencil in compass
Metric compass (found in Appendix C or purchased metric or inch compass)
Logo compass programs for metric and inch

Directions:

1. This program is saved under the name COMPASS
2. Type LOAD   "COMPASS and follow the directions in step 3.

   (If you have no disk, type the following procedures:)

```
TO MC :RADIUS              TO IC :RADIUS
HT DOT [0 0]               HT DOT [0 0]
PU LT 90                   PU LT 90
FD :RADIUS * 10            FD :RADIUS * 10 * 2.54
RT 90 PD                   RT 90 PD
CIRCLER :RADIUS * 10       CIRCLER :RADIUS * 10 * 2.54
END                        END
```

3. Using the metric compass, make circles on wax paper with the radius of:

1 centimeter	4 centimeters
2 centimeters	5 centimeters
3 centimeters	6 centimeters

(*Note:* Mark the center point with the pencil and label each radius.)

4. You are now ready to type directions in the computer program that will replicate the wax paper circles on the screen. If they match, then the compass-drawn circles were measured correctly.

5. *Note:* Because of the elliptical shape of some computer monitors, the top and bottom of some circles may appear to be flattened. However, the outside ring of the right and left circumference should match the right and left side of the compass-drawn circle. You will need to experiment with the monitor you are using.

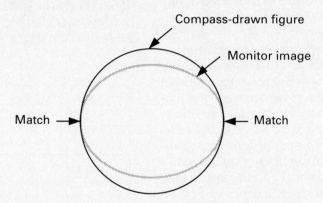

6. Radii of larger than 6 centimeters appear too distorted when checked on the computer monitor and should not be tried with elementary/middle school students.

7. Now type MC 1 (It stands for *metric compass* with a radius of 1.)

8. Hold the 1 cm wax paper circle up to the screen image. If they do not match perfectly or side to side (see step 5), then check your measurement with the compass again.

9. Type CS between each example to clear screen for the next example.

10. Follow the directions in steps 8 and 9 while typing:

MC 2	for the 2 cm radius
MC 3	for the 3 cm radius
MC 4	for the 4 cm radius
MC 5	for the 5 cm radius
MC 6	for the 6 cm radius

11. Variation: Make wax paper models using the inch compass and match them to the Logo inch compass by typing . . .

IC .5	to match at $\frac{1}{2}$-inch radius
IC 1	to match an inch radius
IC 1.5	What will this be?

How it works: Logo uses only metric measurement. One step of the turtle equals one millimeter. So the procedure for the inch compass must

(continued)

find the equivalent metric measure for an inch, which is 2.54 centimeters times 10 to find the equivalent in millimeters, i.e., `FD :RADIUS *` `10 * 2.54` gives the metric equivalent for the inch.

12. Anything more than an inch and a half will be too distorted to use with most computer monitors.

13. *For those who like a challenge:*
    Try other small measures using both the metric and inch compass and check them on the computer.

Solid Geometry

Solid geometry involves three-dimensional shapes, their properties and relationships. Children's early experiences with geometry are centered around objects with three-dimensions—blocks, cans, cones, balls, and boxes. Many children had building blocks at home or at preschool with which they constructed structures and patterns. Collections of empty oatmeal containers, food cans, soda cracker boxes, paper towel rolls, and other various sized boxes provide representations of solids for children to explore, compare, and construct. Physical education teachers talk about personal space, self space, and general space and often provide movement activities to engage children in learning about these concepts. Jensen and Spector (1986) suggest activities to involve children in experiencing solids and their properties. In these teacher-directed activities, the children pretend they are suspended inside various space figures and they describe the figure's properties. For example, "Imagine that you are in something like this round oatmeal box. This is your personal cylinder. Pretend that your fingertips are just touching the inside of your cylinder . . . draw the biggest circles possible by rotating your arms slowly" (Jensen and Spector, 1986, p. 14).

In the classroom, commercial materials should be available for play and exploration. Plastic models and wooden models are needed to provide different perspectives about the figures. The transparent, plastic models help children realize that each figure has an inside and an outside. Ask children to describe and classify the shapes according to their properties. Put the shapes in a container and ask different children to close their eyes, take a shape, and describe it to others who are trying to guess the shape selected. The vocabulary associated with these activities is important. Younger children may call the shapes, "box, ball, can," but as they build greater understanding of the shapes, the proper names will become common, "prism, sphere, cylinder." As children learn to be more precise in describing and classifying solids, have them make a table of the properties:

Solid	Number of edges	Number of faces	Number of vertices

Students need to relate the study of geometry to their environment by finding examples in their homes of objects that look like the solids being studied. They can create wall charts with pictures of the shapes. Have students take a can or box and cut apart all the individual faces that compose that figure. Another activity is to have students build a solid from squares. These experiences help them visualize the individual components of the figure and how a vertex is formed. Students in middle grades often have not had such experiences with informal geometry and need to work with models just as younger children do. Participation in such activities helps build the visualization necessary in working with formulas for solids.

Students can construct poster board or tagboard models from printed patterns. Cutting and assembling the figures help children learn about edges, corners, faces, and other parts of solid figures. With older children, these models can be used to develop understanding of the formulas. The following activity is to illustrate the formula for the volume of a pyramid using cardboard models.

Activity

PROBLEM
SOLVING

Understanding Construction of a Pyramid

Procedure:

1. Cut out one of Figure A and four of Figure B. Enlarge the patterns. Use the patterns in Appendix C to help you.

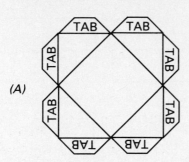

(A)

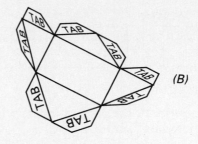

(B)

2. Assemble the five pyramids and tape together.

3. Use tape to hinge together pairs of the smaller pyramids to make three identical pyramids.

4. Line up the pyramids with hinged faces on opposite sides and tape the bases together.

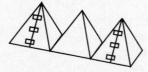 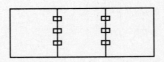

5. Fold the model to produce a rectangular prism.

6. Construct the other solids in Appendix C. Which are prisms and which are pyramids?

A common situation people face is to decide how much wrapping paper is needed to cover a box. This problem entails visualizing the box cut apart and laid flat, a matter of determining the surface area of the shape. A fun and useful experience is to estimate the amount of paper needed to cover various sized and shaped figures. Estimating the capacity of various containers is an interesting and challenging activity. Our perceptions about capacity are usually not accurate because we lack many experiences of this type.

Activity

ESTIMATION

What Is Your Guess?

Procedure:

1. Study the box or rectangular prism below. Decide how much wrapping paper is needed to cover it. Which sheet would you choose?

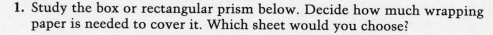
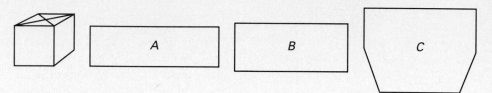

2. Consider the pentominoes. Which of the shapes can be folded into a topless box (without a lid)?

3. The Factory Box Problem (Burns, 1984):
   Someone in a factory bought cardboard that was 5 squares by 4 squares. They figured that each sheet of cardboard could be cut into four pieces so each piece would fold into a topless box. How could the sheet be cut?

4. Have 4–6 empty containers. Label them with alphabet letters. Which one holds the most? Which one holds the least? Which ones hold about the same amount? Test your predictions with rice, milo, salt, or another granular substance.

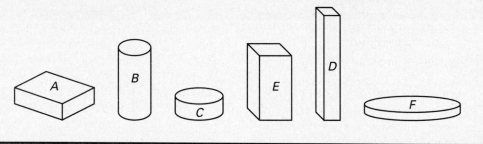

These activities only begin to cover the opportunities to explore ideas about solid geometry. There are a number of resource books on geometry available from the National Council of Teachers of Mathematics and some are listed in the bibliography under these authors: Olson 1975, Pohl 1986, and Wenninger 1975. Extended experiences with a variety of materials allow children to discover, verbalize, and internalize geometric concepts. Children should learn to use mathematics vocabulary and notation in meaningful ways.

Diagnosis and Remediation

Successive Thought Processing

Students who process information successively (from the parts to the whole) perform geoboard activities better if the area of the polygon to be measured is partitioned into segments within the polygon and then all the segments will be added up to find the total area. The left geoboard in Figure 3.16 shows a sample activity as a successive thought processor would perform it. The solid black line identifies the original polygon. The student uses smaller geobands to divide the figure into unit segments.

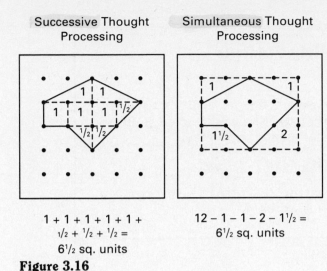

Successive Thought Processing

Simultaneous Thought Processing

$$1 + 1 + 1 + 1 + 1 + \frac{1}{2} + \frac{1}{2} + \frac{1}{2} = 6\frac{1}{2} \text{ sq. units}$$

$$12 - 1 - 1 - 2 - 1\frac{1}{2} = 6\frac{1}{2} \text{ sq. units}$$

Figure 3.16

Simultaneous Thought Processing

Students who process information simultaneously (from the whole to the individual parts) perform geoboard activities better if the polygon to be measured is included in a whole rectangular frame from which the outer segments can be subtracted. The remaining polygon inside the solid black line is the square area measurement. The geoboard on the right of Figure 3.16 shows the action movements as a simultaneous processor would perform them.

Students having trouble obtaining correct answers to area geoboard problems when doing the calculations one of the ways shown in Figure 3.16 should be encouraged to try the other processing style. Frequently, the change helps correct the misconceptions. This seems to hold true for all students—not just elementary ones. The

authors have watched adult learners profit from a change in processing style when errors occurred in geoboard activities. If the geoboard activities in this chapter seemed difficult, try doing them again, using the opposite processing style.

Correcting Common Errors

Angle Identification: Some students do not recognize an angle as the same if it is rotated. Students need many experiences making their own angles and seeing programs like the Logo angle rotation program viewed earlier in this chapter. There are a sizable number of middle school and high school students who do not recognize a right triangle unless the 90 degree angle is in the lower left hand corner of a triangle. The teacher must draw shapes in positions other than the way students are used to seeing them.

Angle Misconceptions with Protractors and Logo: With a protractor, draw several of the shapes mentioned in the review chart in the introductory section of this chapter. It is difficult to know which angle is the significant one that will make the shape when turning the protractor. The same problem occurs in Logo when a person creates a polygon because the turtle must make the exterior angle first and then make the interior angle that gives the polygon its shape.

The Logo program SHAPES can be adapted to show the exterior angle more easily if a straight line is extended from each side of the polygon. Figure 3.17 shows how the extended figure would appear. The Logo program is saved on the Logo disk under the name HELPSHAPES.

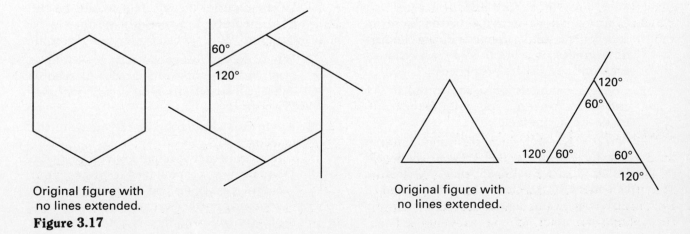

Original figure with no lines extended.

Figure 3.17

Original figure with no lines extended.

Activity

COMPUTERS

Logo Change to Help Students

Directions:

1. For those students with no disk, type the following addition to the repeat line in each polygon except the circle. The modification is underlined here for emphasis—do **not** underline in the program.

   `[FD 80 BK 40 RT]` (angle size remains the same)

2. Type   `ERALL`   and load the program HELPSHAPES Now type `HEX CS TRI CS SQ`   to see how it works.

Students can create a polygon with extended sides using the protractor and then check with the Logo program to make sure the extended sides appear the same. The Logo program becomes a self-check for students.

Summary

This chapter emphasizes hands-on experiences with objects for learning concepts about geometry. An important goal of these informal geometry experiences is that children understand that geometry abounds in their environment. As they move into the formal study of geometry, they will have been exposed to many of the concepts and properties that will lead them to a greater understanding and more positive attitudes. The geometric concepts developed in this chapter will be used to develop mathematical concepts throughout the book. Chapter 4 will use the content of this chapter as numbers and formulas are applied to the geometric concepts.

Exercises

Directions: Read all questions *before* answering any one exercise. Frequently the last question in one category leads to the first question in the next category.

A. Memorization and Comprehension Exercises
Low-Level Thought Activities

1. Examine an elementary textbook series and list the solids covered during the first nine grades of school. Make a sketch of each one and list its properties.

2. Construct models of solids found in A1 and make patterns for your students to make similar models.

3. Piaget's research indicates that children first have a topological view of the world rather than a Euclidean view. What does this mean to the classroom teacher and to the scope and sequence of geometric topics?

4. Explain how the geoboard can be used to teach and show the area formula for a right triangle and for a parallelogram.

B. Application and Analysis Exercises
Middle-Level Thought Activities

1. What material do you feel would be the best medium for teaching perimeter to children? Why do you prefer this material?

2. Where do you feel constructions should be covered in the first nine grades of school? What constructions would you include? Why?

3. Apply this simulation as if you were the teacher:

   It appears that you will not complete the book this year. You need to project your end-of-year plans and discuss them with your principal. The next chapter is geometry followed by a chapter on multiplication with larger numbers. You are certain you cannot cover both chapters in the remaining time. Decide what position you will take—skip geometry or skip the last chapter on multiplication, or invent a creative solution. Defend your position as you would before your principal.

4. Using the evaluation form found in Appendix A, select commercial software for any three topics from this chapter. Include drill and practice, simulation, tutorial, and problem-solving software in the selection. For each software program selected, describe how you would integrate it in a lesson plan.

5. To reinforce math across the curriculum:
   a. List five mathematical questions/problems that come from popular storybooks for children. Show how they are related to the topics of this chapter.
   b. List five ways to use the topics of this chapter when teaching lessons in reading, science, social studies, health, music, art, physical education, or language arts (writing, English grammar, poetry, etc.).

6. Search professional journals for current articles on research findings and teaching ideas with geometry. You may wish to look at the first activity under Synthesis and Evaluation Exercises below and coordinate the activities.

Follow the form below to report on the pertinent findings:

Journal Reviewed: _____

Publication Date: _____ Grade Level: _____

Subject Area: _____

Author(s): _____

_____

Major Finding:

Study or Teaching Procedure Outlined:

Reviewed by: _____

Some journals of note are:
Journal for Research in Mathematics Education
Arithmetic Teacher
Mathematics Teacher
School Science and Mathematics

7. Computer applications for classroom use are found in the journals listed below. They are by no means the only ones. Search the periodical section of the library for more. You may wish to look at the first activity in the Synthesis and Evaluation Exercises below to coordinate activities.
Teaching and Computers
Classroom Computer Learning
Electronic Learning
Computers in the Schools
The Journal of Computers in Mathematics and Science Teaching
The Computing Teacher

C. **Synthesis and Evaluation Exercises**
High-Level Thought Activities

1. Create your own lesson plan for teaching one concept using geometry. Follow these steps:
   a. Look at the Scope and Sequence Chart. Find a grade level and a concept within that grade level. Trace the previous knowledge of the concept taught in earlier grades so you will know where to begin your instruction.
   b. Use current articles from professional journals to help plan the direction of the lesson. Document your sources.
   c. Include at least one computer software program as a part of the lesson. Show how it will be integrated with the rest of the lesson.
   d. Include behavioral objectives, concrete and pictorial materials, procedures, evaluation of student mastery.
   e. Use the research-based format of a) review, b) development, c) controlled practice, and d) seatwork.

2. Write a plan for three activities to teach line of symmetry to fourth graders.

3. Of the three types of motion geometry (translation, reflection, and rotation) which would you think to be the most difficult to teach? Design a lesson plan to teach this concept to sixth graders.

4. Design some activities with Logo to show children that a circle has 360 degrees and that the angles of a quadrilateral have a sum of 360 degrees.

5. Create a computer database (a computerized method of organizing and storing information) on polygons. Include name, number of sides, number of angles, angle size and any other relevant information about the polygon. Prepare a class lesson with appropriate questions using the database.

6. What are the key concepts of geometry that are needed by elementary and middle school students, and how can an innovative teacher develop these concepts in the classroom?

Bibliography

Battista, Michael T. "The Effectiveness of Using LOGO to Teach Geometry to Preservice Elementary Teachers." *School Science and Mathematics* 87 (April 1987): 286–296.

Billstein, Richard, and Johnny W. Lott. "The Turtle Deserves a Star." *Arithmetic Teacher* 33 (March 1986): 14–16.

Burger, William F. "Geometry." *Arithmetic Teacher* 32 (February 1985): 52–56.

Burger, William F., and Michael Shaughnessy. "Characterizing the van Hiele Levels of Development in Geometry." *Journal for Research in Mathematics Education* 17 (January 1986): 31–48.

Burns, Marilyn. *The Math Solution: Teaching for Mastery Through Problem Solving.* Sausalito, Calif.: Marilyn Burns Education Associates, 1984.

Commission on Standards for School Mathematics of the National Council of Teachers of Mathematics. *Curriculum and Evaluation Standards for School Mathematics.* Working Draft. Reston, Va.: The Council, 1987.

Cowan, Richard A. "Pentominoes for Fun and Learning." *Arithmetic Teacher* 24 (March 1977): 188–190.

Craig, Bill. "Polygons, Stars, Circles, and Logo." *Arithmetic Teacher* 39 (May 1986): 6–11.

Crowley, Mary L. "The van Hiele Model of the Development of Geometric Thought." In *Learning and Teaching Geometry, K–12,* 1987 Yearbook of the National Council of Teachers of Mathematics. Reston, Va.: National Council of Teachers of Mathematics, 1987. 1–16.

Edwards, Ronald R. "Discoveries in Geometry by Folding and Cutting." *Arithmetic Teacher* 24 (March 1977): 196–198.

Guay, Roland B., and Ernest D. McDaniel. "The Relationship Between Mathematics Achievement and Spatial Abilities Among Elementary School Children." *Journal for Research in Mathematics Education* 8 (May 1977): 211–215.

Hart, Kathleen. "Which Comes First—Length, Area, or Volume?" *Arithmetic Teacher* 31 (May 1984): 16–18, 26–27.

Jensen, Rosalie, and Deborah C. Spector. "Geometry Links the Two Spheres." *Arithmetic Teacher* 33 (April 1986): 13–16.

Kidder, F. Richard. "Euclidean Transformations: Elementary School Spaceometry." *Arithmetic Teacher* 24 (March 1977): 201–207.

Kouba, Vicky L., Catherine A. Brown, Thomas P. Carpenter, Mary M. Lindquist, Edward A. Silver, and Jane O. Swafford. "Results of the Fourth NAEP Assessment of Mathematics: Measurement, Geometry, Data Interpretation, Attitudes, and Other Topics." *Arithmetic Teacher* 35 (May 1988): 10–16.

Mansfield, Helen. "Projective Geometry in the Elementary School." *Arithmetic Teacher* 32 (March 1985): 15–19.

McKnight, Curtis C., F. Joe Crosswhite, John A. Dossey, Edward Kifer, Jane O. Swafford, Kenneth J. Travers, and Thomas J. Cooney. *The Underachieving Curriculum: Assessing U.S. School Mathematics from an International Perspective.* Champaign, Ill.: Stipes Publishing Co., 1987.

Olson, Alton. *Mathematics Through Paper Folding.* Reston, Va.: National Council of Teachers of Mathematics, 1975.

Pohl, Victoria. *How to Enrich Geometry Using String Designs.* Reston, Va.: National Council of Teachers of Mathematics, 1986.

Resink, Carole J. "Crystals: Through the Looking Glass with Planes, Points, and Rotational Symmetries." *Mathematics Teacher* 80 (May 1987): 377–382.

Renshaw, Barbara S. "Symmetry the Trademark Way." *Arithmetic Teacher* 34 (September 1986): 6–12.

Retzer, Kenneth A., "Inferential Logic in Geometry." *School Science and Mathematics* 84 (April 1984): 277–284.

————. "Logic, Tradition, Truth in Algebra, and Inference for Geometry." *School Science and Mathematics* 84 (March 1984): 181–188.

Thomas, Eleanor M., and Rex A. Thomas. "Exploring Geometry with Logo." *Arithmetic Teacher* 32 (September 1984): 16–18.

Usiskin, Zalman. "Resolving the Continuing Dilemmas in School Geometry." In *Learning and Teaching Geometry, K–12*, 1987 Yearbook of the National Council of Teachers of Mathematics. Reston, Va.: National Council of Teachers of Mathematics, 1987, 17–31.

Wenninger, Magnus J. *Polyhedron Models for the Classroom.* Reston, Va.: National Council of Teachers of Mathematics, 1975.

Young, Jerry L., "Improving Spatial Abilities with Geometric Activities." *Arithmetic Teacher* 30 (September 1982): 38–43.

4

Measurement

Key Questions: *What kinds of activities prepare students to use discrete measurement systems in everyday situations?*

Activities Index

Introduction

In the last chapter, we developed an intuitive understanding of shapes, objects, and their relationships. In this chapter, we will explore the use of numbers and formulas to describe the world around us, and in later chapters, we will use these measurement applications in problem-solving situations. Since time began, people have always measured things in the world around them. Cavemen judged distances by time or eye and they compared sizes by paces or matched objects with trees, stones, or other objects common to their surroundings. There was no accuracy in this world.

As needs grew, greater accuracy was required. In order to measure clothing, weapons, and other items, rough methods of measurement were developed. Outstretched arms, heights, feet, and hands were all used for measuring. These measures differed between people and, therefore, were non-standard. All persons had their own method of measuring. In 6000 B.C., the first known standards of measurement such as stones or feet or hands were established.

Today, measurement is used in many ways in our lives and is vital for communication. The sciences have always used measurement for communication. Most professions require measurement in some way or another. These uses vary in terms of scales, codes, numerals, etc. For example, water hardness is measured in terms of mineral content, earthquake intensity is measured in terms of the Richter scale, and rock hardness is measured by Mohs' Hardness scale. Oven and room temperatures are measured with a thermometer that may be controlled by a thermostat. Daily, people weigh themselves, cut measured material for sewing clothing, panel a room, figure the distance they have jogged, or mark off part of their yard for vegetable planting. So measurement can be different things to different people and professions.

In this book, we are emphasizing the measurement units of volume, capacity, length, mass, area, temperature, and time. Measurement procedures included are direct comparison (2 and 3 objects), serialization, indirect comparisons, repeated measure (object used repeatedly for measurement), and select unit measures. These methods constitute the common elementary school measurement curriculum. In recent years, measurement has become a topic of increasing interest due to the activities generated by the metric system.

Measurement refers to a quality, not a quantity. In mathematics we attach qualities of measurement in order to describe them in one way with some amount of precision and accuracy independent of the particular measurement unit used. Estimation and approximate measurements need to be emphasized. "About," "close to," "nearly," "almost," and "approximately" are common terms used by children when discussing their measurements. My finger is "about 10 centimeters long" or "it is nearly one kilometer to my home." When teaching measurement we usually go through a procedure of comparison, (indirect and direct) vocabulary, instrument use, and reading the instruments. Therefore, no matter what the measurement, we usually begin with direct comparisons. With children we say, "Andrea is taller than Mario"; "Maggie's chalk is longer than Juan's"; "My glass holds more than your glass." These direct comparisons are common at all levels, but especially in the primary grades. Direct comparison is done by comparing like units to one another. This leads to such non-standard measures as this book is ten paper clips long or three chalkboard erasers long.

But soon a need for communication becomes apparent and the non-standard measures are compared to standard measures. An example of this would be how many pencils long is a meter stick or how much clay is as much as one kilogram of butter. Measurement now has gone from general to specific units and the general

mechanics of measurement become apparent. Reading rulers, balances, scales, and containers requires some proficiency on the part of the student. Not only are numbers used but labels of specific measure are also applied. For example, "the book is 15 cm wide" or "the coffee cup holds 250 ml." More generally, "I have a mass of 50 kg" or "I am 100 cm tall." These standard measures are now real to the student. The standard units require an awareness on the part of the student with estimation used for discussions or initial feelings. Proof and verification follow. For example, to prove that five cups of water is more than a liter of water, actually checking the answer can be done to prove the initial estimate or statement. Experimentation and experience make these statements on estimation more accurate. Therefore, students must have frames of reference for many of the standard units of measure. Specifically, the student must be able to discover that it is easier to measure a room in meters than millimeters or kilometers. The tools for measuring should have the same quality as the object being measured. Obviously, liters are not an appropriate tool for measuring distance and grams are not an appropriate tool for measuring area.

Specific types of measurement follow. Since students now have feelings for taller than, shorter than, wider than, or bigger than, application and the language of measurement must be developed. Remember, children need as much help in building the language of measurement as they do in building the concept of counting. Once a general language of measurement has been accomplished, the language is refined to be more specific.

The types of measurement include length, mass (weight), volume, capacity, area, temperature, and time. These are the most common measurement topics in the elementary and middle school. General activities for introducing each of the concepts are essential and numerous approaches are necessary for understanding. Therefore, building language, questioning techniques, and activities for each topic should be established.

There are two common measurement systems in the world. The United States uses the Customary system and some metric while most of the world uses the metric system exclusively. Most elementary textbooks include both systems, with each system taught separately and with no conversion of one system to the other. Consequently, the approach of this book is to treat each system independently of the other.

The Customary System

The Customary System (Table 4.1), often thought of as the English System until England adopted the metric system, has been the basic measurement system of the United States. It is based upon an arbitrary system of measurement. At one time, the inch was defined as the width of a thumb and the yard was the distance from a person's nose to the tip of the outstretched arm. Historically, an acre was measured by the amount of land two oxen could plow in one day. Obviously, this is not a very rational and orderly system. In this system, an ounce of gold weighed more than an ounce of feathers, yet a pound of feathers weighed more than a pound of gold. This brought about the 16 oz pound and the 12 oz troy, which were introduced as being the basic units of weight. The exact weight of a pound was further confused in the Middle Ages in England when the Tower (three-quarters of an ounce heavier than the troy) pound and the avoirdupois pound (16 oz) were invented.

The Metric System

The metric system is used as a standard measuring system in many industries in the United States. Since its development by Gabriel Mouton in 1670, the metric system has been the ideal system to use. The French adopted it in 1790 as part of the Revolution. Since then all but a few small countries and the United States have made the transition to the metric system. In the U.S., early attempts by John Quincy Adams and Thomas Jefferson to convert to the metric system met with defeat. However, Adams's study on the merits of the metric system set the basis for most of the arguments to follow. He main-

Table 4.1 Customary Measurement

Linear Units	Capacity	Area
inch (in.)	gill	square inch (in.^{2})
foot (ft)	pint (pt)	square foot (ft^2)
yard (yd)	quart (qt)	square yard (yd^2)
mile (mi)	gallon (gal)	square mile, acre

Weight	Volume	Temperature
ounce (oz)	cubic inch (in.^{3})	Fahrenheit degrees °F
pound (lb)	cubic foot (ft^3)	
ton	cubic yard (yd^3)	

tained that the main advantage of the metric system was that it uses a natural invariable standard from nature for the base measure. In other words, the physical phenomena of a metal bar used to represent a meter was not a true international standard as its context changed with different temperatures and latitudes. This obviously caused inconsistency. With the metric system, consistency is the goal. Adams also indicated that one basic unit was used for measuring mass and another unit for capacity whether wet or dry. Finally, the system is based on decimals (groups of 10) just like our numeration and monetary systems. Decimal computation is necessary in each and most computation is based on division by tens.

The metric system uses Greek prefixes:

> deka (10^1), hecto (10^2), kilo (10^3), mega (10^6), giga (10^9), and tera (10^{12})

and Latin prefixes:

> deci (10^{-1}), centi (10^{-2}), milli (10^{-3}), micro (10^{-6}), nano (10^{-9}), pico (10^{-12}), femto (10^{-15}), and atto (10^{-18}).

The metric units are shown in Tables 4.2, 4.3, and 4.4.

Activity

COMPUTERS

PROBLEM SOLVING

Metric Table Spreadsheet

Directions: Prepare a metric table using a computer spreadsheet. Make it so you can enter a measurement and it will give you all its corresponding measures. Follow the directions of the user's manual for the spreadsheet you are using (see Appendix E).

| kilo | hecto | deka | unit | deci | centi | milli |

Record what appears on the monitor in the space above.

Table 4.2 Metric Measurement

Name of Unit	Symbol	Value	Exponential Form	Base 10 Equivalent Numerals
kilo	k	thousands	10^3	1000
hecto	h	hundreds	10^2	100
deka	da	tens	10^1	10
base		ones	10^0	1
deci	d	tenths	10^{-1}	$\frac{1}{10}$
centi	c	hundredths	10^{-2}	$\frac{1}{100}$
milli	m	thousandths	10^{-3}	$\frac{1}{1000}$

Table 4.3 Metric Comparisons

Linear Meters	Mass Gram	Capacity Liters	Exponential Form	Base 10 Numeral
kilometers (km)	kilogram (kg)	kiloliter (kL)	10^3	1000
hectometer (hm)	hectogram (hg)	hectoliter (hL)	10^2	100
dekameter (dam)	dekagram (dag)	dekaliter (daL)	10^1	10
meter (m)	gram (g)	liter (L)	10^0	1
decimeter (dm)	decigram (dg)	deciliter (dL)	10^1	0.1 or $\frac{1}{10}$
centimeter (cm)	centigram (cg)	centiliter (cL)	10^{-2}	0.01 or $\frac{1}{100}$
millimeter (mm)	milligram (mg)	milliliter (mL)	10^{-3}	0.001 or $\frac{1}{1000}$

Table 4.4 Most Common Measures

Area (most common)	**Volume** (most common)	**Temperature**	**Capacity**
Square millimeter (mm^2)	Cubic millimeter (mm^3)	Celsius °C	Liter (L)
Square centimeter (cm^2)	Cubic centimeter (cm^3)		
Square meter (m^2)	Cubic meter (m^3)		
Square kilometer (km^2)			
Hectare (ha)			

Metric measure can be facilitated by remembering the following:

Length

10 mm = 1 cm
10 cm = 1 dm
100 mm = 1 dm
10 dm = 1 m
100 cm = 1 m
1000 m = 1 km

Area

100 mm^2 = 1 cm^2
10 000 cm^2 = 1 m^2

10 000 m^2 = 1 ha

Volume

1000 mm = 1 cm^3
1000 cm = 1 dm^3
1 000 000 cm^3 =
    10^6 cm^3 = 1 m^3

Mass

1000 mg = 1 g
1000 g = 1 kg
1000 kg = 1 t

Capacity

1000 mL = 1 L
1000 L = 1 kL

Scope and Sequence

Kindergarten
- Comparison—size, length, capacity, temperature.
- Length—using arbitrary units.
- Capacity—holds more or less.
- Time—duration (more time or less).
  Sequence—first, next, last.
  Night and day
  Clock face
- Calendar
- Money—coin recognition
       (nickel, dime, penny).
- Money—value of pennies.

First Grade
- Length comparison.
- Measuring with arbitrary units.
- Measuring with centimeter ruler and inch ruler.
- Capacity—more, less using liter, cup, quart, pint.
- Mass/weight comparison using kilogram, pound.
- Temperature comparison.
- Time to hour and half hour.
- Digital clocks.
- Calendar—days of week.
- Comparing time duration.
- Money—value of all coins except half dollars.
- Counting collections of coins.

Second Grade
- Linear measurement with centimeter, millimeter, meter, kilometer, inch, and foot.
- Capacity using liter, cup, pint, quart, gallon.
- Mass/weight using kilogram, gram, pound, ounce.
- Temperature comparisons.
- Time to nearest minute, hour and half hour.
- Calendar—days of week, months.
- Digital clocks.
- Money—values of all coins.
- Counting collections of money.
- Adding and subtracting money.
- Compare prices with amount of money.

Third Grade
- Linear—centimeter, meter, millimeter, kilometer.

- Equivalent metric measures.
- Using inch, half-inch, foot, yard, mile.
- Capacity—metric cups, liter, milliliter, cup, pint, quart, gallon.
- Equivalent metric and customary measures.
- Mass/weight—gram, kilogram, ounce, pound.
- Temperature—Celsius and Fahrenheit.
- Choosing sensible metric measures.
- Perimeter and area with grid squares.
- Volume—counting cubes.
- Time—earlier and later.
- Minutes before and after; a.m., p.m.
- Time to nearest minute; elapsed time.
- Calendar concepts.

Fourth Grade
- Linear—centimeter, millimeter, meter, kilometer, inch, half-inch, quarter-inch, eighth-inch, foot, yard, mile.
- Capacity—milliliter, liter, cup, pint, quart, gallon.
- Equivalent metric and customary measures.
- Mass/weight—gram, kilogram, ounce, pound, ton.
- Temperature—Celsius and Fahrenheit.
- Choosing sensible measures.
- Perimeter and area.
- Volume of rectangular prisms.
- Time—a.m. and p.m.
- Using time intervals.
- Days, hours, minutes, seconds.
- Time to nearest minute.
- Money—multiplying, adding, and subtracting.
- Making change.
- Using catalogs, sales, ads.
- Sales tax.

Fifth Grade
- Linear—all previous units.
- Capacity—all previous units and fluid ounce.
- Mass/weight—all previous units and milligrams.
- Temperature.
- Choosing sensible measures.
- Equivalent metric and customary measures.
- Perimeter of polygons.
- Area of rectangles.
- Circumference.
- Surface area.
- Volume of rectangular prisms.

(continued)

- Time zones and changing units.
- Adding and subtracting time.
- Money—all operations.
- Unit price—better buy, discounts.

Sixth Grade

- Linear—all previous units and dekameter, hectometer, sixteenth-inch, miles.
- Capacity—all previous units.
- Mass/weight—all previous units.
- Choosing sensible measures.
- Operating with measures.
- Perimeter and area with formulas including circles, triangles, rectangles, parallelograms.
- Surface area—rectangular prisms, cylinders, pyramids.
- Volume of rectangular prisms, cones, pyramids, cylinders.
- Time—review all previous concepts.
- Money—installment buying, checking accounts, best buy.

Seventh Grade

- All previous units.
- Mass/weight—metric ton.
- Operating with measures.
- Choosing sensible measures.
- Perimeter, circumference, area of trapezoids.
- Surface area—triangular prisms.
- Volume.
- Measuring angles.
- Time—all review.
- Money—using operations and finding unit price, sales tax, discounts, simple interest.

Eighth Grade

- Review of all previous units.
- Review of perimeter, area, volume, surface area, circumference.
- Measuring angles.
- Scale drawings.
- Time—24-hour notation.
- Money—Income tax, sales tax.
- Simple and compound interest.
- Percents and buying on credit.

Teaching Strategies

Linear Measurement

In order to teach linear measurement, activities should begin with concepts familiar to children. Direct comparison activities on height, width, and length can be achieved with their bodies, pencils, and miscellaneous objects. Comparison of lengths such as one object is longer, shorter, or the same as another is essential. Comparisons should be followed by non-standard measures such as strides, digits, palms, or paces. The children may compare their own strides with those of others in the class. Results should indicate to the children a need for a standard measure. The non-standard units can be coordinated with standard measures by discovering how many pennies long a meter stick is or how many paper clips wide a book is.

Children eventually find that the use of some unit of measure helps them in making comparisons. Before children enter school, they begin using body parts as measuring units.

Activity

ESTIMATION

Different Ways to Measure Length and Width

How many different ways can you measure the length and width of the table using your body? List at least five ways.

At first children use units that are larger than the objects being compared or measured and later move to smaller units, counting how many smaller units are in the object(s) being measured. Finally, although the body is still indirectly involved, children begin using various objects as the unit of measure.

Activity

PROBLEM
SOLVING

ESTIMATION

Guessing Metric Measurements

Directions: How good are you at guessing your metric measurements? For this activity you will need a meterstick and a piece of string.

Procedure:

1. Using string, have a partner cut a piece as long as you are tall. How does that length compare to your arm span?

2. Count how many times the string can be wrapped around your head, waist, ankles, wrist, neck, and thigh.

3. Cut a piece of string that you guess would go around your waist. Don't measure until you have cut the string. Now try it around your waist. How many centimeters off were you? _____ cm

   Your waist is _____ cm.

4. Repeat this activity for other parts of your body.

   Your neck = _____ cm. Answers will vary.

   Your wrist = _____ cm.

   Your thumb = _____ cm.

5. Your waist measures about _____ times your neck.

   Your waist measures about _____ times your wrist.

   Your waist measures about _____ times your thumb.

6. Make a bar graph representing the different lengths of each body part.

Working with parts of units is preliminary to working with standard units, such as inches, feet, and yards. The more ways in which children learn to divide non-standard unit measurements into parts, the better prepared they will be to understand how to use standard units of measure.

After working with non-standard units of measure, they soon find that the same type unit differs in length (people's feet differ), and consequently different people measuring the same object obtain different results. As the need to interpret and communicate measurements arises, the usefulness of these conventional standard units becomes clear. We now use a ruler marked in standard units to measure length.

The next step is teaching the children to use the ruler (Thompson and Van de Walle, 1985). The 1986 National Assessment of Educational Progress for Mathematics(NAEP) indicates that students' ability to work with measurement instruments may be limited to simple situations as suggested by the results of two items involving pictures of rulers (Kouba et al., 1988). Little difficulty was noticed when the ruler was aligned correctly. However, when the situation required a mental transformation to determine the length of the segment being measured, few third-graders and only about one-half of the seventh-graders could choose the correct length. Students need to be instructed to use the ruler correctly by carefully aligning the beginning location and reading the length. First, they learn to read centimeters on the ruler. To practice this, the children could measure the paper to find its length and width in centimeters. Larger objects requiring meter measurements should then be introduced as well as perimeter problems. Situations where the millimeter and kilometer would be used should also be introduced. The children should be made aware through illustration that the approximate measure depends upon the unit used. As an example, one does not measure the size of a book in kilometers or the distance from Phoenix to Los Angeles in meters. Frames of reference should be established for each unit such as "my waist is 1 m high"; "a dime is 1 mm thick"; "it is 1 km across the school yard"; and "my little fingernail is 1 cm wide." These benchmarks are essential for establishing a comfortable feeling and a means of communication for distance measures.

Activity

ESTIMATION

Estimation with Large Measures

Directions:

1. Have the class line up on a start line. At the signal to begin, all participants should go forward to their best estimate of a distance of 50 meters. Measure the results to determine the first three places based upon those students nearest to 50 meters.

2. Metric Olympics could be planned for an entire class with activities such as a 25-meter measure, a 100-meter measure, or create a 1-kg clay ball.

3. Have teams of students each estimate one hectare. Verify the estimations by measuring each other's estimate. Rank order the groups by best estimate.

When students have gained facility with using standard units of measure, they will again begin finding discrepancies in measurements. They may even come to the conclusion that measurements can never be exact. Discussions and activities surrounding a theme of measurement accuracy are a natural. The following two activities are examples of ways to blend several topics together in a busy classroom.

Activity

MATH
ACROSS THE
CURRICULUM

Map Scales in Social Studies

Directions:

1. Find a highway map of the state in which you live.

2. Using the legend for the map scale, measure the distances via air and highway.

3. Compute the miles or kilometers to and from all major cities in the state.

4. Compare results to actual distances recorded by air maps and highway department maps.

5. *Extension Activities:*
   a. Ask an architect for old blueprints to see the scale on paper that produces a room of actual size.
   b. Map the classroom using a similar scale to the architect's map.

Activity

CALCULATORS

PROBLEM
SOLVING

Metric Measurements

Directions:

1. Find the thickness of one page of a book. You will need a metric ruler, thick book, and a calculator.

2. Open the book to page 1 and to the last page. This will actually be how many pieces of paper?

3. With your ruler, carefully measure the thickness of this part of the book in millimeters.

4. To find the thickness of one page, divide your measurement by the number of pieces of paper.

   Example: 300 pages measured 25 mm. Divide the 25 mm by 150, using a calculator.

5. How thick is the book?

Another aspect of teaching measurement involves geometric measurement. There are many terms that label parts of shapes. Length and width are designated as the sides of rectangles. The circumference is the distance or perimeter of a circle. The diameter is the length of a straight line drawn from side to side through the center of the circle. The radius is the distance from the center to a side. The formula for circumference is:

$$C = \pi d \quad \text{or} \quad C = 2\pi r$$

where d is the diameter and r is the radius.

Activity

CALCULATORS

PROBLEM
SOLVING

Exploring Circumference and Diameter

Procedure:

1. Use your calculator for the computations. Measure the following items and find the circumference and diameter in centimeters. What patterns or relationships do you notice?

Object	C	D	C + D	C / D	C – D	C × D
Soda can						
Coffee can						
Plate						
Glass						
Waste basket						

Total = _____ _____ _____ _____

(continued)

2. Which column shows a constant relationship between circumference and diameter?

3. What is the average value of circumference divided by diameter?

4. What can you say about the ratio between circumference and diameter?

Mass (Weight)

Mass and weight are often used interchangeably, but their meanings differ. Weight is related to the gravitational pull of the earth. It relates the action of a force (gravity, for example) upon a mass. Mass remains the same regardless of location.

Research (Bitter, Mikesell and Maurdoff, 1976) has shown that children find measuring mass a more difficult concept to undertake than measuring distance. Therefore, the teaching of mass is usually accomplished through indirect comparisons, although direct comparisons are equally important. Directly comparing trucks, cars, dolls, blocks, and miscellaneous toys by hand comparison is common. However, most comparisons are with the use of a balance where object A is compared to the known mass piece and then object B is compared to the known mass piece. The end result is a statement that object A is more or less massive than object B and also that object A has a greater or smaller mass than the mass piece. This direct comparison involves knowledge of a standard unit which should be introduced after non-standard units. Some non-standard materials that can be used in conjunction with a balance for units of mass (weight) are nails, pennies, interlocking cubes, and clay.

Activity

ESTIMATION

MATH ACROSS THE CURRICULUM

Finding Mass with a Balance in Science

Directions:

1. Build a balance to find the mass of objects using pennies, interlocking cubes, or clay, and make comparisons to determine objects with the greater mass.

For example: 5 nails have a greater mass than 2 pennies or 3 pennies have the same mass as 10 interlocking cubes.

2. Estimate and record the answers before finding the actual mass.

3. Record keeping should indicate the status of the estimates. Hopefully, these estimates improve with experience.

After comparison of masses (weights), students will discover that a more standard measure is needed. Each measurement unit should be introduced in terms of a frame of reference. A milligram has about the same mass as a drop of water. A nickel has a mass of about 5 grams. A kilogram is about the mass of this book and a tonne is used for heavy measures. As these frames of reference are developed through experimentation, ample time and instruction should be provided for students to read a balance as well as a scale. Obviously, the kilogram will mean more to students if they can record their own masses in kilograms. Smaller items, such as paper clips, nickels, and raisins can be measured in grams. The milligram and tonne are more difficult to introduce, but awareness can be realized by magazine articles, pictures, and stories indicating their use and establishing a sense of what they are.

In an attempt to improve estimation, the teacher may give statements that contain limits.

Such statements might be: it has a mass of more than one gram but less than 5 grams, or the student's mass is between 50 and 75 kilograms. Have several students ask which objects are about equal in mass, have the greater mass, or have the least mass. The statements may be verified by actually finding each object's mass and recording it on a graph. Graph results should be discussed and the original statements should be validated. These final procedures should help in understanding the measuring of mass.

One of children's first encounters with the concept of mass (weight) measurement is comparing objects in their environment to see which is heavier. While the "hold one in each hand" method works well for objects that are quite different in mass (weight), more sophisticated methods must be found for objects that are quite close to the same mass (weight). Exploring and experimenting with the balance will eventually lead children to an understanding of heavier and lighter in terms of the balance. An improvised teeter-totter (a plank over a brick) can provide many meaningful experiences.

The use of rubber bands will probably be the child's first experience with spring scales. Light and heavy objects should be suspended by elastic bands and their length noted. Similar experiences can be provided with springs of all kinds: fragile and strong, long and short, extension and compression.

Eventually, the need to interpret and communicate weights leads to the desire for standard weights. Many schools have sets of standard weights so students may feel and experiment with them. Students should experience activities with the metric system as well as with the customary system.

Activity

CALCULATORS

Mass with Silver Dollars

Directions:

PROBLEM SOLVING

1. If a silver dollar has a mass of 18 g, use a calculator to determine how much money you would have if you had a kilogram of silver dollars.

2. What would be the mass of 1000 silver dollars?

3. Use a calculator to determine how many silver dollars you are worth based on your mass.

4. Do the steps above using a nickel as the unit of measure.

Area

The concept of area can be established through the use of manipulatives such as the geoboard, colored rods, or blocks. Using the geoboard, the establishment of a non-standard square unit can lead to finding the area of numerous regions as in Figure 4.1.

As the student constructs the square unit on each figure, there is the realization that a square

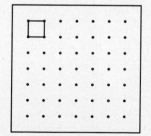
Area = 1 sq. unit   Area = 2 sq. units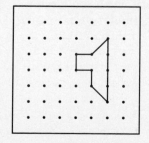
Area = 4 sq. units

Figure 4.1

unit takes up a certain amount of area regardless of the shape of the original figure. A vocabulary and a frame of reference still need to be established. Once a unit of area, such as a square, a hand, or a triangle, is established, the area of larger items like books and tables can be found. However, for communication of this unit of area, a standardized unit is required. The direct comparison of which book has more area is easily accomplished without the standardized unit by placing one on top of another. As they become more nearly the same in area, measurement becomes necessary. In order to investigate which desk has the most area, newspapers, books, or bodies can all be used as measures for indirect measurement. An intriguing question might be to discover whether the round or the square table has greater area. How can this be solved? This activity leads to perimeter measure as students will probably measure the distance around. By estimating how many units will fit on each table, a specific area measure can be used to determine which has the larger area. This activity can be reinforced by having children trace hands on squared paper and then count the number of squares covered. Estimation will be required to make whole squares. Understanding the area concept can be reinforced by giving each student five tagboard squares, 5 cm on a side. Have each student make a design. Ask what is the area of each design. Will the area be the same for any two designs? Likewise, perimeter measures should be taken of each design. Are the perimeters of each design the same? Why or why not?

The standard unit of area for small surfaces such as a book, notebook, or desk is the square centimeter. The square meter is appropriate for determining the area of rooms in a home or school. For example, finding how much carpeting is needed to cover a certain number of square meters or how much wallpaper is needed to cover a certain number of square meters. Frames of reference can be established for the square centimeter and square meter through experimentation.

Hectare is the unit used for small land measure while the square kilometer is used for measuring the areas of large land and oceans. A region of land will be needed to establish a frame of reference for the square kilometer and hectare.

Results of the fourth NAEP mathematics assessment report that students appear more familiar with smaller units of measure than with larger units (Kouba et al., 1988). Both third-grade and seventh-grade students have more experience measuring with smaller units like centimeters which they can relate to the physical world. Assessment items with pictures furnished visual clues that helped students.

One of children's first encounters with the concept of area measurement is in measuring body parts. Children have fun with activities of this nature. A similar activity is to trace around the child's body as the child lies on butcher paper. In addition to comparing their bodies, children might want to see how many children it would take to cover a wall. When children assume different action positions (running, jumping, crouching, etc.), some interesting tracings result, offering a unique opportunity for art experiences as well.

Activity

With Centimeter Graph Paper

Directions:

1. Place your hand with the fingers closed on centimeter graph paper. Draw around your hand.

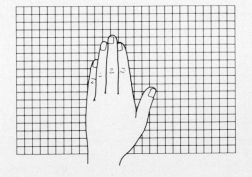

2. Guess the area of your hand in square centimeters. _____ cm²

3. Count the square units to find the area of your hand. You should count a square if more than $\frac{1}{2}$ of it is inside the outline of your hand.

4. The area of my hand is about _____ cm².

5. Now draw your foot on centimeter graph paper. Try to make a better guess.

6. Guess the area of your foot in square centimeters. _____ cm²
   Now count the square units to find the area of your foot.

7. The area of my foot is about _____ cm².

Children eventually find that the use of some unit of measure helps them in making comparisons. They most naturally begin by using various parts of the body as possible measuring units. Children may see how many hands or feet, for example, it takes to cover their body tracing. Eventually, however, children begin to use units not so directly related to their bodies. As the need to interpret and communicate measurements arises, the need for standard units becomes apparent.

Activity

ESTIMATION

PROBLEM SOLVING

Using a Table and Square Centimeters

Directions: Complete the table by estimating and measuring area in square centimeters.

Object	Estimate Area in cm²	Measure Area — cm²
a quarter		
your closed hand		
bottom of a 1 lb coffee tin		
3 x 5 index card		
heel on your shoe		
a picture of a lake		
surface area of a block		
surface area of a pyramid		

The unit of area measure is a square unit which is a square with each side equal to one unit. Although the square inch, square foot, and square yard are the most common units of area measure in the United States, most of the world uses the metric system. Hence, children should experience activities with the square meter as well as the usual standard units. The activities should lead to discovery or reinforcement of the area formula of the square (s^2), rectangle ($L \times W$) or circle ($\pi\, r^2$). The square unit has each side of length, s, with the area formula of the square (s^2).

Activity

ESTIMATION

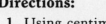

PROBLEM SOLVING

Estimation with Graph Paper

Directions:

1. Using centimeter graph paper (centimeter squares), draw possible rectangles with a constant perimeter of 20 cm.

2. Record the dimensions of each rectangle and its area.

3. Which rectangle has the greatest area?

4. Using scissors, cut out the rectangles and arrange them in some order. Do you see any patterns? Describe them.

5. Try rearranging them to form other patterns.

Activity

MANIPULATIVES

ESTIMATION

Metric Measurement

Directions:

1. Estimate the area of your room in square meters.

2. Construct a square meter from cardboard or wrapping paper.

3. Use your square meter to again estimate the area of your room.

4. Discuss the differences in your estimates.

5. Experiment to come up with a formula for finding the area of a room.

Activity

COMPUTERS

PROBLEM SOLVING

Calculating with a Spreadsheet

Directions:

1. Prepare a spreadsheet for the following chart.

2. Have the spreadsheet calculate any missing information.

Area	Radius	Diameter
379.98 cm^2		
256.78 cm^2		
		625.47 cm
385 cm^2		
3215.36 cm^2		
		47.5 cm
	8.4 cm	

3. Make any generalizations you can from the data.

Here is another example of the possibilities of spreadsheets when studying the measurement of circles.

Activity

COMPUTERS

Exploring Circles with Spreadsheets

Directions:

1. Problem to explore: The WKYK television station can transmit broadcasts as much as 50 km away. How large an area is within the listening area of WKYK station?

2. Draw a circle on the board to represent the problem with the midpoint labeled WKYK. Draw a radius and label it 50 km. How do you find the area of the circle?

3. What other information about a circle can be determined by knowing the value of the radius?

4. Use the spreadsheet to experiment with different radii to determine the corresponding values of the diameter, circumference, and area. How are these four circle measurements related?

Capacity and Volume

Capacity and volume are referred to as synonymous terms in most elementary programs. These two terms have different dictionary meanings. Volume is defined as how much space a region takes up, while capacity is how much a container will hold. Therefore, anything that can be poured is usually measured in capacity units while the space in a room is measured in volume units.

In working with young children, the terms can be used interchangeably, but middle grade students will need to distinguish between them. To begin working with these units, estimating how much a jar holds or which container holds more is necessary for a grasp of the concept. Often children confuse the measurement of mass and capacity; therefore, different size and shape containers are needed for estimation practice. In order to make the distinction, several different sized containers can be labeled and sand or cereal can be available to fill them. Have the children estimate which container holds the most, the least, or about the same as an identified one. Next, have them verify their guesses. The children may also order the containers on the basis of how much they will hold, their heights, widths, or top circumference. Teams could be selected to perform these activities. This gives the children a feeling of capacity depending upon size and shape, not just size, as is often the case. Data from the fourth NAEP mathematics assessment suggest that third-grade students appear to have an intuitive notion for volume as related to the shape of an object (Kouba et al., 1988). Almost 90 percent of them correctly identified which vase held the most water.

This activity can be followed with the question, "How much does each container hold?" The children must select a non-standard unit to measure its capacity. They could give their answers in paper cups full, styrofoam cups full, or whatever units are selected. More specific activities would be to guess which holds more than a liter, less than a liter, more than a milliliter, and so on. The answers should be proven using the standard measures of milliliter or liter. Graduated cylinders and beakers are convenient to carry out the standard measure applications. Finally, a frame of reference can be established, such as a can of soda holds _____ mL. Familiarity with the unit has been established with the standard measure so the student would know when to use milliliters or liters. Frames of reference could be 5 mL = 1 teaspoon, _____ liters = amount of water you drink a day. Discuss what units are used to measure milk, medicine, or soda pop.

For volume measurement, the preceding procedures are appropriate with the exception of the units. Remember, volume is the space that is occupied. The volume of the room is the number of cubic meters that will fit into the room. This includes length, width, and depth. For small measures such as the volume of a shoebox, use the cubic centimeter. Construct a cubic centimeter and have the children estimate how many will fit into a shoebox. Likewise, construct a cubic meter and have the children estimate how many cubic meters would fit into the classroom.

Computing the volume of the shoe box in cubic centimeters and the classroom in cubic meters using length, width, and depth (height) measurements would help establish the formula for volume as length × width × height. The es-

timates could then be compared with the actual computed measurements for establishing frames of reference for cubic meter and cubic centimeter. Which unit will be used for finding the volume of a basement or the volume of a dresser drawer?

In summary, the liter and milliliter are normally used for capacity unit measures while the cubic centimeter and the cubic meter are common measurement units for volume. In the metric system, the volume, capacity, and mass measurements of water have a commonality: a cubic centimeter of pure water has a mass of one gram and a capacity of one milliliter.

Activity _____

MANIPULATIVES

Cubic Centimeters

Directions:

1. Construct a cubic centimeter from heavy cardboard or with any other non-porous substance.

2. Fill with one milliliter of water.

3. Now set up an experiment to verify that it has a mass of one gram.

4. Experiments can also verify this for larger units.

Since volume and capacity are much more sophisticated concepts than either length or area, children usually gain an understanding of volume later than length or area. Children should begin their work with volume and capacity by freely playing and experimenting with pouring sand and water into containers. In so

doing, they are dealing with three-dimensional space and begin establishing basic notions about it. Although they may not wonder about how many cups of sand or how much water they have for some time, they will begin to wonder which of two containers holds more or how much water they need to fill an aquarium.

Activity _____

PROBLEM SOLVING

Comparing Volumes and Capacities

Materials Needed: A piece of construction paper or clear acetate, scissors, tape, some beans or other dry material, and a pan.

Directions:

1. Cut the construction paper or acetate in half.

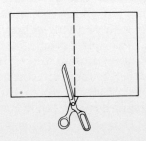

2. Make a round tube by rolling one piece of the paper the long way. Put one edge of the paper over the other and tape them together.

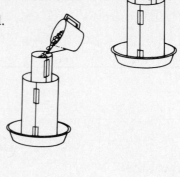

3. Make another round tube, but roll the piece of paper the other way to make a shorter, fatter tube. Put one edge of the paper over the other and tape them together.

4. Place the shorter tube in the pan. Put the longer tube inside the shorter tube.

5. Fill the long tube with the dry material.

6. Mark where you think the dry material will come on the shorter tube if you pull the long tube out.

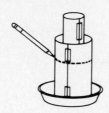

7. Now pull the long tube out. How close is your guess?

8. Does the answer change if different dry materials are used such as sand or popcorn?

9. Verify the outcome by finding the volume of each using the formula of a cylinder ($\pi r^2 h$).

The strategy of comparing volumes and capacities by pouring water from one container to another is not easy for children to grasp. They find it difficult to believe that the quantity of the contents remains the same even though the shape of the container changes. Piaget claims that most children learn to conserve volume at about age eleven (Labinowicz, 1985). Sometimes the water or sand that fills one container does not fill another. Is the first container bigger because it is full, or is the second one bigger because more could be added to it? What if the water or sand overflows? Which container is larger then?

Children must work through these puzzling questions in their own way and in their own time before they can be expected to make comparisons and measurements with any degree of understanding. Since there are so many factors to consider, volume and capacity concepts naturally evolve later than length and area concepts. Once children have begun making volume comparison, they may begin considering how much one container holds.

When sand or water is placed in many containers it simply runs out. After some experimentation children find that blocks or sugar cubes will better account for the space in such containers. Thus, they have begun to get the idea of the cubic unit of measure.

As the need to interpret and communicate measurements arises, the need for standard units becomes apparent. Although the inch-cube, foot-cube, yard-cube and gallon are the most common units of volume measure in the United States, most of the world uses the metric system. Hence, children should experience activities with the meter-cube and liter as well as the usual standard units.

Activity

Building Cubes

Directions:

1. Use a centimeter ruler to measure the length, width, and height of a block.
2. Build larger and larger cubes using blocks.
3. Record the number of blocks on any edge and the volume for each of these larger cubes.
4. What patterns do you observe?
5. Generalize to a formula for volume of a cube.

Activity

Using Volume

Directions:

1. What is the volume of your body?
2. Since your body is almost entirely water and water has a constant weight, you may find this to be an easier problem than you thought.
3. Use a calculator to compute the volume of each person in your group and record this data.
4. Generalize the results.

When students have gained facility using standard units, they will begin finding discrepancies in measurements. How accurate are the measurements? Think about what degree of accuracy is needed for real-life situations.

Activity

Using Formulas

Directions:

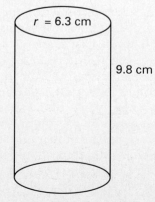

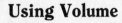

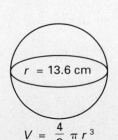

$r = 6.3$ cm

9.8 cm

17 cm

$r = 8.5$ cm

$r = 13.6$ cm

V = Area of base x height $V = \frac{1}{3}$ Area of base x height $V = \frac{4}{3} \pi r^3$

1. Use the given formulas and dimensions, and a calculator to find the volume of the figures shown.

2. Use the cylinder shown to answer the following questions. Using a calculator, find the volume of the cylinder if:
   a. The height is doubled.
   b. The radius is doubled.
   c. The radius and the height are doubled.
   d. The height is halved.
   e. The height and the radius are halved.
   f. The radius and height are multiplied by 5.

3. Follow the same procedures for formulas using the cone and sphere.

Time

Time is another aspect of measurement and is usually introduced with clocks and calendars. A clock measures the passage of time for each 24-hour period. Early experiences that are indirectly and unintentionally given to young children do not help them develop a clear feeling about the passage of time or the meaning of a minute or a second. For example, many adults and parents are guilty of saying phrases like:

"I'll be back in just a minute," or "Just a second and I will help you." These inadvertent comments distort the child's understanding of how long a period of time is represented by the terms "minute" and "second."

Some of the children's first encounters with the concept of time measurement are general comparisons of which person or task took the least amount of time. Like other areas of measurement, since you cannot see time, you must infer it from observations of other things.

Activity

ESTIMATION

PROBLEM
SOLVING

Simultaneous Comparisons

Directions:

1. Drop two different size balls (without a push) simultaneously from the same height.

2. Which one takes the least amount of time to reach the floor? How can you tell?

Eventually, the need to communicate and interpret time leads to the desire for standard units of time: second, minute, hour. The teacher should give children opportunities to experience one-minute intervals to help them gain a better perception of the passing of time. To help children develop a sense of time, have them do an activity, such as counting, bouncing a ball, listening to a story, or walking about the room, for a minute. Later, have children estimate one-minute intervals. Have them close their eyes for about how long they feel a minute lasts and

when they feel a minute has elapsed, indicate with a "thumbs-up" sign. Generally, there is great variation among a group of children, which may be due to a lack of an intuitive feeling about the basic unit of time, a minute. Help children become more aware of time intervals such as one minute, five minutes, ten minutes, and longer. This experience will clarify notions children have about the duration of time. The following activity stresses durations that go beyond one minute.

Activity

MATH
ACROSS THE
CURRICULUM

Use of the Egg Timer in Science

Directions:

1. Using an egg timer, record how many times you can do the following activities before the sand runs out:

    touch your toes, bounce a ball, play a chord on the piano, or hop in place.

2. Record how many seconds it takes for all the sand to run out.

3. Design a timer of your own, i.e., water dripping from a can. Explain how the pioneers may have used your timer.

To understand the concept of the passage of time, experiences should include activities to develop a sense of yesterday, today, and tomorrow along with the idea of the continuation of time through months. Daily calendar activities provide an excellent base. Children want to know how many more days until a special event such as a vacation, a holiday, or a school happening. With a calendar available as a reference point, the child builds some clear sense of the passage of time. The calendar shown in Figure 4.2 illustrates the many concepts included: yesterday, today, tomorrow, days of the week, days in a week, days in the current month, name of the current month, special days that month. In the primary grades, including a pattern for the days adds interest, builds an awareness of the previous month and the coming month to see the continuation of time, and reinforces previous patterning skills. During the calendar time of each school day, the teacher can include discussion about the weather and temperature, which can be excellent graphing activities (covered in Chapter 12), events which will occur that day, and any sharing children might want to do.

Understanding time sequence is another aspect of developing an awareness of time apart from the simple reading of a clock. Have children arrange pictures of various events to indicate the order in which the events happen. For

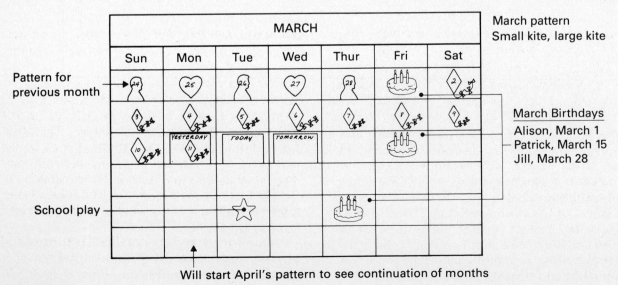

Figure 4.2

example, the pictures might include: a child brushing his or her teeth, eating breakfast, just waking up, and leaving for school; another might be: a child with a deflated balloon, with a large inflated balloon, the popped balloon, and the child just beginning to blow up the balloon. The task is to arrange these pictures in a logical time sequence. First experiences should involve comparing events during longer time periods such as events in their day: eat breakfast, go to school, play after school, go to bed. Many commercially made time sequence cards are available. In the classroom, many opportunities are present to discuss time sequence. For instance, the teacher might say, "First I want you to get

your crayons, then color the worksheet, cut out the pictures, and put them in the collection tray."

Children also need to develop a sense of the duration of various events and decide which event takes longer. Have children think about how long it takes to do certain things. An activity might be to look at pictures of pairs of events and decide which takes longer: brushing your teeth or taking a bath; walking to school or time spent in school; giving a dog a bath or playing a soccer game. The problem of comparing events which cannot occur simultaneously or of indicating "how long" for a single event leads children to the use of various units of time measure.

Activity

Using a Pendulum

Directions:

1. Set up a pendulum with a string and any object attached to the end of the string.

2. Can you change the beat of a pendulum? In what way(s) can you do this?

3. Make two pendulums that swing the same. Then alter one so that it swings twice as often as the other.

4. Use the pendulum to compute how long it takes to walk the length of the room.

Once the children are ready to begin telling time, clock reading skills can be introduced. Children must learn to read a variety of clock faces with hands that have different sizes, colors, and lengths. Some clocks have a second hand which may add confusion. Although digital clocks are popular and easier to read initially, they do not allow children to see the relative position of the displayed time to the next hour or to the times that come before or after it.

To simplify the task of learning to tell time, one approach is to begin with only an hour face clock. Using only an hour hand allows children to read the time as soon as the numerals 1 to 12 are recognized. The hour hand can be moved to a position a little past the hour and the child can learn to interpret this as "a little after"

and when the hour hand is almost to a number the child can confidently call this "almost _____, o'clock."

When children have success with the hour hand and feel comfortable with it, the minute hand can be introduced. This usually begins with the hour and half-hour times. The child should find the shorter hand first, read the hour, then look at the longer hand. A large wooden clock with gears is helpful for children to see the hands moving together. Allow each child many opportunities to manipulate the clock to explore its features.

When children are ready to tell time to the nearest five minutes (usually in second grade), a prerequisite is being able to count by fives to 60. A paper plate clock may be made for the child to practice counting by fives as each numeral on the face is touched (Fig. 4.3). The teacher should emphasize that these numbers indicate minutes after on-the-hour times. To simplify the task of understanding time, we suggest that children learn only to give the minutes after the hour. This is in agreement with reading time on digital clocks and reduces teaching the concept of "minutes before the hour."

Children need practice seeing how the time represented on a digital clock would look on an analog clock. They can begin with the digital time and use Logo to see the analog time. The following activity is adapted from Babbie (1984, p.327).

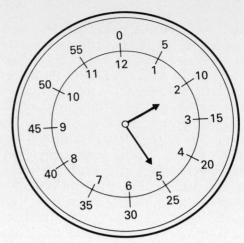

Figure 4.3

Activity

COMPUTERS

Analog Clock in Logo

Directions:

1. This program is saved under the name CLOCK on the Logo disk.

2. Type LOAD "CLOCK and follow the directions in step 3.

   (If you have no disk, type the following procedures:)

   ```
   TO MINUTE :M
   RT 360 * (:M/60) FD 45
   END

   TO HOUR :H :M
   RT 360 * ((:H + (:M/60))/12) FD 30
   END

   TO CLOCK :H :M
   PE LT 90 FD 55 RT 90 PD
   CIRCLER 55
   PU HOME PD
   HOUR :H :M
   PU HOME PD
   MINUTE :M
   END
   ```

3. To see 8:30 on the clock, type:

   CLOCK 8 30   (Notice that the colon is left out because Logo reserves the colon for variable designations in procedures.)

   Try several more.
   (*Remember* to CS (clear screen) between examples.)

4. Comparisons can be made from one clock face to another if the CS command is left out. Then one clock will be drawn over the other. Change the color between entries to distinguish one time from the other.

For example, type:

```
CLOCK 3 20
SETPC 2
CLOCK 3 25        (This program shows comparisons
SETPC 3            between every 5 minutes over
CLOCK 3 30         a 15-minute period.)
SETPC 4
CLOCK 3 35
```

5. *For those who like a challenge:*
  a. What would you change to see 10-minute intervals over a 50-minute period?
  b. ... 7-minute intervals over 20-minute period?
  c. ... 20-minute intervals over 65-minute period?
  Test your predictions with the Logo program.
  d. Make up other variations. Keep a list and share it with others.

Another recommendation is to refrain from teaching the terms "quarter after," "quarter to," and "half past." These are outdated terms and add more confusion since the terms "quarter" and "half" are associated with 25 and 50 in money and now indicate 15 and 30 in time. It is better to reserve these terms for money.

Money

Children enter school with a wide range of competencies in knowing the names and values of the coins. This variation is due to the experiences they have had handling money. Often the underprivileged child has a clearer conception about money than the privileged one who has had limited opportunities to handle money.

Unfortunately, many employers comment and complain about young people's lack of competency in money skills. Although the technological advances in cash registers, calculators, and computers have altered the business scene, there is still an urgent need to better prepare students to face the demands of the job market and the world of the consumer.

Teaching money is far more effective if real coins are used. This may not be a wise situation all the time, but many activities should involve the manipulation of actual coins. Play coins should be selected carefully. Some coins are made from inexpensive plastic with pictures that do not closely resemble the actual heads and tails of real coins. The cardboard punchout coins found in the back of students' textbooks are often good representations. Have the student put these punchouts in an envelope marked with the child's name and keep the envelope in a central location for frequent use. Working in cooperative learning groups proves to be an effective method for students to practice counting coins and making change. Many commercial games and materials are available to provide practice in pretend buying and selling. Coin recognition usually begins in kindergarten, and in first grade heavy emphasis is placed on coin recognition and counting collections of coins. "Counting on" as a method for determining change is usually introduced in third grade and continues throughout the intermediate grade textbooks.

One of the problems associated with learning about coins is the lack of a physical relationship among the coins. Two nickels cannot be shown to be equal in physical value to a dime. Textbooks do not contain sufficient pages to develop coin equivalencies, yet this is a key concept in the study of money. One concrete aid that is available in many schools is Cuisenaire rods. These rods can show the numerical relationships among all coins and the dollar. Figure 4.4 shows how the rods can model money. Bradford (1980) suggests a teaching sequence to be used with these models if children have not worked with Cuisenaire rods. Some activities using proportional models to teach money concepts are included in this section. *A note of caution:* Do not say, "How many pennies are in a nickel?" In determining equivalencies, it is the relative value of the coins and not the physical relationship that is being established. The question should be, "How many pennies does it take to equal the value of a nickel?"

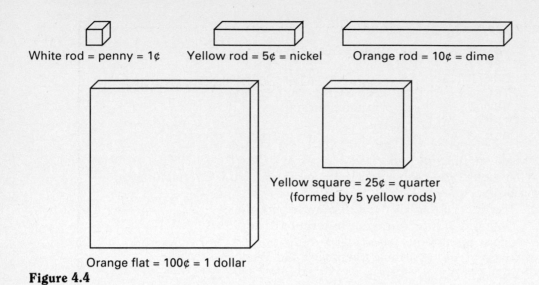

White rod = penny = 1¢   Yellow rod = 5¢ = nickel   Orange rod = 10¢ = dime

Yellow square = 25¢ = quarter
(formed by 5 yellow rods)

Orange flat = 100¢ = 1 dollar

Figure 4.4

Activity

MANIPULATIVES

Understanding Coin Equivalencies with Cuisenaire Rods

Procedure:

1. How many different ways can you cover a "quarter"?

2. How many different ways can you cover a "dollar"?

3. Exchange 4 dimes and 2 quarters for the fewest coins (rods) possible.

4. Exchange 3 nickels and 7 pennies for the fewest coins (rods) possible.

5. How much money is this: 6 pennies, 2 dimes, and 1 quarter?

6. What are the fewest coins you can use to make 43 cents?

Money reinforces place value concepts since it is composed of a base 10 system. In counting numbers, exchanges of 10 individual units for a set of ten can be compared to exchanging 10 pennies for 1 dime. An important aspect of numeration is knowing how to continue a given counting sequence. This skill is needed in counting coins. The child must be able to begin a counting sequence by some value (tens) and switch to other values (fives or ones). This skill is also important in knowing how to "count on" to determine change.

The newspaper and classroom store have long been recognized as useful instructional aids to develop money concepts. The newspaper can provide comparative shopping and estimation activities, including conducting a shopping spree with a given amount of money, figuring percentage discounts on items, saving money with coupons, and monthly installment buying, and discounts for cash. The classroom store offers a direct experience with buying items, paying for them, and deciding the amount of change. Children can serve as the customer or the merchant. Older children can set up and operate a school supply store to sell items such as pencils, notebooks, and school T-shirts. Purchasing imaginary stock and watching the stock market provides another valuable experience with money.

Activity

Customary System of Weight with Money

Directions: Use a calculator and solve the following money problems.

1. Bananas 68¢ per lb.

> 10 bananas in a lb. What is the average cost per banana?

> How much will 3½ lbs. cost?

2. Cookies 49¢ per lb.

3. What is the cost for a 3-mile taxi ride?

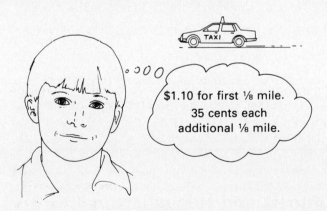

> $1.10 for first ⅛ mile.
> 35 cents each additional ⅛ mile.

4. Using a menu from a local restaurant, plan a meal to feed each member of your class a hamburger and a soft drink. What is the total cost?

5. Using a grocery store ad from a local newspaper, create a budget for a class party.

Temperature

Label a thermometer with a common event for certain temperatures. This helps the child build a frame of reference to relate the number to real events. Careful planning must be included in developing an ability for the child to read the thermometer. Introduce the concept of temperature into the curriculum with concrete experiences and establish frames of reference such as 32°F is freezing and 212°F is the boiling point of water, while in the metric system, water freezes at 0°C and boils at 100°C.

Activity

ESTIMATION

Experiments with Temperature

Directions: Do experiments to fill in the blanks.

Customary System
Body temperature _____ degrees F
Freezing point of water _____ degrees F
Room temperature _____ degrees F
Boiling point of water _____ degrees F

Metric System
Body temperature _____ degrees C
Freezing point of water _____ degrees C
Room temperature _____ degrees C
Boiling point of water _____ degrees C

Activity

MATH ACROSS THE CURRICULUM

Social Studies and Science

Directions:

1. Sketch a map of the United States. Include 25 major cities.

2. Estimate and label the average temperature in degrees for each city.

3. Compare the results to a U.S. weather map.

4. Repeat the activity with your estimate of average winter temperatures.

5. Repeat the activity for average spring and fall temperatures but use degrees F.

Diagnosis and Remediation

Correcting Common Errors With the Concept of Time

Deficiencies in telling time may relate to insufficient real-life experiences reading time on a clock face. Today's child has quick access to a digital clock, which cannot give the needed exposure to telling time in a meaningful way. A child may be able to read "9:42" but will often not know that this time comes between 9:00 and 10:00. Time elapse becomes difficult to relate for the child. These children need experiences focused on constant reading of a clock face. They need to read the time every 5-minutes to actually witness the meaning of an hour and to see how the hour hand moves only a small distance during each 5-minute interval.

Since the minute hand is constantly changing position, some children have trouble seeing time elapse. When the previous minute hand remains in view as the count continues, children can see how many minutes have passed while doing an activity. Thus, the Logo clock minute hands, which remain in sight, provide a benefit over the regular clock.

Activity

Keep Time with a Logo Clock

Directions:

1. This program is saved under the name TIMECLOCK

2. Type LOAD "TIMECLOCK and follow the directions in step 3.

   (If you have no disk, type the following procedures:)

```
TO MINUTECOUNT
CLOCK :H :M
WAIT 3600
MAKE "M :M + 1
MINUTECOUNT
END

TO TIMECLOCK :H :M
MINUTECOUNT
END
```

   Copy the procedures, MINUTE, HOUR, and CLOCK from the previous Logo activity.

3. *What the program does:*
   The previous procedure CLOCK is used with a minute counting command to move the minute hand once every minute from whatever time you choose to start as your beginning time.
   When the next minute comes around, the turtle makes a circle and draws the next minute hand and continues to do so until *CONTROL key* and the *G key* are pressed at the same time. This stops the program. Children can count the number of minute hands to see how many minutes have passed since the last minute.

4. Type TIMECLOCK 3 30 (to start the clock at 3:30). See what happens.

5. Children enjoy taking timed tests with a friend watching the monitor to tell them when their time is up. Academic relay races can be played with two students keeping track of two teams' time on two separate computers.

6. If children have trouble with other elapsed times, the program can be adapted to fit the needed time period.

7. *For those who like a challenge:*
   Change the MINUTECOUNT procedure to count at another time interval.

Two common errors in reading time are reversing the hour hand and the minute hand and reading the actual numbers indicated rather than assigning the minute hand appropriate time values. An example of the first error is the child reading 10:20 as 4:10 and an example of the second error is the child reading 10:20 as 10:04. See Figure 4.5.

To remedy these problems, children should be given a progression of clock faces, each one a little more detailed than the preceding one. This procedure allows children to focus on each part of the clock independently, first working with hour-face clocks then minute-face clocks. A teaching sequence is offered by Horak and Horak (1983).

Another aspect of understanding telling time is to relate the numbers 1–12 with count-

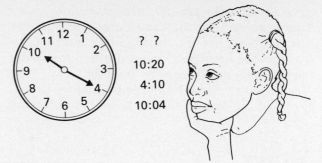

Figure 4.5

ing by intervals of five. A flexible number line might help see this relationship. An activity is provided to show a teaching aid that can be made for the children's use.

Activity

MANIPULATIVES

Making Clocks

Materials: Paper plates
Brad fastener
Grosgrain ribbon or adding machine tape
Permanent marking pen

Procedure: Divide the paper plate into twelve congruent regions and label with the numbers 1–12 as on a clock face. Cut two clock hands from another piece of cardboard or poster board and attach in the center with a brad fastener. Take the grosgrain ribbon and stretch around the circumference of the plate. Cut the ribbon and tape around the plate. Mark the ribbon at each place that matches with the clock numbers. Untape the ribbon and divide each segment into five equal regions. This will look like a flexible number line with the numbers marked in fives to 60 and with the increments by ones noted between the fives.

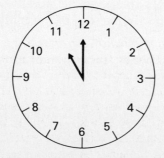

Ribbon with markings

Correcting Common Errors with the Concept of Money

A common error in money skills is the inability to count collections of coins to determine how much money in all. One problem may be the lack of adequate counting skills to count by ones, fives, and tens beginning at various numbers. Children cannot shift from counting by tens to counting by fives and then to counting by ones. The hardest sequence is to start with quarters and then count by fives or tens. Figure 4.6 shows a chart to help children gain flexibility in changing counting sequences. The coins on the chart have been cut and pasted from first or second grade workbooks and laminated onto the chart. When counting a collection of coins, children should first sort them into like coins, then begin counting with the largest value to the least value. In this manner, the child will not have to regroup numbers as often.

Another difficulty with money may be the confusion of the coins themselves. A common problem is distinguishing between a quarter and a nickel. If the teacher concentrates only on size, the child is confused when the nickel is on the page without a quarter as a reference. The nickel is larger in size than the dime or penny, but is it the largest in value? Without the quarter as a reference point, it may be difficult to decide. Also, there is the issue of heads or tails—which is easier for children to identify with the coin? Which face means which coin value? Some educators suggest that instruction should begin with the tails of the coins, however most textbooks introduce coins with the heads. Most teachers feel more comfortable teaching coins with the heads

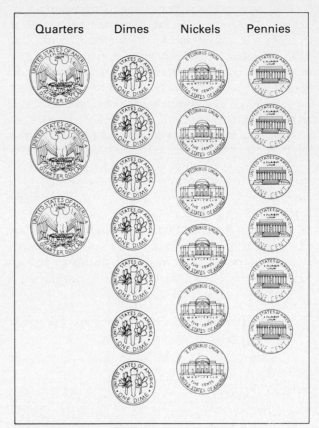

| Quarters | Dimes | Nickels | Pennies |

Figure 4.6

first as they consider this the front side of the coin. Think about this question—heads or tails? This issue deserves your consideration when buying ancillary materials, in evaluating the texts of a mathematics program, as well as when you are walking about the room listening to children count collections of coins.

Summary

Measurement is an often overlooked topic in the elementary school. With the introduction of the metric system, our numeration system, and our monetary system, the measurement system can be taught as an integrated curriculum with each complementing the other. Measuring encourages children's active involvement in solving problems and discussing mathematics. The National Council of Teachers of Mathematics (NCTM) curriculum standards (Commission on Standards for School Mathematics, 1987) state that measurement is central to the curriculum because of its power to help children see that mathematics is useful in everyday situations.

The metric system uses meters, grams, and liters as the base units. Prefixes are uniform throughout the system with milli-, centi-, deci-, milli-, deka-, hecto-, and kilo- being the most common. Celsius is the unit of temperature measure. Volume is measured in cubic units and whole area is measured in square units. Students are encouraged to establish a frame of reference to get an intuitive feeling for the units.

The customary system uses inches, feet, yards, quarts, ounces, gallons, and miles as the most common units. Conversion from one unit to the next varies by unit. Fahrenheit is the unit of temperature measure. Cubic and square inches, feet, and yards are used to measure volume and capacity. The United States is one of the last five countries in the world that continues to use the customary system, although many major U.S. corporations have switched to metric measure.

In the next chapter we will begin the development of the concept of number, known as number readiness. The attributes of the geometry chapter and the understanding of number, one-to-one correspondence, and counting are the thrust of preschool and kindergarten mathematics. We have included geometry and measurement content before the chapter on number readiness so you can use the ideas to develop the understanding of number, number systems, and the use of them.

Exercises

Directions: Read all questions *before* answering any one exercise. Frequently the last question in one category leads to the first question in the next category.

A. Memorization and Comprehension Exercises
Low-Level Thought Activities

1. Compare mass, volume, and capacity of the following

   1 liter = _____ = _____

2. List advantages and disadvantages of the customary system.

3. Discuss the difference between non-standard and standard units. Give examples of each.

4. Describe and give examples of direct and indirect measurement.

5. Describe a series of experiments similar to the measurement model.

6. List the advantages and disadvantages of the metric system.

7. Obtain a copy of the Scope and Sequence of mathematics for your state. Are topics for metric education included? List how your state uses the metric system.

8. Review a current elementary school textbook series for its measurement content. List the topics and concepts emphasized at each grade level.

B. Application and Analysis Exercises
Middle-Level Thought Activities

1. Research the different types of measurement to learn more about them:
   Water hardness
   Mohs' hardness scale
   Richter earthquake scale

2. Set up five events that take one minute; one hour.

3. Develop a classroom thermometer and have labels of similar events at various temperatures.

4. Write a report on the history of the metric system.

5. Write a paper on what measurement is.

6. Suggest some activities in linear measurement for kindergarten and the primary grades.

7. Preview some filmstrips, films, video tapes, video disks, or computer software on the teaching of measurement in the elementary school and write a summary of their strengths and weaknesses. You may want to modify the software evaluation form found in Appendix A for use with films, filmstrips, etc.

8. Write a summary of a research article pertaining to the teaching of measurement in the elementary school.

9. Select a grade level of your choice. Compare a pupil text published since 1986 with one published prior to 1970. Compare the units on measurement. How do they differ in content? How do they differ in process? Prepare a brief, written report.

10. To reinforce math across the curriculum:
    a. List five mathematical questions/problems that come from popular storybooks for children. Show how they are related to the topics of this chapter.
    b. List five ways to use the topics of this chapter when teaching lessons in reading, science, social studies, health, music, art, physical education, or language arts (writing, English grammar, poetry, etc.).

C. Synthesis and Evaluation Exercises
High-Level Thought Activities

1. Develop a computer- or calculator-based unit showing the relationship of the metric system, monetary system, and the base 10 numeration system.

2. Develop a grade level measurement curriculum and outline what measurement objectives will be a part of the curriculum.

3. Develop a unit of 6–12 lessons for teaching at a specific grade level as seen on the Scope and Sequence Chart:

   Linear measure   Mass
   Capacity   Temperature
   Area   Time
   Volume

4. Design a lesson illustrating the relationship between metric measure, our numeration system, and our monetary system.

5. Devise an original measurement project from the classroom environment that calls for determining volume.

6. Plan a metric field day or Olympics using estimation activities as found in this chapter.

7. Develop a measurement lesson based on a commercial software program.

8. Develop several area class activities using the calculator.

9. The database in Table 4.5 was prepared from the National Center for Health Sta-

Table 4.5 Growing Up Differences

Age	Sex	Shortest	Tallest	Difference	Lightest	Heaviest	Difference
2	B	82.5	94.4	11.9	10.49	15.50	5.01
3	B	89	102	13.0	12.05	17.77	5.72
4	B	95.8	109.9	14.1	13.64	20.77	7.13
5	B	102	117	15.0	15.27	23.09	7.82
6	B	107.7	123.5	15.8	16.93	26.34	9.41
7	B	113	129.7	16.7	18.64	30.12	11.48
8	B	118.1	135.7	17.6	20.40	34.51	14.11
9	B	122.9	141.8	18.9	22.25	39.58	17.33
10	B	127.7	148.1	20.4	24.33	45.27	20.94
11	B	132.6	154.9	22.3	26.8	51.47	24.67
12	B	137.6	162.3	24.7	29.85	58.09	28.24
13	B	142.9	169.8	26.9	33.64	65.02	31.38
14	B	148.8	176.7	27.9	38.22	72.13	33.91
15	B	155.2	181.9	26.7	43.11	79.12	36.01
16	B	161.1	185.4	24.3	47.74	85.62	37.88
17	B	164.9	187.3	22.4	51.5	91.31	39.81
18	B	165.7	187.6	21.9	53.97	95.76	41.79
2	G	81.6	93.6	12.0	9.95	14.15	4.20
3	G	88.3	100.6	12.3	11.61	17.22	5.61
4	G	95	108.3	13.3	13.11	19.91	6.80
5	G	101.1	115.6	14.5	14.55	22.62	8.07
6	G	106.6	122.7	16.1	16.05	25.75	9.70
7	G	111.8	129.5	17.7	17.71	29.68	11.97
8	G	116.9	132.6	15.7	19.62	34.71	15.09
9	G	122.1	142.9	20.8	21.82	40.64	18.82
10	G	127.5	149.5	22.0	24.36	47.17	22.81
11	G	133.5	156.2	22.7	27.24	54	26.76
12	G	139.8	162.7	22.9	30.52	60.81	30.29
13	G	145.2	168.1	22.9	34.14	67.3	33.16
14	G	148.7	171.3	22.6	37.76	73.08	35.32
15	G	150.5	172.3	21.8	40.99	77.78	36.79
16	G	151.6	173.3	21.7	43.41	80.99	37.58
17	G	152.7	173.5	20.8	44.74	82.46	37.72
18	G	153.6	173.6	20.0	45.26	82.47	37.21

tistics, U.S. Department of Health and Human Services, 1982.

Prepare a lesson using the database with appropriate questions about measurement, including collecting related class measurement data and comparing it to the national average.

Prepare a lesson that has students create a computerized database with this data and add their class measurement data. Construct problem-solving activities using the database.

10. What kinds of activities prepare students to use discrete measurement systems in everyday situations?

Bibliography

Alexander, F. D. "The Metric System—Let's Emphasize its Use in Mathematics." *Arithmetic Teacher* 20 (May 1973): 395–396.

Ashlock, Robert B. "Introducing Decimal Fractions with the Meterstick." *Arithmetic Teacher* 23 (March 1976): 201–206.

Babbie, Earl. *Apple LOGO for Teachers.* Belmont, Calif.: Wadsworth Publishing Co., 1984.

Bitter, Gary G. *Discovering Metric Measure.* New York: McGraw-Hill, 1975.

————. *Investigating Metric Measure.* New York: Webster/McGraw-Hill, 1975.

————. "Measuring Metrically Pleasurably." *Educating Children* 19 (Summer 1974): 11–14.

————. "Metric Estimation Olympics." *Teacher* 94 (November 1976): 118.

————. *Teacher's Handbook of Metric Activities.* Boston: Allyn & Bacon, 1978. (Duplicator masters and Task cards).

Bitter, Gary G., and Charles Geer. *Materials for Metric Instruction.* Columbus, Ohio: ERIC, 1975.

Bitter, Gary G., and Thomas Metos. *Exploring with Metrics.* New York: Julian Messner, 1975.

Bitter, Gary G., Jerald L. Mikesell, and Kathryn Maurdoff. *Activities Handbook for Teaching the Metric System.* Boston: Allyn & Bacon, 1976.

Bradford, John W. "Making Sense Out of Dollars and Cents." *Arithmetic Teacher* 27 (March 1980): 44–46.

Clason, Robert G. "When the U.S. Accepted the Metric System." *Arithmetic Teacher* 24 (January 1977): 56–62.

Clements, Douglas, and Michael Battista. "Geometry and Geometric Measurement." *Arithmetic Teacher* 33 (February 1986): 29–32.

Commission on Standards for School Mathematics of the National Council of Teachers of Mathematics. *Curriculum and Evaluation Standards for School Mathematics.* Working draft. Reston, Va.: The Council, 1987.

Hallerberg, Arthur E. "The Metric System: Past, Present—Future?" *Arithmetic Teacher* 19 (April 1973): 247–255.

Hart, Kathleen. "Which Comes First—Length, Area, or Volume?" *Arithmetic Teacher* 31 (May 1984): 16–18, 26–27.

Horak, Virginia M., and Willis J. Horak. "Teaching Time with Slit Clocks." *Arithmetic Teacher* 30 (January 1983): 8–12.

Inskeep, James E., Jr., "Teaching Measurement to Elementary School Children." In *Measurement in School Mathematics.* Yearbook of the National Council of Teachers of Mathematics Vol. 38. Reston, Va.: National Council of Teachers of Mathematics, 1976: 60–86.

Kouba, Vicky L., Catherine A. Brown, Thomas P. Carpenter, Mary M. Lindquist, Edward A. Silver, and Jane O. Swafford. "Results of the Fourth NAEP Assessment of Mathematics: Measurement, Geometry, Data Interpretation, Attitudes and Other Topics." *Arithmetic Teacher* 35 (May 1988):10–16.

Labinowicz, Ed. *Learning from Children: New Beginning for Teaching Numerical Thinking, a Piagetian Approach.* Menlo Park, Calif.: Addison-Wesley, 1985.

Mullen, Gail S. "How Do You Measure Up?" *Arithmetic Teacher* 33 (October 1985): 16–21.

Nelson, Glenn. "Teaching Time-Telling." *Arithmetic Teacher* 29 (May 1982): 311–334.

Spitler, Gail. "The Shear Joy of Area." *Arithmetic Teacher* 29 (April 1982): 36–38.

Szetela, Walter, and Douglas T. Owens. "Finding the Area of a Circle: Use a Cake Pan and Leave Out Pi." *Arithmetic Teacher* 33 (May 1986): 12–18.

Thompson, Charles, and John Van de Walle. "Learning about Rulers and Measuring." *Arithmetic Teacher* 32 (April 1985): 8–12.

Walter, Marion. "A Common Misconception about Area." *Arithmetic Teacher* 17 (April 1970): 286–289.

Yvon, Bernard R., John W. Butzow, and Gregory W. Marshall. "Training Metric Leaders." *Arithmetic Teacher* 19 (April 1982): 43–47.

5

Number Readiness–
Early Primary Mathematics

Key Question: *How much should a teacher expect preschoolers and kindergarteners to know about the concept of numbers, and what activities are developmentally sound to use with young children?*

Activities Index

Introduction

Before children can use numbers with meaning, they must have a firm understanding about what numbers are and how to name them. This development period is between the ages of 5 and 7, and is the age span this chapter covers. Young children develop a sense of number through kinesthetic experiences. Objects need to be matched, sorted, grouped, counted, and compared. In order to best understand the importance of these activities, learning theory will be discussed in terms of how children learn mathematics. Piaget's work with young children of-fers the greatest insight into their thinking and developmental stages (refer to Chapter 2).

There are some prenumber concepts that children should acquire before formal work with numbers. Teachers should be able to assess a child's readiness for number by determining what abilities the child has acquired and what perceptions the child has about the world. Prenumber concepts from Piaget's theory of cognitive development will be discussed in the next section: classification, class inclusion, number conservation, seriation, and set equivalence.

Scope and Sequence

Kindergarten
- Identify positional words—top, bottom, left, right, above, below.
- Classify objects by attributes of size, shape, color, kind.
- Compare groups as same, more, less, most, least.
- Recognize, copy, extend, and create patterns.
- Continue a counting pattern.
- Identify groups 0 through 10.
- Write numbers 0 through 10.
- Count groups through 10.
- Order numbers 0 through 10.
- Associate symbols to sets.
- Readiness for addition.
- Readiness for subtraction.

First Grade
- Identify groups 0 through 10.
- Write numbers 0 through 10.
- Order numbers 0 through 10.
- Compare numbers through 10.
- Identify ordinal numbers through tenth.
- Demonstrate operations of addition and subtraction.
- Understand positional words.
- Recognize, copy, extend, and create patterns.

A Child's Understanding of Number

Although Piaget's work concentrated on how children learn, it provides valuable information on understanding how children deal with mathematical concepts and offers ways to determine if a child is ready for certain instructional topics in mathematics. The young child between the ages of 5 and 7 is in the stage of preoperational thought. This means the child's attention is centered on a limited perceptual aspect of an object or event. The child relies on perceptual evaluations in relating to the world. These are some areas of prenumber concepts that reveal how a child perceives number: classification, class inclusion, number conservation, seriation, and equivalence of sets. The child's development in terms of acquiring these skills greatly influences performance level in mathematics. Early primary teachers must have background on how to conduct interviews with children on these Piagetian tasks and how to interpret the findings in terms of curriculum considerations.

might include size, color, shape, thickness, texture, function, or any combinations of these. As for example in Figure 5.1, a collection of three shapes in two sizes might be shown to a child. The child is asked to put the shapes into piles so that all of the objects in each pile are alike in some way. Classification is the earliest stage of logical thinking and is the foundation for graphing. The preoperational child will not determine a classification scheme but will begin with one plan and change as another feature of the material becomes obvious and important. There is no

Task: Classification

Classification experiences involve making decisions about certain attributes of objects and sorting them based on that classification. Attributes

Sort into two groups:

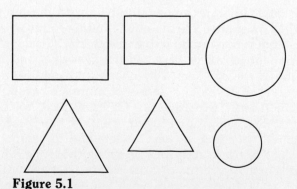

Figure 5.1

consistent thinking strategy for sorting the material. Preschool and kindergarten children sort objects first by color as it is the most salient feature and with maturity and experiences, sorting will be done by shape and size.

Teaching Ideas: Since classifying is a fundamental task in life and appears across the curriculum, forming classes and determining relations needs to be encouraged through many activities. Sorting can be done with shapes, dyed pasta, bottle caps, plastic animals, items from nature (shells, seeds, pine cones, leaves, nuts), anything! The materials should be varied and interesting. Collections of fabric, pictures from magazines, junk such as old keys or nuts, bolts, screws, and buttons offer many varied opportunities for classification. Encourage children to describe and name their sets. Sorting should be done using a sorting area with a defined boundary such as a chalked circle, yarn hoops, or a sorting mat. The teacher should circulate among the children asking key questions: How are these the same? How are these different? Why does this belong in your group? What is your sorting rule? It is important to sort materials many different ways (using different sorting rules). Children need to be encouraged to use different sorting criterion and be flexible.

Associated with sorting is the idea of noting similarities and differences. This involves logical thinking. Children need a wide variety of experiences. The teacher can group children from the class into two sets—long sleeves or short sleeves; with glasses or without glasses; wearing tennis shoes or other types; shirts with logo and shirts without; or harder ones—girls with pierced ears and girls without; tennis shoes with Velcro and those with ties. This can be an exciting way to explore materials in greater detail as well as to prepare children to think logically and draw conclusions. It also leads into graphing as a visual way to describe the results.

Activity

MATH
ACROSS THE
CURRICULUM

Sorting Projects

Directions:

1. Use the list of class members names and have the children sort the names by a given letter.

2. Say a list of short words to the students and have them indicate all those that rhyme and those that do not rhyme.

3. In social studies, have children sort animals as farm or zoo.

4. In science, have children sort objects into those that sink and those that float or into those that a magnet attracts and those that are not attracted. Collect seeds or leaves and sort by type.

Task: Class Inclusion

A part of classification is the ability to see relationships between groups at different levels in the classification system. Grouping on the basis of class is a basis for classifying objects in the physical world. It is related to logical reasoning. To classify objects, their relation to other objects must be known. The idea is that all of one group can be part of another group at the same time (e.g., The group of boys is part of the group of children). The Piagetian task to test class inclusion is discussed first, followed by some additional questions that prove insightful in a child's acquisition of this concept.

Show the child a box containing 20 yellow plastic beads and 7 blue plastic beads. Discuss the properties of the beads with the child—plastic, a hole through it, uneven or bumpy surface, colored. Ask, "Are there more yellow beads or more plastic beads?" The characteristic response of younger children is to answer more yellow beads. Children have problems seeing the relationship of the two classes and end up basing their responses on appearances. This indicates an inability to consider the quantity of plastic beads (the larger and general class) because the answer is based on appearances and the visible, bigger set is the yellow beads.

Additional experiences should include questions about children (are there more boys or more children), animals (are there more cows or more animals), fruit (are there more apples or more fruit). When assessing young children who do not have the class inclusion concept, take the questioning one step further (Fig. 5.2). Suppose you give the child a collection of 5 plastic cows and 3 horses and ask, "Show me all the cows. Show me all the horses. Show me all the animals. Are there more cows or more animals?" The four- and five-year-old child answers, "More cows," as the whole group does not exist. Ask, "Than what," and the typical response is, "Than horses." The child cannot consider the whole because the two parts are horses and cows (what is seen). The child does not have the mental structures to form such classes.

Most children younger than seven have difficulty seeing that all of one group can be some or part of another group. In terms of mathematical logic, a class may be considered in terms of its parts of partial classes. This means that number can be related to logic. Piaget maintains "that both class (in logic) and number result from the same operational mechanism of grouping and that one cannot be fully understood without the other." (Copeland, p. 112). This is directly related to understanding the meaning of addition or number inclusion. Since addition is putting two sets together and naming it as a single number, the child who does not have the ability to place numbers into a mental relationship will have difficulty with the addition principle. The "counting on" strategy for addition

(Chapter 7) holds limited meaning for the pre-operational child and probably should not be attempted. Piaget (1952) contends that children at this stage of development are unable to have reversibility in thinking about whole to part and back to the whole again. Until these concepts are developed, which is around seven years of age (second grade), Piaget concludes: "In a word, it seems clear at this stage (stages 1 and 2) that the child is still incapable of additive composition of classes, i.e., of logical addition or subtraction" (Piaget, p. 174).

A quick assessment can be done by asking a child to count a collection of objects. Then add two more to the group and ask how many things there are now. Many first graders will count the entire group beginning at number 1 rather than arriving at the solution by thinking of the relation of five to adjacent numbers. The unnecessary counting is a result of being unable to count on from the first set and arrive at the solution sooner.

Teaching Ideas: Children need many varied experiences of putting sets together to make the whole. They can make up number stories with counters, connecting cubes, or painted lima beans. Pattern block designs offer opportunities to explore the class inclusion concept. The *Sesame Street* television program has a song, "One of These Things is Not Like the Others." In this activity, children must form mental relationships between objects to determine which one does not belong. Children can compare seasonal pictures by attributes that they have in common (are alike) and not in common (are different). For example, in February pictures of Lincoln and Washington can be compared or in October, compare two large poster pumpkins that are different in several ways. These activities lead to more sophisticated mathematical logic in later grades such as Venn diagrams. Children need to explore relationships according to quantifiers such as "all," "some," and "none."

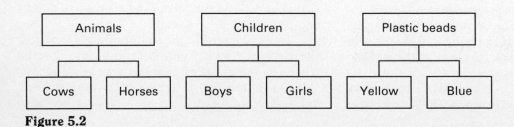

Figure 5.2

Task: Number Conservation

Conservation of number shows how the child perceives number invariance and the degree to which the child is tied to perceptual cues. Reversibility of thought is part of this task. Can the arrangement of a constant number of objects be changed without changing the number? The ability to maintain the equivalence of sets despite their arrangement is developmental and is acquired gradually.

Place a row of eight colored chips (blue for this example) before the child. Piaget recommends at least eight objects be used for the task. Otherwise, the child might know the number perceptually without the use of logic. Beside this row, have the child form an equivalent set with chips of a different color (yellow, for example). See Figure 5.3. This task also tests the child's ability to establish one-to-one correspondence.

Point to the first row. Ask, "Are there more chips in this row or in this row (point to the second row), or are there the same number of chips?" Child responds. Ask, "How do you know that?" Child responds. Now spread out the second row of yellow chips so that the length has been extended. Ask the same question about equivalence. Ask, "Why do you think so?" Additional chips may be added to both rows or the second row may be grouped in a stacked column form or bunched together. Repeat the same questions with justification of the answers.

The child who is a nonconserver cannot maintain the equivalence of number due to changes in length that are irrelevant to number. The real issue is that the child cannot reverse the line of reasoning back to where it started. Preoperational children are so focused on the perceptual aspects of the task that a lasting equivalence or number invariance is not possible.

Conservers are not persuaded by changes in configurations or countersuggestions. They can justify their answers of "the same" by explaining that no chips were added or taken away, or that the chips could be arranged in the original

position, or that the chips are spread out so the rows are longer but the number is the same.

When we relate this to mathematics instruction, Piaget claims that without reversibility of thought and number conservation, the additive property cannot be understood. The child cannot understand that five will remain five if grouped as four and one, one and four, three and two, or two and three. The number in the set is still five. "Fiveness" has no meaning for the child.

Other educators or philosophers have studied Piaget's findings and have repeated his interviews to find the same results. Direct instruction and schooling on conservation skills show that conservation cannot be taught with any lasting, permanent effects. Conservation abilities evolve from maturation due to experiences children have. Piagetian theory says that conservation abilities will not emerge until cognitive schemata, or the logico-mathematical structure, are in place. Number is something each human being constructs from within, and not something that is socially transmitted (Kamii, 1985).

Research (Heibert, Carpenter, and Moser, 1983; 1982) has shown that ability to do Piagetian tasks in conservation, class inclusion, and transitivity may not be needed to solve simple addition and subtraction problems. When the same arithmetic problems were compared to a task assessing a child's information-processing ability, a small but consistent correlation was seen. However, some students scoring low on the information-processing task could solve the arithmetic problems. The findings would seem to indicate that the information-processing task and the three Piagetian tasks may not be prerequisites for applying simple strategies to solve early number facts.

Teaching Ideas: Children need experiences exploring various numbers in different arrangements. Sets must be constructed and compared by children so that lasting equivalence can be

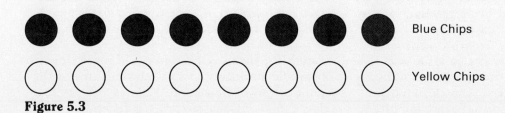

Blue Chips

Yellow Chips

Figure 5.3

achieved no matter what the configuration. Number fans and bead cards (Fig. 5.4) are teaching devices that can be used to build understanding of relationships needed for conservation. A game with painted lima beans or two-sided colored counters develops number invariance (Fig. 5.5) and allows many repetitions without boredom. Connecting cubes can be joined together to represent the same sum with various configurations and color patterns.

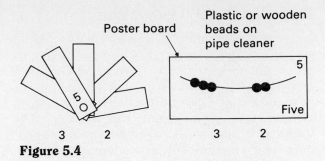

Figure 5.4

Activity

MANIPULATIVES

Spill the Beans

Materials: Lima beans painted on one side or both
Colored counters
Small cup

Procedure: Put a given number of beans in the cup. Shake and spill the beans. Group those beans with the painted side showing and count for the first addend. Group the beans with the white side showing and count for the second addend.

(*Note:* There is no need to count the total amount as that should have been determined at the beginning and will not change.)

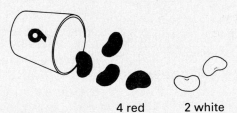

4 red   2 white

Figure 5.5

Task: Seriation

The ability to seriate involves the ordering of objects and events, for example the smallest blocks to the largest blocks. To seriate correctly, the child must make comparisons and make decisions about differences. Preoperational children have no overall plan for arranging things in a sequential order such as by length and cannot coordinate the relationships. According to Piaget, the ability to seriate is vital to the child's understanding of number. It also leads into the child's understanding of the relationship of cardinal and ordinal numbers. The thinking processes necessary to seriation skills are important also in learning science, social studies, and language arts.

To assess this concept, the child might be given a set of ten sticks of graduated lengths and asked to arrange them from longest to shortest. The child could be given a set of pictures of graduated sizes. Since it is difficult for young children to establish a baseline for making comparisons, a ruler could be used as a straightedge or the pictures could be stood in a chalktray.

Teaching Ideas: Children could place rings on a spindle according to size relationships or they could arrange Cuisenaire rods into a staircase (Fig. 5.6). (Cuisenaire rods are different colored

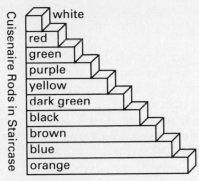

Cuisenaire Rods in Staircase

white
red
green
purple
yellow
dark green
black
brown
blue
orange

Figure 5.6

Child's set   Examiner's set

Figure 5.7

rods in graduated lengths from 1 cm to 10 cm. Patterns for the rods are found in Appendix C.) Classmates can be placed in order of height. A set of shapes of a given region, rectangles for example, can be made in graduated sizes from heavy posterboard for the child to arrange in serial order.

Task: Equivalence of Sets

A task associated with understanding number is equivalence of sets. Here the child must form an equivalent set and be able to match sets for equivalence. Perceptual cues may interfere with the young child's understanding of this aspect of number. The tasks below describe how to assess set equivalence.

Give the child a pile of lima beans. Make a set of five beans. Ask the child to make a set on a margarine lid (or any specific place to identify the set) that has the same number of beans. Ask, "Why do you think your set has the same number of beans as my set?" A child can have this understanding of one-to-one correspondence without being able to count. Another task is to make a line of counters. Below it start a second line of counters but stop before the lines are equal in number. Have the child continue the second set until it equals the numbers in the first set.

The next task tests set equivalence as well as how tied the child is to perceptions. The teacher and child each have a paper cup or small container (Fig. 5.7). The child is given a group (around 10–12) of large lima beans or large counters. The teacher has an equivalent number of small lima beans (counters). The task is to simultaneously drop a bean into each of their cups. Because children love to race on this task, it might be advisable to say "drop" each time to

keep the pace together. After all beans have been dropped into the cups ask, "Are there the same number of beans in each cup or do you have more or do I have more?" After the child responds, show the cups with the beans inside and repeat the question. The child may change the response because the size difference in the beans give perceptual cues that may be interpreted as associated with the number. Ask for justification of each answer to gain valuable insight about how the child is thinking.

The same thing holds true if the child is given two glasses of different sizes but the same size beads (Fig. 5.8). Repeat the activity but now the size of the container affects the child's perception of quantity even though a bead was placed simultaneously into each container.

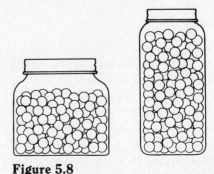

Figure 5.8

Teaching Ideas: Children need experience in naming the number when given a representative set and in constructing sets of a specific amount. Have children work in pairs with junk material. Each child, taking turns, makes a set and the other child makes a set that is equal. Use yarn to establish one-to-one matching between sets and help the child see set equivalence. Relying on counting is sometimes insufficient for the young child. In the task with the beans, it is helpful for the developmentally young child who makes decisions about number from perception to take the beans from the cups, place them in matched pairs to see equivalency, and discuss differences. Set equivalence and one-to-one matching can be done with cups and saucers, juice cans or milk and straws, toy babies and bottles, jars and lids, and plastic flowers and vases (Fig. 5.9).

Figure 5.9

If these Piagetian ideas are applied to the learning of mathematics, there are several implications for the curriculum. First, prenumber concepts should be developed prior to introducing the child to abstract symbols. Second, teachers must develop diagnostic skills described in this chapter to assess a child's logico-mathematical knowledge and developmental level. Third, children need a learning environment that permits free exploration with concrete materials. Since most of these concepts are attained around the age of six or seven, the kindergarten year and a good portion of the first grade year should be spent enjoying informal explorations with number.

With this knowledge of Piaget's observations and theory, compare traditional early primary classrooms. Children may lack the logical operations necessary for understanding number, yet they are introduced to basic facts, counting on strategies, comparing numbers using greater than or less than symbols, missing addends, and many more concepts. Textbooks present number concepts through pictorial representations followed by abstract symbolism. Rather than permitting children to construct meaning about mathematics through creating and coordinating relationships, learning becomes rote memory of rules and procedures. In Piaget's words, "The true cause of failures in formal education is therefore essentially the fact that one begins with language (accompanied by drawings, fictional or narrated actions, etc.) instead of beginning with material action." (Labinowicz, p. 167)

The remainder of this chapter discusses the additional developmental needs of the young child and presents activities to match the child's thinking and to build mathematical understanding.

Building the Concept of Number

Patterns

Patterns are inherent in mathematics and the skill of recognizing patterns is valuable. Pattern recognition means identifying the repetitive nature of something. The pattern may be of a visual, auditory, or physical dimension. Discovering a pattern requires detecting differences and similarities between the elements in the pattern. Patterns should be experienced visually, auditorially, and kinesthetically. Burton states, "Engaging students in a search for patterns is both a pedagogically rewarding occupation and a good preparation for later mathematical development." (1985, p. 49)

Pattern should begin with an AB repetition. This means two elements will be alternated to produce the AB pattern. For example, red square, blue square, red square, blue square or circle, square, circle, square, circle. When presenting patterns to children, it is important to include at least three repetitions. Once children have explored many varieties of this form, other simple patterns may be introduced: ABB, AAB, AABB, ABC, ABCD (Fig. 5.10).

Children need to identify, analyze, copy, extend, and create many different patterns. Children need to read the pattern aloud to help focus their attention. As expected, patterns should be presented to children in physical ways. They can form patterns with other classmates: boy, girl; sitting, standing; hands on head, hands on hips; short pants, long pants. Body movements can be used to make patterns: snap fingers, clap hands; touch your knees, touch your toes; step, hop. Patterns may be created using manipulatives:

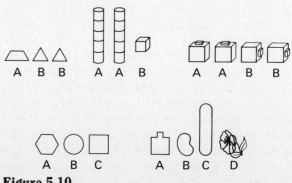

Figure 5.10

pattern blocks, color tiles, peg boards, connecting cubes, dyed pasta, or whatever is available. The teacher creates a pattern such as snap fingers, clap hands and students can translate the pattern with the materials on hand (red circle, green circle; toothpick, macaroni; bottle cap, milk cap). When they are comfortable with the concept of patterns, have them record their pattern by gluing construction paper shapes like the pattern blocks or connecting cubes, using stickers, or gluing the actual objects, if not intended for use again.

Activity

MATH ACROSS THE CURRICULUM

Finding Patterns in Our Lives

Procedure:

1. Take a "pattern walk" to investigate where patterns occur in our everyday lives. (Social Studies)

2. Listen to songs or music that has a repeated pattern. (Music)

3. Listen to stories that have a repeated pattern or a broken pattern. (Reading)

4. Create a pattern that can be used in gym class with exercises such as hop, jump, hop, jump; touch toes, hands above head, touch toes, hands above head. (Physical Education)

5. Do an art project where pattern is reflected: make strings of construction loops; string plastic or wooden beads on yarn; make a potato print pattern; paint a pattern. (Art)

Counting

One of the first skills children learn in mathematics is rote counting. They count one, two, three . . . ten. Some children do this at an early age and it is assumed by adults that the child will be a good math student. Rote counting is saying the number names in isolation without actually counting anything. It can be introduced from other children, parents, television, books, rhymes, games, and finger plays. A detailed collection of finger plays, verses, songs, and books for counting experiences can be found in *Towards a Good Beginning: Teaching Early Childhood Mathematics* (Burton, 1985). Rote counting has no conceptual understanding of the numbers associated with it. It can be nothing more than nonsense names to some children. But as the rote counting skill is investigated for understanding, it is obvious that the child has little or no understanding of what the number name means or represents. Rote counting is similar to learning the alphabet but having no idea of the names of the individual letters. Research (Fuson and Hall, 1982) indicates the three-and-

a-half-to-four-year-old group can count to 13 on the average and the five-and-a-half-to-six-year old group can count to 51 on the average. Ordering the decades presented the most difficulty to the counting sequence.

A simple activity to check the rote counting skill for understanding is to provide a container with 10 objects in it. Ask the child to give you four of the objects. If the child can correctly select four objects, it indicates more of an understanding than just rote counting. This is called rational counting. It is the ability to assign a counting name or number to a group of objects. One-to-one matching is necessary for the child to be able to accurately count the number of objects or people. Textbooks have pictures of sets and the child is to circle the number name for the number of objects in the set. Other pages may show the set of objects and the child is to write the number for the set (Fig. 5.11).

Because several skills are required for this task, children can rote count much farther than they are able to rationally count. It proves more difficult for a child to successfully count objects in random array while objects in a linear ar-

Write the number.   Write the number.

Figure 5.11

rangement may pose no problem (Fig. 5.12). Generally, the child physically touches the objects while counting and gradually will rely on visual counting. Counting does not become meaningful until about six years of age, when the child has a mental structure of number.

How many are there?

Where do I begin counting?

Figure 5.12

Involve the class in many counting activities to reinforce the one-to-one matching concept. Encourage children to keep records of objects by tallying. For example, have the children draw a tally for each boy that you ask to stand. Have them tally for you the number of children who are eating bag lunches.

The final and most difficult counting stage is that of counting with meaning. At this point, a number meaning is assigned to the counting words. This stage develops slowly and is associated with developmental progress in the child as well as opportunities to explore the invariance or "manyness" of number. Research (Fuson and Hall, 1982) indicates that even though young children can extend their conventional string of counting words, the development of number meaning comes much later. This means that a child may know "twenty-five" in the counting sequence but may not have a number-structure meaning of two tens and five ones.

It is difficult for many adults to remember how difficult and abstract the idea of number is for children to grasp. Imagine a new counting system with nonsense number names that you have to remember *in order and with meaning* and you can identify better with the frustrations of the young child.

As children become aware that the quantity in a set they counted is named by a specific name, they begin to associate meaning to numbers. When symbols are introduced, they hold no meaning to the child. Directed activities must be provided over a long period of time in the early primary program. Richardson (1984) devotes a chapter to developing beginning number concepts with connecting cubes. Using connecting cubes and counting boards, she shows how number can be developed from the concrete level to the symbolic level.

Additional books with activities to develop number concepts are *Mathematics Their Way* (Baratta-Lorton, 1976) and *Developing Number Concepts Using Unifix Cubes* (Richardson, 1984). Detailed explanations and pictures are provided for setting up number stations. Children freely explore numbers through a variety of materials and activities that focus on the process rather than on the answer. They construct arrangements of objects (for numbers they can count) to develop numerosity, conservation, counting skills, number sequence, number combinations (Fig. 5.13).

The design of *Mathematics Their Way* allows for the gradual evolution of number from

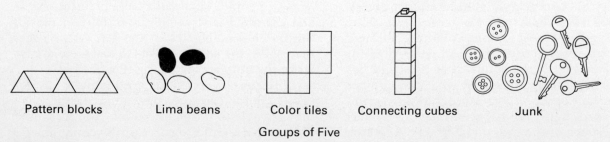

Pattern blocks   Lima beans   Color tiles   Connecting cubes   Junk

Groups of Five

Figure 5.13

the intuitive concept level to the connecting level and finally to the symbolic level. This developmental sequence of activities takes into account the intellectual capacities and maturation of a child. There must be a one-to-one correspondence between each number name and the objects being counted in the group for the child to truly understand the number concept being studied. The teacher's role is to provide ways to help children see the relationship between activities and the traditional math symbols. The teacher must also be sensitive to know when to schedule activities at the next level by assessing when each child has acquired *real understanding* of the quantity of a given number. The assessment strategy is described in Figure 5.14.

Activity

MANIPULATIVES

Assessment of Number Acquisition
(Baratta-Lorton, 1976, p. 287)

Situation 1

Teacher: Count five blocks into my hand.
How many blocks do I have?

Child: Five (without counting).
(Shows "fiveness" and understanding of number invariance.)

Teacher: Divides set of blocks between hands. Opens one hand and shows the blocks. (3) I have 3 blocks and how many blocks are hiding in my other hand?

Child: Two. (Answers immediately without counting or guessing.)

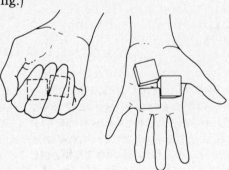

Figure 5.14

This procedure is repeated several times with different combinations for the number. After many successful times, the child is ready to move to number 6.

Situation 2

Teacher: Count five blocks into my hand.
How many blocks do I have in my hand?

Child: (Counts the blocks again.) Five.

The rest of the procedure is continued and the child needs to count on or randomly guess to give answer. This means the child has not mastered an understanding of five and needs to have additional experiences with combinations for fives.

Writing Numerals

Recognizing numerals and writing numerals involve different skills. Children are asked to match numerals to sets before writing them. The skill of associating a symbol with an amount is different from that of writing numerals. The physical skills of writing numerals include small muscle control as well as copying skills. Educators use various techniques to introduce the numerals 0–9. Normally, the order found in textbooks is 1–5, 0, 6–9. Experienced teachers claim other sequences are more appropriate based on the difficulty of the specific numeral. Regardless of the order of presentation, associating a verbal sequential order with writing the numeral provides structure that helps recall. Some teachers use a color sequence for writing the numbers. The first segment is assigned a certain color and the second segment is assigned a different color. The color sequence is the same for all the numbers.

Children may experience reversed numerals. Many different problems exist. Some children have certain numerals they consistently reverse and others reverse numerals when they hurry. A great number of errors are made in writing the "teens." Many first graders write "71" for "seventeen." This error is made because children write the numbers according to how they hear the number spoken. This may also be a problem later in place value understanding. Since children often write numerals before they receive formal instruction at school, teachers see many strange procedures that are difficult to stop. For example, the numeral nine is formed starting at the baseline with a straight line and the top loop added last. To help establish the form of each numeral in the mind, teachers should provide tactile experiences accompanied with verbal structure of the sequential order of the writing. Make numerals from sandpaper or highly textured materials and have children trace the numerals. Fill a small cake pan with salt and have the child trace the numeral, shake the pan and practice again. Care should be taken to have easy-to-see models available for copying. Tracing in the air, on each other's back, and on "magic slates" or individual chalkboards provide a variety of practice activities. Furnish opportunities for writing the numerals in association with a matching set rather than in isolation.

In addition, students can name the words as they are counted for relating reading to a math lesson. When students have a clear understanding of the numerals and their matching sets, the numeral names can be introduced. This should be done when the child has constructed a set rather than with the typical procedure followed in textbooks whenever possible. Build number charts that list the items there are "one" of in the classroom. Post the number charts and add to the lists as children discover more items. Create number charts for other numbers (Fig. 5.15).

In our class we have . . .

One 1	Two 2	Zero 0
1 teacher	2 cabinets	0 zebras
1 flag	2 big desks	0 dogs
1 sink	2 big chairs	0 cars

Figure 5.15

The relation of the number in a set, the number name, and the written symbol must be well understood. An activity to develop this relationship for numbers through 10 is illustrated in Figure 5.16. The child is given cards with the individual numerals 0–10 on them, pictures of sets from 0–10, and the number names. The child's task is to put appropriate cards together. The numeral-set associations should form a large part of an early primary mathematics program.

Figure 5.16

Various arrangements of patterns for a number provide flexibility and greater transfer later. Children need to be able to recognize any sets that represent a number, regardless of the direction or composition of the members of the set. For example, 8 may be associated with each of the sets in Figure 5.17.

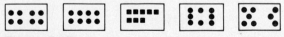

Figure 5.17

Another important consideration in early number development is the concept of more and less. As mentioned earlier, the young child who does not have class inclusion will not perform well on these tasks. This child does not understand relative relationships of number, namely that seven can be more (more than six) and also less (less than nine). The concept of number must be firmly established before the child is ready to make comparisons between numbers. One device that teachers have found helpful is a walk-on number line (the counting numbers beginning at 0). The child can step as the numbers are counted and relative position is sometimes easier to understand. Experiences to compare quantities are required also to build understanding. Only the concept of more and less should be the focus at first, then as children develop the number relationships, how many more or less can be addressed.

Activity

MANIPULATIVES

Developing More and Less

Materials: Connecting cubes
Number cube with 0–6 on faces

Procedure: First player rolls the cube. Get connecting cubes and make a train that long. Next player rolls and builds a train according to the number rolled. Compare trains and decide who has more (or who has less). A spinner divided into two equal regions with one-half as the "more side" and the other half as the "less side" can be used if desired. After comparing the two trains, one child spins and the pointer tells who is the winner (Fig. 5.18).

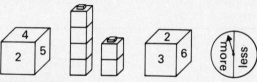

Figure 5.18

Materials: Connecting cubes
More/less spinner

Procedure: Each child makes a cube train out of the other's sight. They show their trains, compare them, and one child spins. The winner is the child whose train matched the spinner. *Example:* One player made a train of six; other player made a train of three. Spinner points to "less." The child with three wins.

Collecting data by tallying and counting is a popular activity. The ability to associate the numeral with the tally can lead into real-world use of the numbers. Databases can be used for surveys and the organization of data.

Activity

COMPUTERS

Using a Database for a Survey

1. Create a database that can be used to record results of a survey of favorite foods.

2. As a class, determine which favorite food choices should be included in the survey.

3. Collect the results from at least 20 people.

(continued)

4. Enter the information into the database.
The database record could look like this:

Name	Favorite Food	Age	Sex
Gary	Tacos	12	M
Mary	Cake	13	F
Nancy	Ice Cream	10	F

5. Have the computer sort by favorite food.
Which food is the favorite?
Which food was the least favorite?

6. Have the computer sort by age.
Are there any conclusions about age and favorite foods?

7. Have the computer sort by sex.
What are the findings?

8. Have the class decide if there is any other information that can be concluded from the survey.

9. Have cooperative student groups write a newspaper story on their survey findings. Be sure to include an appropriate headline.

Readiness for Addition and Subtraction

In this section, the readiness activities for addition and subtraction will be discussed—the "real mathematics" to many parents and teachers. As discussed earlier in this chapter, the ability to add and subtract with understanding develops much later than many adults think. Textbook practice pages of basic facts in first grade may be inappropriate for many children who are not developmentally ready for the abstractions. The child needs many experiences with number and symbols before being introduced to signs (+ and −) for the operations. The child may be able to respond to questions about answers for basic facts, but often this is simply good recall rather than solid understanding.

As children develop the mental structures to deal with numbers, they can explore the operations at a concrete level. To gain understanding of the operations of addition and subtraction, children should manipulate materials and discuss and model addition and subtraction in problem-solving situations. They need to verbalize their actions and internalize the concepts before performing written work with the symbols. Activities are illustrated in Figure 5.19. After participating in and describing concrete experiences, the connecting level with symbols is modeled. When a solid understanding of the language and action is shown, the symbolic level is introduced.

Although typical scope and sequence of textbooks presents addition to 6, subtraction to 6, addition to 10, subtraction to 10, educators continue to debate whether addition and subtraction should be taught together or separately. A great deal of research has been conducted to investigate this question without any firm conclusions. Whichever approach is followed, children should be comfortable expressing the operations verbally and in print to ensure understanding. Through these situations, children will gain an intuitive awareness that addition is joining and subtraction is separating or comparing objects.

Using Technology

In 1984, the National Council of Teachers of Mathematics issued a benchmark report entitled, "The Impact of Computing Technology on School Mathematics." One of the recommendations was that calculators should routinely be available to all students in all activities, beginning with kindergarten. Calculators have a place in number explorations in early primary mathematics programs. Some calculators can be used

 Tree has 4 green apples and 2 red. Child models with felt apples on tree.

 Same situation but child has number cards to show with apples.

4   2

 Same situation with equation cards to show symbolic level and child gives answer.

4 + 2 = 6

Concrete Level   Connecting Level   Symbolic Level

Figure 5.19

for counting by having the child count and push the equal key to see if the counting number matches the calculator display. Counting by multiples of a number follows naturally by having the first number in the series change to reflect the counting pattern. The calculator's constant feature causes the number automatically to keep increasing or decreasing by a given amount. Counting backwards is another skill that can be done on the calculator. This skill is necessary when using the "counting back" strategy for subtraction.

Activity

Number Sequence on a Calculator

Materials: Calculator

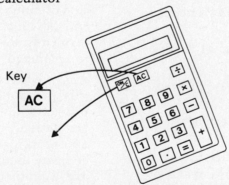

Key

AC

Procedure: Count "one more" beginning at 0. Press "+" key, then "1," and "=" key. Display reads: 1. Press "=" key again (2), again (3), again (4), and continue. This means the calculator is counting by ones or showing "one more."

Count "one less" beginning at 20. Start with 20. Press "−" key, then "1" key, and "=" key. Display reads: 19. Press "=" key again (18), again (17), and continue with each number showing "one less" or how to count backwards.

Counting by other multiples is equally easy to show. Count by twos. Start with 0. Press "+," "2," and "=." Continue to press "=" to see multiples of two.

NOTE: Some calculators work differently than the one described here.

The early primary program should provide opportunities for estimation skills in various dimensions—duration of time, number, length, temperature; should include geometry explorations with pattern blocks and geoboards, and; should allow explorations with numeration concepts. Each of these areas will be discussed in later chapters in this text.

Summary

The early primary curriculum for mathematics must take into account the aspects of active, direct participation by the child. The environment must be child-oriented with time provided for investigations as well as for interactions with other students and the teacher. Biggs and MacLean (1969) comment, "The pupil of the electronic age requires an educational environment that allows him maximum participation in discovery; schools must relate and synthesize rather than linearly fragment knowledge; they must provide a multiplicity of stimuli that will spark the child's curiosity and engender a continuing desire to learn." (p. 4) As we keep the child's needs in mind, the primary grade experiences must be structured to nurture young children and guide them toward logical, operational thinking about the world of mathematics.

Exercises

Directions: Read all questions *before* answering any one exercise. Frequently the last question in one category leads to the first question in the next question.

A. **Memorization and Comprehension Exercises**
Low-Level Thought Activities

1. Name the earliest stage of logical thinking; it is the foundation for graphing.

2. Name the types of counting and give an example of each one.

3. Name four manipulatives that can be used to teach children patterns.

4. Suppose there are several children in your first grade classroom who are making number reversals. Describe the remedial activities you would use to help.

5. You are in a first-grade classroom with almost no manipulatives. The children are not understanding the operation of subtraction. The pictures in the textbook of some birds on a limb and some birds flying away hold no meaning. What techniques would you use to help?

6. In view of the discussion in this chapter on Piagetian tasks and how children learn, discuss how this compares to the way you experienced numbers early in school.

B. **Application and Analysis Exercises**
Middle-Level Thought Activities

1. Collect items to form a diagnostic kit to test these prenumber concepts: conservation of number, set equivalence, seriation, and class inclusion. Apply what you have learned to make a "manual" of questions to accompany these materials.

2. Apply what you have learned to handle this parent-teacher simulation: You are using a child-oriented program with number stations, manipulative materials, children working together and talking as they compare findings (cooperative learning), and a mother who is serving as an aide during math period challenges you that the children are "just playing" rather than "doing math." How would you respond?

3. Analyze the following experience to see what understanding a child has for the concept of number:
   A child is counting a collection of shells and begins by touching each shell as the number is said. After eight, the child continues to count but faster so the counting is not in a matched relationship with the shells. What remedial techniques would you use to help?

4. Examine a first grade textbook where numerals and their corresponding sets are shown. Comment about the approach used, the teaching suggestions offered in the teacher's manual, the prerequisite skills you feel are needed to successfully complete the page, and any alternations you would make in the teaching sequence.

5. Apply what you have learned in this chapter to handle the following simulation:

    You are a kindergarten teacher using a child-oriented program for teaching mathematics. You have some parents who are pressuring you to give their children traditional worksheets and practice dittos so they have a firm evidence that mathematics is being taught. How would you respond?

6. Using the evaluation form found in Appendix A, select commercial software for any three topics from this chapter. In the selection, include drill and practice, simulation, tutorial, and problem-solving software. For each software program selected, describe how you would integrate it in a lesson plan.

7. To reinforce math across the curriculum:
    a. List five mathematical questions/problems that come from popular storybooks for children. Show how they are related to the topics of this chapter.
    b. List five ways to use the topics of this chapter when teaching lessons in reading, science, social studies, health, music, art, physical education, or language arts (writing, English grammar, poetry, etc.).

C. Synthesis and Evaluation Exercises
High-Level Thought Activities

1. Develop five activities, not seen in this chapter, for teaching the concept of number. Look at the Scope and Sequence chart in this chapter to help you choose concepts for a desired grade level. Find computer programs to help young children. The following explorations will help you find appropriate activities:
    a. Search professional journals for current articles on number readiness teaching ideas and research findings.

    Follow this form to report on the pertinent findings:

Journal Reviewed: _____

Publication Date: _____ Grade Level: _____

Subject Area: _____

Author(s): _____
_____

Major Finding: _____

Study or Teaching Procedure Outlined: _____

Reviewed by: _____

Some journals of note are:
Journal for Research in Mathematics Education
Arithmetic Teacher
Mathematics Teacher
School Science and Mathematics

    b. Computer applications for classroom use are found in the journals listed below. These programs are by no means the only ones. Search the periodical section of your library for more.
Teaching and Computers
Classroom Computer Learning
Electronic Education
Computers in the Schools
The Journal of Computers in Mathematics and Science Teaching
The Computing Teacher

2. Develop five activities for teaching one-to-one correspondence to young children. Follow the same steps as you did in C1.

3. Devise an evaluation plan for determining if a child can count rationally.

4. Write a proposal to your principal or supervisor describing manipulative materials you wish to buy and/or have resources to equip your early primary classroom for teaching mathematics. Include a rationale and budget. Remember that many manipulatives can be made inexpensively. Some materials to consider are felt, various colors of plastic canvas (sold for needlepointing in craft stores), sponges, strips of balsa wood, popsicle sticks, free cardboard meat containers to make Cuisenaire rods, pattern blocks, and other manipulatives men-

tioned in this chapter. The patterns are found in Appendix C.

5. Read the NCTM Curriculum and Evaluation Standards with regard to the developmental aspects of early childhood education. Write a position paper concerning whether the standards have addressed the mathematical needs of the young child as well as the developmental needs.

6. How much should a teacher expect preschoolers and kindergarteners to know about the concept of numbers, and what activities are developmentally sound to use with young children?

Bibliography

Baratta-Lorton, Mary. *Mathematics Their Way.* Menlo Park, Calif.: Addison-Wesley, 1976.

———. *Workjobs II.* Menlo Park, Calif.: Addison-Wesley, 1978.

Biggs, Edith E., and James R. MacLean. *Freedom to Learn.* Ontario, Canada: Addison-Wesley, 1969.

Burton, Grace. *Towards a Good Beginning: Teaching Early Childhood Mathematics.* Menlo Park, Calif.: Addison-Wesley, 1985.

Commission on Standards for School Mathematics of the National Council of Teachers of Mathematics. *Curriculum and Evaluation Standards for School Mathematics.* Working Draft. Reston, Va.: The Council, 1987.

Copeland, Richard W. *How Children Learn Mathematics.* New York: Macmillan, 1974.

Fuson, Karen, and James W. Hall. "The Acquisition of Early Number Word Meanings: A Conceptual Analysis and Review." In *The Development of Mathematical Thinking,* edited by H. P. Ginsburg. New York: Academic Press, 1983.

Hiebert, James, Thomas P. Carpenter, and James M. Moser. "Cognitive Development and Children's Solutions to Verbal Arithmetic Problems." *Journal for Research in Mathematics Education* 13 (March 1982): 83–98.

———. "Cognitive Skills and Arithmetic Performance: A Reply to Steffe and Cobb." *Journal for Research in Mathematics Education* 14 (January 1983): 77–79.

Kamii, Constance. *Young Children Reinvent Arithmetic: Implications of Piaget's Theory.* New York: Teachers College Press, 1985.

Kouba, Vicky L., Catherine A. Brown, Thomas P. Carpenter, Mary M. Lindquist, Edward A. Silver, and Jane O. Swafford. "Results of the Fourth NAEP Assessment of Mathematics: Measurement, Geometry, Data Interpretation, Attitude and Other Topics." *Arithmetic Teacher* 35 (May 1988): 10–16.

Labinowicz, Ed. *The Piaget Primer: Thinking, Learning, Teaching.* Menlo Park, Calif.: Addison-Wesley, 1980.

———. *Learning from Children: New Beginnings for Teaching Numerical Thinking, a Piagetian Approach.* Menlo Park, Calif.: Addison-Wesley, 1985.

Mueller, Delbert W. "Building a Scope and Sequence for Early Childhood Mathematics." *Arithmetic Teacher* 33 (October 1985): 8–11.

National Council of Teachers of Mathematics. "The Impact of Computing Technology on School Mathematics: Report of an NCTM Conference." Reston, Va., March 1984.

Piaget, Jean. *The Child's Conception of Number.* New York: Humanities Press, 1952.

Richardson, Kathy. *Developing Number Concepts Using Unifix Cubes.* Menlo Park, Calif.: Addison-Wesley, 1984.

Secada, Walter G., Karen C. Fuson, and James W. Hall. "The Transition from Counting-All to Counting-On in Addition." *Journal for Research in Mathematics Education* 14 (January 1983): 47–57.

Steffe, Leslie P. "Differential Performance of First-Grade Children When Solving Arithmetic Addition Problems." *Journal for Research in Mathematics Education* 1 (1970): 144–161.

6

Numeration

Key Question: *What are the unique qualities of our base 10 numeration system, and how can they be shown as the basic building blocks of many number concepts to come?*

Introduction

Development of Numeration Systems

The development of numeration systems has two major characteristics of note which have fascinated historians as well as mathematicians over the centuries. One is the concept of regrouping. The other is the existence of numeration systems as a mark of intelligent societies. One of the most fascinating facts of civilization is that all highly developed cultures created some kind of numeration system at times when many cultures did not even know of each other's existence. Whether using concrete objects or tally marks (strokes) to represent objects, the cultures came to the realization that an efficient way of counting beyond one-to-one correspondence was needed. The answer was to regroup, to arrange a certain set of symbols (numerals) in varying positions or patterns to denote larger and larger numbers.

There are several definitions that will be helpful in the discussion of numeration throughout the chapter.

139

- *Numeration System:* The process of writing or stating numbers in their natural order; a way of regrouping numerals to represent numbers.
- *Number System:* A mathematical system consisting of a set of objects called numbers, a set of axioms (basic propositions), and operations that act on the numbers. It is a way to classify numbers, their properties, and operations which exist no matter what numeration system is used in various cultures.
- *Base of a Number System:* The number of units in a given digit's place that must be taken to denote 1 in the next higher place. In the decimal (10) base, ten in the units place is denoted by 1 in the next higher place, the tens place. In base four, four in the units place is denoted by 1 in the next higher place, the fours place.

For the purpose of this discussion, two kinds of numeration systems will be explored, name-value systems and place-value systems along with some of the ancient civilizations who used them. Some systems have name-value characteristics, meaning the name of the regroupings tells the pattern. Some systems have position or place-value characteristics, meaning the position of the regroupings tells the pattern.

Our numeration system has both characteristics. The written symbols (numerals) are place-value based (examples: 1, 10, 101, 1011), but our number words are name-value based (examples:

22 means two tens and two; 47 means four tens and seven). The combination of both characteristics may cause confusion when young children are learning our numeration system. For example, some children write 407 for forty-seven.

Learning about ancient numeration systems helps children acquire an understanding and appreciation of how the systems developed. The Romans used the main base of ten and the sub-base of five for regrouping.

I II III IV V VI VII VIII
IX X XI XII XIII XIV XV

with

L = 50, C = 100, D = 500, and M = 1000

Once you know the pattern, you can continue writing any numeral.

The Mayans used the following regrouping pattern:

. .. . — ⋅— ⋅⋅— ⋅⋅⋅— ⋅⋅⋅⋅— —
1 2 3   4   5   6   7   8   9   10

Can you see a main base and a sub-base? After the number 19, the Mayans switched to a base of twenty.

The Babylonians regrouped with a base of sixty. In their early numeration system, the size of the symbol indicated how large the number represented. These early numbers indicate traces of notch recording.

Activity

COMPUTERS

MATH ACROSS THE CURRICULUM

Early Babylonian Numerals in Logo

1. A Logo program that creates early Babylonian numerals is shown in Appendix D. It has been placed in the book to show what upper elementary or middle school students are able to create if they have had exposure to Logo since the early primary grades.

2. The program may be used as a history assignment, blending a knowledge of early civilizations and the development of mathematics.

3. The following are early Babylonian numerals and their counterparts in our numeration system. These figures were drawn by the program in Appendix D.

216000 36000 3600 600 60 10 1

After looking at early Babylonian numerals, you can see how difficult it would be to construct a system on the size of a numeral. As the Babylonian culture matured, the *position* of the numeral became important. Hence, it became a place-value system. A bar was placed over the symbols written from right to left to indicate the value of each position. The Babylonian numerals of the early and later periods are seen in Figure 6.1. Both represent the same amount and are equal to the numeral 72 in our system.

Our system of numeration is based on regrouping at ten and has its roots in the Hindu-Arabic system. An early recording of the numerals looked like those in Figure 6.2.

Early
Babylonian
Numerals

Later
Babylonian
Numerals

Figure 6.1

Figure 6.2

Activity

Different Numeration Systems

From the introduction on numeration, you can solve these problems by discerning the regrouping pattern. Give the corresponding numeral in the base 10 numeration system.

Name of the Numeration System Seen Below: _____

LXVII = _____

CLII = _____

Make some new ones of your own:

Name of the Numeration System Seen Below: _____

‥ = _____   ⋯ = _____

The fact that humankind developed efficient ways of counting and created systematic patterns of regrouping symbols is a mark of higher-ordered intelligence. Numeration systems did not occur overnight; the systems and patterns developed over centuries. It is essentially such number patterns that we are sending into space as radio waves. What we are saying to the universe is that we are a people of intelligence, that we have produced something that does not occur in nature. If these patterns are understood by others in the universe, they too would have had to create numeration systems much as we have done. What scientists are listening for is a new pattern of regroupings to come back to us, indicating that other intelligent beings are alive in the universe. Numeration systems represent one of the most important creations of humankind.

Structure of Numeration Systems

Exploring the structure of a numeration system generates an appreciation and understanding of how efficiently the system operates. Numeration systems have a specific structure from which emerges place-value notations, powers, and exponents. The configuration of the base 10 system is shown in Figure 6.3. Notice that the system begins with a single basic building unit, often referred to as a UNIT. From there, each regrouping is done at ten. The models shown are known as the base 10 blocks (also referred to as place-value blocks, multi-base blocks, Dienes blocks, powers of ten blocks).

All units are single ones until the first regrouping. The configuration of the first regrouping looks like a long set of units fused together; hence, the name LONG. The 10 means one set of ten and no single units present except for those accounted for in the LONG.

The second type of regrouping occurs when there are ten LONGS (or ten groups of ten). The configuration looks like a flat with its ten LONGS fused together, hence, the name, FLAT. There are ten times ten sets of the basic building unit that can be seen in the FLAT or one hundred pieces of the basic UNIT. The numeral 100 means one set in the second regrouping with no sets of ten and no single units present except for those already accounted for in the FLAT.

The third type of regrouping occurs when there are ten FLATS (or ten groups of hundreds). The configuration looks like a block with its ten FLATS fused together, hence the name, BLOCK.

There are ten times a hundred sets of the basic building unit that can be seen in the block or one thousand pieces of the basic UNIT. Therefore, 1000 means one set in the third regrouping with no sets of hundreds, no sets of tens, and no units except for those already accounted for in the BLOCK.

Exponents are known as powers of ten when the base is 10. Exponents are introduced to students in the sixth or seventh grades in most textbooks. The exponent is just another way of indicating which type of regrouping is being considered. Notice that:

$$10^2 \quad \text{is read as "ten squared."}$$

It is not by chance that the configuration for "ten squared" is in the shape of a square (the FLAT), nor is it by chance that:

$$10^3 \quad \text{is read as "ten cubed."}$$

As Figure 6.3 shows, the shape is a cube (BLOCK). Many college students are unaware that any number that is squared or cubed becomes the shape of a square or cube, when using base models. When middle grade students are first introduced to exponents, the configuration and meaning of the base 10 system (as seen in Figure 6.3) should be shown to them.

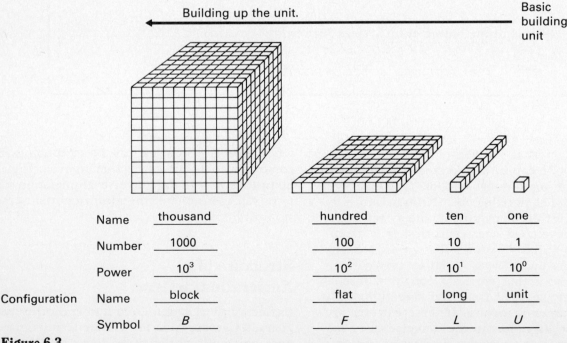

Building up the unit. ←   Basic building unit

Configuration		thousand	hundred	ten	one
	Name	thousand	hundred	ten	one
	Number	1000	100	10	1
	Power	10^3	10^2	10^1	10^0
	Name	block	flat	long	unit
	Symbol	B	F	L	U

Figure 6.3

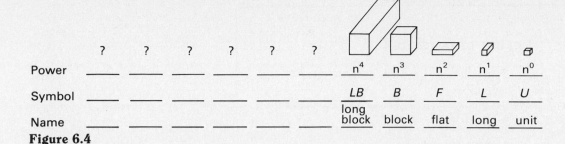

	?	?	?	?	?	?	n^4	n^3	n^2	n^1	n^0
Power	___	___	___	___	___	___	n^4	n^3	n^2	n^1	n^0
Symbol	___	___	___	___	___	___	LB	B	F	L	U
Name	___	___	___	___	___	___	long block	block	flat	long	unit

Figure 6.4

It is in the middle school that students learn (and quickly forget) that any number to the zero power is one. For example:

$$395^0 = 1$$

Any number to the zero power means that no regrouping has taken place. The number is still being counted as a "one" until we know where the first regrouping occurs. An informal proof for 395 to the zero power would be:

$$1 = \frac{395^1}{395^1} = 395^{1-1} = 395^0$$

This is a common question on college aptitude tests and it is one of the most frequently missed questions. We believe that such questions may be missed because upper grade teachers neglect to show models of the numeration system in a concrete or pictorial form so students can see the logical sequence of the system's development.

Although work with different number bases has been deemphasized in current student textbooks, teachers need to see how the base 10 system fits into the scheme of things from a larger point of view. Figure 6.4 shows the configuration for any number base, n. Notice that there are no lines drawn on the configurations after the first regrouping because these "multi-base blocks" apply to a multitude of bases. When the blocks are used in the base 10 system, they are simply called base 10 blocks.

Activity

PROBLEM SOLVING

Using Multi-Base Blocks in Any Base

Study the configuration patterns in Figure 6.4 and fill in the missing information below each shape. (*Note:* The key word is PATTERN.)

See if you can draw the rest of the six shapes. You have all the information needed to do it. They form a pattern also. Since the shapes get progressively larger, you will need an extra piece of paper to accommodate the drawings. Most people have never seen a picture of the numeration system going to the ninth regrouping, yet they are able to answer all these questions. Why?

Activities like the preceding one were found in the upper grade math textbooks of the 1960s during the period in mathematics history known as the "New Math" era. Working with different number bases was considered quite new and revolutionary by many parents and teachers. Figure

398　　　SCALES OF NOTATION.

numbers in other scales should be read by naming the number of units of each order.

Thus, 342 in the *quinary* scale should be read : *quinary scale, 3 units of the third order, 4 of the second, and 2 of the first.*

WRITTEN EXERCISES.

540. **1.** Write in the quinary scale the numbers corresponding to the numbers from 1 to 13 in the common or decimal scale.

EXPLANATION. — Since the radix of the scale is 5, the characters employed are 1, 2, 3, 4, 0.

Since 5 units of any order are equal to 1 of the next higher order, the numbers including 5 will be expressed by 1, 2, 3, 4, 10.

Since 6 is equal to 1 unit of the second order and 1 of the first order it is written 11 ; since 7 is equal to 1 unit of the second order and 2 of the first it is written 12.

Expressing the numbers from 1 to 13 in accordance with the law just illustrated, they are 1, 2, 3, 4, 10, 11, 12, 13, 14, 20, 21, 22, 23.

Write the numbers corresponding to the numbers from 1 to 20 in the common or decimal scale :

2. In the quaternary scale.　　**6.** In the nonary scale.

3. In the octary scale.　　**7.** In the undenary scale.

4. In the senary scale.　　**8.** In the septenary scale.

5. In the ternary scale.　　**9.** In the binary scale.

541. **To change from the decimal to another scale.**

1. Change 58375 from the decimal to the senary scale.

```
6 | 58375
6 |  9729 + 1
6 |  1621 + 3
6 |   270 + 1
6 |    45 + 0
6 |     7 + 3
        1 + 1
```

EXPLANATION. — By dividing by 6, we obtain the number of units of the second order and the number of units of the first order remaining.

By continuing to divide by 6, the number of units in the successive orders is obtained and the number of units remaining after division. It is thus found that 58375 when expressed in the *senary* scale contains 1 unit of the *seventh* order, 1 of the *sixth*, 3 of the *fifth*, etc., or the number is expressed by 1130131_6.

For convenience in notation the radix of the scale is indicated by a small subscript figure.

Figure 6.5

6.5 shows an exercise from an upper elementary textbook published in 1892 (*Standard Arithmetic,* by William J. Milne, p. 398). If the word "scale" is replaced with the word "base," it becomes apparent that different number bases have been a part of the curriculum off and on for nearly 100 years.

The following activities will reconstruct numerals in different number bases using pictures of the multi-base blocks as the manipulatives.

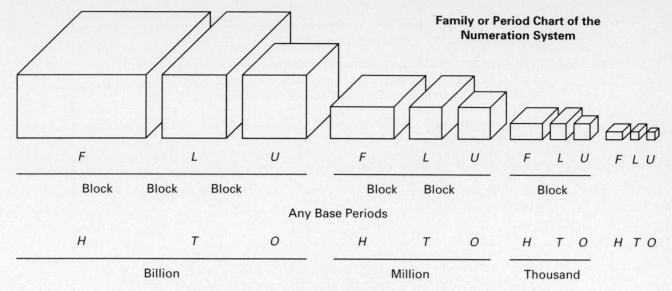

Family or Period Chart of the Numeration System

F	L	U	F	L	U	F	L	U	F L U
Block	Block	Block	Block	Block		Block			

Any Base Periods

H	T	O	H	T	O	H	T	O	H T O
Billion			Million			Thousand			

Base Ten Periods

Key to Chart

U = unit	O = one
L = long	T = ten
F = flat	H = hundred

Figure 6.6 is just like Figure 6.4, with several more regroupings drawn to the left. Looking from right to left, the same three configurations keep recurring. They are the cube shape (BLOCK), the long shape (LONG), and the square shape (FLAT). Our system reads these sets of three as families or periods and the names are consistent with the configurations.

Other bases have the same configurations and the numerals can be read as if in families or periods as well. The names BLOCK, LONG, and FLAT, are used because they apply to the configurations (shapes) which appear in any number base. The family or period names can be generic too, so they can apply to any number base. This chart answers the questions first posed in the preceding problem-solving activity. How close did you come to filling in the chart correctly?

Figure 6.6

Activity

Different Bases in Logo

Directions:

1. This program is saved on the Logo disk under the name BASES

2. Type   LOAD "BASES   to work with the program, and go to step 3.

   (If you have no disk, type the following procedures)

```
TO UNIT
REPEAT 4 [FD 15 RT 90] FD 15 RT 45 FD 5
RT 45 FD 15 RT 135 FD 5 BK 5 LT 45 FD 15
RT 45 FD 5 LT 135 BK 15 LT 90
END

TO LONG :BASE
REPEAT :BASE [UNIT FD 15]
BK :BASE * 15
END
```

(continued)

3. Type the word UNIT and press return. Now you see the basic building unit upon which any base is built.

4. Type CS to clear the screen (remember to do that between all examples) and then type the following commands. Remember to space just the way it is shown here. Press RETURN after typing each line:

```
UNIT FD 15
UNIT FD 15
UNIT FD 15
```

You have just made a LONG in base 3. Another way to do the same action is to type the line:

```
REPEAT 3 [UNIT FD 15]
```

Try it and see how it works. The commands in the brackets [ ] are repeated the number of times seen after the repeat command.

5. Using the repeat command, predict how you would make a LONG in base 4. Type it and see if your prediction was correct. Check to see if you were right by typing:

```
LONG 4
```

Were you correct? If not, can you analyze (problem solve) what you need to do to correct your original prediction?

6. Look at the following procedure printed below (the TO in front of a procedure name tells the computer that what follows is a special set of commands to be done together every time you type the name).

```
TO LONG :BASE
REPEAT :BASE [UNIT FD 15]
BK :BASE * 15
END
```

The :BASE after the name of the procedure tells the computer that :BASE is a variable that can have any number placed there and the computer will respond that many times to repeat the LONG. Type LONG 5 and see what happens. Now type LONG 8 and see what happens.

7. Since FLATS are the LONGS regrouped at the base number in the second regrouping, we can use the LONG procedure to make FLATS. Type

```
REPEAT 3 [LONG 3 RT 90 FD 15 LT 90]
```

and see what happens.

8. Predict what the Logo commands would be for FLAT 5.

9. A FLAT for any base would look like the following procedure:

```
TO FLAT :BASE
REPEAT :BASE [LONG :BASE RT 90 FD 15 LT 90]
END
```

10. Look at the procedure printed in step 9 as you watch the turtle make a FLAT in base 6 and one in base 3. See if you can tell which command is

being followed all the way through the procedure. Type `FLAT 6 CS FLAT 3`

11. Blocks can be made from the FLAT procedure with modifications for depth. If you have the Logo disk, type `ED "BLOCK` to see the commands that will make a BLOCK in any base. Remember to hold down the control key and press G to get out of the edit mode. Within the superprocedure named BLOCK, there are several subprocedures that help move the FLAT into a three-dimensional shape and place a top on the FLAT so it looks like a real BLOCK.

12. Try making some BLOCKS in several bases. Try these:

    `BLOCK 6 CS BLOCK 2`

13. You have just constructed the numeration system just as early mathematicians did . . . only you did it with a computer.

The previous programs in Logo can be used to show the multi-base blocks for any number in any base. The numbers have been limited to lower bases so that the pictures of the blocks can fit on the computer monitor.

Activity

COMPUTERS

PROBLEM SOLVING

Building Numbers in Any Base with Logo

Directions:

1. Load the program BASES if you have turned the computer off since doing the preceding activity.

2. The turtle will draw the BLOCKS for different numbers in different bases. You will need to leave a space between each formation to see each BLOCK clearly. Type the letters `SP` between each block you want to make. (If you have no disk, type the procedures for the space and the beginning movements of each number:)

```
TO SP
PU RT 90 FD 30 LT 90 PD
END

TO STARTNUMBER
PU LT 90 FD 140 RT 90 PD
END
```

The STARTNUMBER procedure moves the turtle to the left side of the screen so you won't run out of space when making large numbers.

3. To see the number 123 in base 4, you would have 1 FLAT, 2 LONGS, and 3 UNITS with all the BLOCKS regrouped at base 4. To see the number, type: `STARTNUMBER FLAT 4 SP LONG 4 SP LONG 4 SP UNIT SP UNIT SP UNIT`

4. Look at the following commands and draw what you think they will make. Check your prediction by typing these commands in the computer:

(continued)

```
STARTNUMBER BLOCK 3 SP FLAT 3 SP LONG 3 SP LONG 3
SP UNIT
```

You have just made the numeral 1121 in base 3.

5. Now make the number 1011 in base 2. What commands would you type in?

6. What are the numbers represented by these Logo drawings? See if you were correct by typing the commands that would duplicate the drawings.

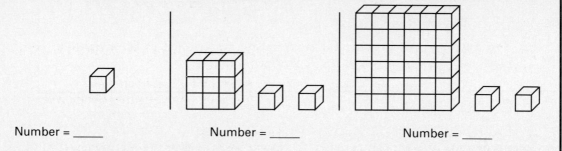

Number = _____   Number = _____   Number = _____

7. Make up more numbers in different bases. Have a partner see if he or she can figure out the number you had in mind from the drawings on the monitor. This is an example of the way cooperative learning can occur in classroom settings.

Activity

PROBLEM
SOLVING

Base 3 Numeration System

Directions:

1. Sketch the configuration of the base 3 numeration system to the third regrouping.

MANIPULATIVES

2. Now count by units in base 3 using the number line that appears below the multi-base shapes. The first four have been done for you. Label each with the corresponding numeral.

To facilitate drawing, let

⬜ = Unit   ⬜ = Flat

⬜ = Long   ⬜ = Block

Multibase block			
Numeral	0	1	2	10

Scope and Sequence

First Grade
- Ordinal numbers through 12th.
- Place value—ones and tens.
- Identify and write numbers through 100.
- Order numbers through 100.
- Compare numbers through 100.
- Count by ones, twos, fives and tens.

Second Grade
- Ordinal numbers through 20th.
- Place value—ones, tens, hundreds.
- Write numbers through 1000 in standard form.
- Identify place value of a digit through hundreds.
- Count by ones, fives, tens, and hundreds.
- Use > and < symbols to compare.
- Compare and order numbers through hundreds.
- Rename tens as ones and ones as tens.

Third Grade
- Write whole numbers through hundred-thousands.
- Identify the place value of a digit through hundred-thousands.
- Compare and order numbers through hundred-thousands.
- Round numbers to nearest ten or hundred.
- Rename through thousands place.
- Continue counting patterns through thousands.

Fourth Grade
- Write numbers through hundred-millions.
- Interpret numbers in word form.
- Write number in expanded form.
- Identify place value of a digit through hundred-millions.
- Compare and order whole numbers through hundred-millions.

- Round numbers to nearest ten, hundred, or thousand.

Fifth Grade
- Write numbers through hundred-billions in standard form.
- Write numbers in expanded form.
- Write numbers from word form.
- Identify place value of a digit through hundred-billions.
- Compare and order numbers through billions.
- Round whole numbers.
- Interpret numbers in Roman system.

Sixth Grade
- Write numbers in exponential form.
- Write numbers through hundred-billions.
- Identify the value of a digit through hundred-billions.
- Compare and order numbers through hundred-billions.
- Round whole numbers.
- Understand other bases.
- Interpret numbers in other numeration systems.

Seventh Grade
- Write, interpret, compare, and order numbers through trillions.
- Identify the value of a digit through trillions.
- Round numbers to the nearest designated place.
- Interpret numbers in other bases, other numeration systems, and in exponential form.

Eighth Grade
- Review all skills from seventh grade.

Teaching Strategies

Understanding the place-value features of our number system is of critical importance in learning algorithms and estimation skills. The National Council of Teachers of Mathematics curriculum standards (Commission on Standards for School Mathematics, 1987) for grades 5–8 include a standard on developing an understanding of number systems that focuses on the underlying structure of the arithmetic of fractions, decimals, and integers. In this way, students can develop a number sense for the whole number system that extends to the rational system and beyond.

In the upper grades, numeration often occurs in the first chapter of the student's textbook. Unfortunately, inadequate coverage is frequently seen and insufficient skill maintenance is offered. Our number system is efficient with certain properties that need to be fully developed in students' minds. The first property of the system is the nature of place value.

The Nature of Place Value

Place value means that the position of a digit represents the value of the digit. In our base 10 system, there are ten digits: 0,1,2,3,4,5,6,7,8,9. All other numbers are composed of these digits. For example, the numeral 5 looks the same in any place, but the value is different depending on the position.

Another aspect of our number system is the number's total value. *Total value refers to the additive property of the system.* This means that 358 represents the number 300 + 50 + 8 and 358 is the total value. *Expanded notation refers to the actual value of each digit* such as 412 = 400 + 10 + 2, and *standard form refers to writing the compact number* such as 412. During the modern math movement, heavy emphasis was placed on showing expanded forms of numbers $(4 \times 100) + (1 \times 10) + (2 \times 1)$, but much less time is currently devoted to this concept. Perhaps the reduced coverage on this topic in textbooks may prove to be a limitation on a child's understanding of the system. Here are some activities that focus on reinforcing this skill and developing understanding.

Activity

MENTAL MATH

Expanded Notation Aid to Memory

Materials: Poster board or construction paper cut into 3x3-inch squares. Write the digits 0–9, one per piece, in large form. From another color of paper, cut 3x6-inch rectangles and space in the left 3x3-inch square of the rectangular region the numbers 1–9. On the right 3x3-inch square of the region write 0. This will form all the decade numbers. Cut 3x9-inch rectangles from a third color of paper. Think of these rectangles as three 3x3-inch square sections. In the left section, write the digits 1–9 in large form. In the middle section, write 0, and in the far right section, write 0. This forms all the hundreds (300, 500, 200).

Procedure: The teacher or groups of children may use these cards to form any three-digit number. The teacher or group leader says a number. The cards are selected that name that number in expanded form. Stack the cards aligning the right edges to see the standard numeral and separate to see the expanded form. The number can be recorded with the plus sign between each card to illustrate the additive property. To include four-digit numbers, create cards that are 3x12-inches and construct cards for the thousands in a similar fashion.

9"

| 4 | 0 | 0 | 3"

| 6 | 0 | 3"
6"

| 7 | 3"
3"

| 4 | 0 | 0 | + | 6 | 0 | + | 7 | = | 4 | 6 | 7 |

Activity

MENTAL MATH

Place Value Strips—An Aid to Mental Computation

Materials: Felt tip pens or dri-mark pens
Laminated construction paper strips 1x9-inches long

Procedure: The teacher says a three-digit number and the students write the number on their strip in the rectangular region at the far left. After the equal sign, students place the single digit beside each appropriate place-value name. In folded compact form, the strip should read 258 = 258. In unfolded, expanded form, the strip should read 258 = 2 hundreds + 5 tens + 8 ones. Various folds can be opened to expose different forms of the number such as:

258 = 25 tens + 8 ones
258 = 258 ones
258 = 2 hundreds + 58 ones

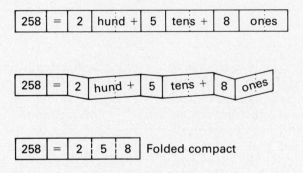

Activity

MENTAL MATH

Aid to Quick Memory of Notation

Materials: 48 cards—4 forms for 12 numbers
Samples:
600 = 60 tens = 600 ones = six hundred
8326 = 8000 + 300 + 20 + 6 = 83 hundreds twenty-six
= eight thousand three hundred twenty-six

Procedure: Play in groups of 4. Determine who goes first. Deal all cards to players. The object is to get books (total sets of 4 forms) of equivalent names for a number. When a book is formed, the player may lay down the cards as a set. When play begins, players may trade any cards to the player on the left. The game pauses for players to check on equivalent numbers and form books. Play continues until one player has succeeded to make all the cards into books.

Students can profit from experiences with the calculator when dealing with large multi-digit numbers. A quick counting activity is to use the constant feature of the calculator and input 1+ = = =. As the student continues pushing the equals key, the counting sequence of the numbers is displayed. Students can see the changing ones column and the 0–9 sequence recurring.

Activity

CALCULATORS

Using Calculators with the Expanded Form

Materials: Set of number cards
Calculator

Procedure: Read the number on the card. Express the number in expanded form (such as 3000 + 40 + 6). Enter the expanded form of the number into the calculator in any order. Did the same compact form of the number occur as the final number? This shows the additive property as well as the commutative property for addition.

Base 10 blocks serve as excellent models for place-value concepts especially in establishing an understanding of ones, tens, and hundreds. Keep in mind that numbers are abstractions and children need concrete materials to construct an understanding of number and place-value notation. Interviews with children by Labinowicz (1985) provide ample evidence that many third-grade children do not relate to the thousand block because they do not conserve volume until age 11. They focus on a surface area strategy and fail to consider the inside of the cube. Therefore, when asked how much the thousand block represents, the non-conserving child replies, "600," counting six sides of 100 each.

Generally, the feeling is that it is important to allow children opportunities to explore the structure of other number systems by performing activities in other number bases. This is usually done in upper elementary and middle school; however, base work can be done with younger children using recording charts and counters. The emphasis should be on the trading between groups. Trading or grouping rules apply to all number bases. If the base is 4, then trades are four for one. If the base is 7, then trades are seven for one. Children may attach fear and failure to exploring patterns in base 10 because they may have been rushed to the symbolic level before real understanding of the system has occurred. Instruction with physical ma-

terials offers children a chance to investigate the patterns in our number system and internalize them.

Both proportional and non-proportional models may be used to work in other bases. When proportional models are used, the same configuration that indicates grouping occurs. The units are individual cubes, the first regrouping results in a LONG, the next regrouping forms a FLAT, and the next regrouping produces a cube or BLOCK. This pattern continues as regroupings continue. The commercial base 2 or base 4 blocks are excellent materials to show the place-value concepts in other bases. They can relate directly to the base 10 blocks found in many classrooms.

The pictures and descriptions in Figure 6.7 of place-value models may help to demonstrate the differences between proportional and non-proportional models. They may also inspire ideas for appropriate teaching strategies for each one.

Do you see which models provide a one-to-one correspondence with the number represented and which models do not? Some models attach irrelevant attributes to the place-value positions such as size or color. Do you feel there might be problems with such models? Why?

If base activities are done with proportional models, such as the multi-base blocks described earlier in the chapter, it is clear to see the defi-

Proportional Models

Nonproportional Models

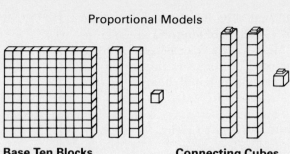

Base Ten Blocks
Purchased commercially or made out of meat trays; plastic canvas; scored with a pen or pencil

Connecting Cubes
Purchased from stores to allow for perfect connections

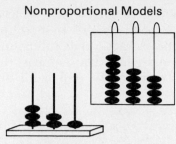

Abacus
Made from wooden blocks, dowels, pipe cleaners, or block of cardboard

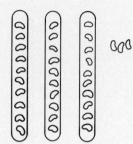

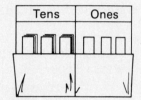

Tens	Ones

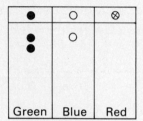

●	○	⊗
●	○	
●		
Green	Blue	Red

Beansticks
Tongue depressors with 10 dried lima beans pasted on each stick

Pocket Chart
Tag board folded over and stapled to make a pocket, paper strips are bundled in sets of ten

Chip Trading
See Appendix C for patterns to make materials out of heavy paper or plastic containers

Figure 6.7

nite configurations indicate regroupings. This makes it easy for children to see the consistency between the various bases. The ability to visualize a size increase per column helps children feel comfortable with large numbers. If a nonproportional model is desired, chip trading (different colored chips) materials work nicely. The key feature is that once trading rules are defined, all activities follow these rules. If trades are five chips of one color for one chip of another color, base 5 is used. Trades are made for the name of the base. If trades of seven of one color for one of another color are made, then base 7 is being used. Follow these steps to see how chip trading works.

Activity

MANIPULATIVES

Chip Trading

Materials: Chips of three colors
Chip trading mat with three columns color coded as the chips
Number die

Procedure: The game mat and the chips are color coded, e.g., the far right column is red, the middle column is blue, and the left column is green. The

(continued)

trades will be five for one, so base 5 will be used. The object of the activity is to be the first player to regroup chips until a trade is made for a green chip. This means that 5 reds can be traded for 1 blue, 5 blues can be traded for 1 green and when that happens, that player wins. The focus of the activity is on making trades of five for one. Sample plays are listed below for one player:

Player rolls—3. Put 3 red chips in column farthest to right.

Player rolls—3. Put 3 red chips on mat. Since a total of 6, trade 5 reds for 1 blue. This leaves 1 red.

Player rolls—2. Put 2 red chips on mat.

Player rolls—4. Put 4 red chips on mat. This makes a total of 7 red chips, so trade 5 reds for 1 blue.

Player rolls—3. Put 3 red chips on mat. Make trade of 5 reds for 1 blue.

Player rolls—4. Put 4 red chips on mat.

Play continues until a trade of 5 blues for 1 green occurs.

This player is the winner, because a green was obtained.

Early activities with number bases for children can be done using nonsense names for the base names. Since a four does not exist in base 4 (rather it is a 10), a nonsense name such as "frump" can be used. Therefore, the digits are 0, 1, 2, 3, and trades are made whenever there is "frump" in a group. To count, we would say, "Zero, one, two, three, frump, frumpty-one, frumpty-two, frumpty-three, frumpty-frump." Now we need to create another new name for the former term "hundred." This pattern continues.

Names of the place-value columns in base 4: (Theoretically, these names for the numbers do not exist in base 4 but we use the base 10 numbers to help us understand the other bases.)

Two hundred fifty-sixs
$4 \times 4 \times 4 \times 4$

Sixty-fours
$4 \times 4 \times 4$

Sixteens
4×4

Fours
4

Units
4^0

Base 4
Digits
0, 1, 2, 3,
Trades are 4 for 1

Base 7
Digits
0, 1, 2, 3, 4, 5, 6,
Trades are 7 for 1

In any base the pattern is:

$$\text{base} \times \text{base} \times \text{base} \times \text{base}$$
$$B^4$$

$$\text{base} \times \text{base} \times \text{base}$$
$$B^3$$

$$\text{base} \times \text{base}$$
$$B^2$$

$$\text{base}$$
$$B^1$$

$$\text{units}$$
$$B^0$$

Many educators feel that students should be comfortable with other bases and these activities greatly enhance and strengthen base 10 ideas. As Burns asserts, "The goal for place-value instruction is to give children a concrete understanding of how our base 10 number system works. However, in order to achieve this goal, the children benefit from exploring groupings in bases other than 10 and examining the patterns that emerge. If enough time is spent with these smaller groupings, children can readily make the transfer to the patterns in base 10. In this way, base 10 evolves as part of a general structure which the children have experienced concretely." (p. 151) Mathematically capable students can be given computational problems in other bases, such as:

$$
\begin{array}{r}
13 \text{ (four)}\\
+\ 12 \text{ (four)}\\
\hline
31 \text{ (four)}
\end{array}
$$
Add 3 and 2 and get 5, but in base 4 it will be 11.

An addition table could be constructed as in Table 6.1.

Table 6.1 Addition Table

+	0	1	2	3
0	0	1	2	3
1	1	2	3	10
2	2	3	10	11
3	3	10	11	12

The Logo activity found in the introduction to this chapter can be used with problems like the preceding one. Students can make up many numbers using the blocks in the different base numeration systems.

Counting Activities

Counting activities in base 10 are valuable for discovering the many patterns in our system as well as for understanding the number sequence. Recording the numbers on graph paper cut into three or four columns will help clarify some patterns. Study the numbers in Figure 6.8 and determine some number patterns. Can you see a rule that could be used for counting in other bases? If so, try another base and see if the rule holds true for it.

	4	3				0
	4	4				1
1	0	0				2
1	0	1			1	0
1	0	2			1	1
1	0	3			1	2
1	0	4		1	2	0
1	1	0		1	2	1
1	1	1		1	2	2
1	1	2		2	0	0
1	1	3				
1	1	4				
1	2	0				

Figure 6.8

Counting by ones, tens, hundreds, or whatever is an important activity to allow children a sense of how and when numbers change. For example, counting by ones even with six- or seven-digit numbers, only the digit in the ones column changes or increases. Some children who have not had opportunities to explore this concept will change several digits in a large number when asked to write the number that is one more or one less.

An excellent device to illustrate this concept is an odometer. If you can secure an old one from a junkyard or used car parts shop, it will demonstrate to children how the gears work together so that when the tenths dial is at 9, one rotation will cause the tenths dial to turn to 0 and the column on the left, the ones, will increase by one. Likewise, if the tenths and ones have a 9, the next rotation will cause both 9s to become 0 and the column on the left, the tens, to increase by one.

A homemade odometer can be constructed to illustrate such counting patterns and the student can control the moving of the places. Observing such counting patterns helps solidify how trading rules operate.

Activity

MATH ACROSS THE CURRICULUM

Odometer Mathematics

Materials: Oak tag—cut in strips 3.8 cm wide
Aluminum soft drink can or cardboard roll from paper towels
Matte board or heavy cardboard 16 cm × 7.5 cm
Exacto knife and scissors

Construction: Cut three strips from oak tag of equal width and length to cover aluminum can. Tape on overlap. Make an opening or "window" in the matte board that is large enough to expose one number from each strip. Cut with exacto knife and pull tabs to cover ends of can. Cut circles from oak tag to cover the ends of can and tape tabs and circles to can.

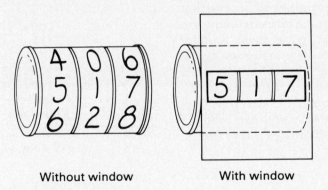

Without window   With window

Procedure: The three strips (or more if using the inner roll from paper towels) can be rotated to expose the digits on the face of the odometer through the "window." Rotate each strip to show children how the digits 0–9 form all the numbers through 999. Some activities to do with the odometer:

1. Teacher or group leader dictates a three-digit number. The student forms that number on the odometer. Check with others in the group.

2. Teacher or group leader announces which counting pattern is being used. The student shows the next three numbers in that counting sequence.

Some additional games or activities, described below, provide more experiences with counting patterns. Remember that numeration occurs at the beginning of the students' textbook so many maintenance activities are needed to ensure firm mastery of these concepts.

Activity

MENTAL MATH

Count on Five—Game 1

Materials: Number die with two sides marked 1, 10, 100
Cards about 5 × 8 cm with a number on each card—limit
size as you desire

Procedure: The deck of cards is face down. Determine first player. The
first player rolls the number die and draws a card from the deck. The object
is to correctly name the next five consecutive numbers counting by the number face up on the number die. Next player rolls the die and draws a card to
determine from which number counting will begin. Example: Roll is 10.
Card is 47. Next five counting numbers beginning at 47 and counting by tens
are: 57, 67, 77, 87, 97.

Activity

MENTAL MATH

Count on Five—Game 2

Materials: Number die with two sides marked as 2, 5, and 10
Three decks of numeral cards, each deck with numbers for 0 to
9, that are color coded for the three counting patterns

Procedure: The stacked decks are placed face down. The first player rolls
the die and draws the numeral card from that particular deck. The player
names the next five consecutive numbers. Second player continues in a similar manner. To distinguish the three decks (besides the color), mark the top
side of the cards with the counting pattern (2, 5, or 10). Example: Roll 5.
Card from fives deck is 165. The next five consecutive counting numbers
from 165 by fives are: 170, 175, 180, 185, 190.

Activity

MENTAL MATH

Place Value Bingo

Materials: Bingo boards
Popcorn, lima beans, or any small markers
Number list for caller

PROBLEM
SOLVING

Procedure: Leader or teacher is the caller who reads statements from the
number list. Anyone finding that number on the board may cover it with a
marker. Winner is first one with five markers in a row, column, or diagonal.
The number list may be written to fit the needs of students playing the game.
A sample list is shown below:

The number that is one more than 34.

The number that is before 58.

The number that is ten more than 12.

A number that has three tens.

Activity

Problem Solving
Through Patterns in Digit Hunt

Materials: Game board with numbers in each "step" of the board
Spinner with place value names that agree with board numbers
Spinner with numbers 0–9
Game board markers

Procedure: Determine who goes first by who spins the largest number. The first player spins the two spinners. The player's marker may be moved from start to the first space on the board that has that digit in the place noted on the place-value spinner. Players cannot occupy the same spot. In that case, the second player to land on that spot must return to start. Players must go to the *nearest* space that has the number in the place, even if it means a move backwards. The first player to reach the end (finish) is the winner.

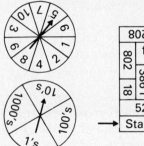

Understanding and Interpreting Large Numbers

Another aspect of numeration is the need to acquire a sense of the relative magnitude of numbers. Using manipulative models with smaller numbers will hopefully create an understanding of the trading process as well as how the powers of ten function. Then, extending these concepts into larger numbers, using proportional and nonproportional models is essential in order to conceptualize large numbers.

It is essential for children to understand numbers from physical situations. This requires additional time, because number meanings develop gradually. Intuition about number relationships helps children decide about the reasonableness of computational answers. Understanding place value is a crucial link to ac-

quiring number sense. As number relationships are developed, children acquire an understanding of the relative magnitude of numbers.

We live in a world that inundates us with large numbers. Any newspaper quickly illustrates this point: $41.2 million trade between two corporations, $3.23 million in shares traded, 2.4 billion cans of soda sold last year, $1.8 trillion national debt, to name just a few examples. Since these numbers are generally written in shortened form, a mystery is created about how much is represented by large numbers. People tend to ignore these large numbers and do not generally relate to them. Clearly, this situation is dangerous for society. If we look to most middle school textbooks for instructional techniques to interpret these numbers in standard form, we find limited coverage of this important skill. Hofstadter (1982) says we cannot have such "number numbness" or "innumeracy." People must be able to make sense of the large numbers that run their lives. Findings from numerous interviews with adults and children indicate that few people know the difference between a million and a billion (in terms of their relationship to one another) and few understand how to translate the shortened form for large numbers into standard number forms.

A book for children by Schwartz (1985) called *How Much Is a Million?* provides some visualization of large numbers. The following activity also offers many exercises that help children get some sense about the size of large numbers. Just ask children or even senior high school students what comes after billions or trillions and some interesting answers are given.

Activity

PROBLEM
SOLVING

Interpreting Large Numbers

1. Ask children what is the largest number in the world and the smallest number in the world.

2. Is it possible to write the largest number? Why or why not?

3. What patterns do you notice in the names of the periods (families)?

4. Why do you think this happened?

5. Is there such a number as a "zillion"? Why do you think this term got invented?

6. How long do you think it would take you to count to a million? A billion? A trillion?

7. What are some things in our environment of which we could quickly find a million? A billion?

8. If you were to put dots on a piece of paper, how many pages would it take for you to make a million dots?

9. Why do you think newspapers use the shortened form of large numbers?

10. Write this number in standard form: 3.45 million.

11. Do you think you were living a million minutes ago? Figure this date to an approximate month and year. What about a million seconds ago?

Activity

ESTIMATION

COMPUTERS

How Big Is a Million?
Find Out Using BASIC

How long do you predict it will take the computer to count to a million? Use this BASIC program and time how long it takes. This program is saved under the name MILLION. Type LOAD MILLION and then type RUN to see the program in action. When the program asks to what number you want to count, type 1000000. BASIC does not recognize commas in numerals. You may want to estimate how long it will take by stopping the computer after ten minutes. Look at which number the program stopped by pressing the CONTROL key and the S key at the same time. How many more minutes/hours will it take to get to a million?

(continued)

```
10 GOTO 80
20 FOR I=1 TO A
30 PRINT I
40 NEXT I
50 PRINT "PRESS SPACE BAR TO COUNT AGAIN."
60 GET B$:IF B$=""THEN 80
70 GOTO 100
80 INPUT "TO WHAT NUMBER WOULD YOU LIKE ME TO
COUNT?";A
90 GOTO 20
100 END
```

Extend the activity to any number. This activity would help children understand the odometer activity seen earlier in the chapter. They can watch the numerals change in a predictable pattern.

A fun, meaningful way to explore large numbers is to take a fact and build visualizations about that number so it becomes meaningful. For example, the National Archives in Washington, D.C. contains 91 million feet of movie film. Think about how far this would stretch across the United States. Figure this distance in terms of well-known cities or nearby cities. It is easier to relate to large distances in the preceding way. Schwartz (1985) offers an example of a million kids standing on each other's shoulders. This column would reach farther up than airplanes can fly. One billion kids standing on each other's shoulders would reach past the moon. One trillion kids standing on each other's shoulders would reach as far as Saturn.

Here is another activity to help see the patterning of "large numbers." This activity involves using a computer spreadsheet.

Activity

COMPUTERS

PROBLEM
SOLVING

Patterns in the Millions

Directions:

1. Prepare a new file using your computer spreadsheet; name it Millions.
2. Type in the formulas as value statements in the places designated in the table below. Remember to move the arrow after you have typed an entry to see the result of your actions.
3. Type in the spreadsheet format below:

	A	B	C	D	E
1		Thousand	Ten Thou	Hun Thou	Million
2		1000	10000	100000	1000000
3	21	+A3*B2	+A3*C2	+A3*D2	+A3*E2
4	121	+A4*B3	+A4*C3	+A4*D3	+A4*E3

4. You will need to widen column E to do the last number. The mark, ## ##### will appear if the number is too large for the width of the column.
5. Do you see what large numbers look like? Are there patterns to be seen? Children can change one or two numbers and see what happens to the whole table.
6. Change 1000 to 1003 in B2. Watch what happens.
7. Change 10000 to 100030 in C2.
8. Now change 100030 in C2 to 100003. What happens?
9. Did you get 254107623 in E4?

10. Now change a million to 3 million, 9 million. How would you read this number?

11. Look at E3. Do you have 189000000? What are some of the differences between E3 and E4? If you only wanted a 7-digit number in E4, what options would you have for numbers in A3?

12. Try some of your numbers to see if your prediction is correct.

13. Choose other numbers in place of 21 and 121. See what changes it makes in the table.

14. Add more columns to the table and go to a hundred-trillion.

15. Experiment with numbers to see what patterns you find. The spreadsheet page will continue over for many more columns even though the first few columns are off the screen. A left arrow will restore you to the first part of the table.

There are several skills involved under the term "reading numbers." Numbers need to be literally read aloud in numeral form (not given adequate time in most classrooms), read in written name form and translated into numeral form, seen in numeral form and written in name form, read in numeral form and modeled with manipulatives, and read in expanded form or exponential form and translated into standard form.

The teen numbers seem to present many problems for some children as they hear the ones digit first and then the tens ("teen"). Many number reversals occur in the teens because upon hearing "seventeen", the child writes 7 and then 1. The child might read it back to you as 17 or, given some time away from that number, the name 71 might be given to the number. It is important to have many practice opportunities for modeling the teen numbers.

A word of caution: never read whole numbers with the word "and" between the hundreds and tens. For example, 259 is read "two hundred fifty-nine" not "two hundred *and* fifty-nine." The word "and" indicates a decimal point in the number separating the whole number from its fractional (decimal) components. We also must be careful and read numbers in complete word form and not simply the single digits. Adults often read multi-digit numbers such as 34,528 as "three four five two eight" to be efficient and faster. This method is a better way to read large numbers when comparing answers or reading answers from a calculator, but it can be destructive to children who need to read, hear, and interpret numbers with place-value labels. Writing checks is an application of this skill and one which students soon see a need to learn.

When reading large, multi-digit numbers, it is important for students to use commas to cluster the periods together. Then it becomes a matter of grouping the digits into groups of three and reading these numbers together followed by the period name (represented by the comma). Many metric groups advocate using a space instead of a comma. However most elementary textbooks use the comma. Each period contains a group of hundreds, tens, and ones—another pattern. The period names also follow a pattern, which is a fun thing to explore and about which students can speculate. Study the period names and see what patterns you notice, reading from left to right:

thousands, millions, billions, trillions, quadrillion, quintillion, sextillion, septillion, octillion, nonillion, decillion, undecillion, duodecillion, tredecillion, quattuordecillion.

Numeration cannot be taught without considering the importance of rounding skills. Most textbooks begin simple rounding to tens around third grade and may include rounding to hundreds. Generally, rounding is introduced with number lines (Fig. 6.9) so students can see the position of the number to be rounded in relation to other numbers (multiples of 10 or 100, etc.) depending on the place to which the number is being rounded. This visual aid helps children see the relative position of the number to each other. However, the rule for rounding seems to be introduced quickly and often prevents greater understanding about why the rule works. The rounding rule used in most textbooks is:

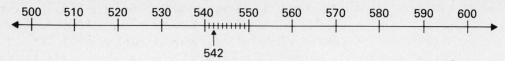

542

Round 542 to the nearest hundred.

Figure 6.9

1. Find the place-value position being rounded.
2. Look at the digit to the right of the place being rounded.
3. Decide if the digit is greater than, less than, or equal to 5.
4. If the digit is equal to or greater than 5, round up (add 1 to the digit to the left).
5. If the digit is less than 5, round down (keep the digit to the left the same).

The term "round down" may be confusing as it implies decreasing the digit. Even though teacher's manuals include this term, using the words in parentheses may be better to use with children.

Many times students do not see a purpose for rounding numbers because the final product is to work the problem using the actual numbers. Unfortunately, teachers often assign problems to be worked both ways—estimated by rounding and actual numbers—and this causes a feeling in students that rounding is extra work that doesn't really count. Estimation is a useful, important skill for determining the reasonableness of your answer and estimation requires an understanding of rounding. Somehow when teachers assign problems to be worked both ways, the importance of rounding and estimating is not instilled in students.

Activity

ESTIMATION

Rounding Large Numbers

Materials: A die or number cube marked with 10 on two sides, 100 on two sides, and 1000 on two sides.

Numeral cards that include four- and five-digit numbers (or larger to match skills for desired grade level).

Procedure: The students decide who begins. The first player rolls the die. The number on the face showing (10, 100, or 1000) represents the place to which the number is to be rounded. The first player draws a numeral card. This is the number being rounded. For example, roll lands on 100 and the card drawn is 4,324. The player is to round that number to the nearest 100 which is 4,300. The deck may be used over and over as the die determines the rounding place each time. The players can receive a point, token, or tally for each correct answer. The player with the most points wins. A game board can also be used. In this case, a correct answer allows the player to have a move on the board. The first player to reach the end is the winner.

Diagnosis and Remediation

Simultaneous Thought Processing

Students who process information simultaneously (from the whole to the part,) frequently do well with activities such as the computer spreadsheet model where the table changes all at once when a number is generated. They can analyze patterns from the whole table seen simultaneously. They also perform well with materials like the place-value strips because everything is already there and just needs to be folded from the expanded notation to the compact form. The base 10 blocks can be used successfully in such activities as "trading down," where the base numbers are exchanged from a large set of base 10 blocks already assembled.

Successive Thought Processing

Students who process information successively (from the parts to the whole) frequently benefit from the use of the abacus and chip trading because both materials require a detailed build-up of the place values from ones to tens to hundreds and so on. Such students may be more comfortable with the Logo computer program that progresses from one number to the next rather than using the computer spreadsheet approach. The base 10 blocks can be used successfully, especially in activities like "trading up" where the student builds large numbers from exchanging smaller place value units to make larger and larger numbers from the smaller blocks.

If materials like the spreadsheet and place-value strips are used, consider pairing a successive learner with a simultaneous learner. It is a cooperative learning effort where both students can learn from watching the other if they talk about what they are doing as they work.

Correcting Common Errors

One common error is that children write too many zeros in the number because they are relating to the number words. For example, two hundred forty-five is written as "20045". The first three digits are recorded for the 200 and then the 45. The remediation technique to use is to ask the child to model the number with base 10 blocks. Also, additional experiences with the place-value strips seen earlier in the chapter would be a valuable connecting level experience.

Another error is developmental in nature—when children do not conserve volume. Since children who do not conserve volume may consider the thousand block as "600," it may be wise to construct 1000 using 10 FLATS (hundreds blocks). Students using the 10 hundreds blocks can easily remove a layer at a time to recall there are more units inside that need to be counted. When they mistake 1000 for 600, let them count by 100s to a 1000, seeing that there are more blocks than 600.

Another common error in writing numerals is just the opposite problem—not using zeros as place holders in numerals where needed. In this case, since the child does not hear or read any numbers for that place value, it is simply ignored. For example, in the number "three thousand fourteen," the child would write, "3,14". Again concrete models such as base 10 blocks with a place-value mat or spiked abacus (Fig. 6.10) help to visualize all the places in numbers and where zeros are needed as place holders.

These activities offer some games and teaching strategies to help develop numeration skills.

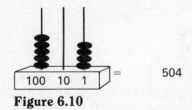

100   10   1   =   504

Figure 6.10

Activity

MANIPULATIVES

Abacus

Materials: Abacus—board 1x2x10-inches with 5 or 7 finishing nails evenly spaced across the top (with flat heads—no sharp ends) Small washers or plastic discs with hole in center

Procedure: The washers are placed on the nails for the place value positions represented. If the teacher reads a number such as "three hundred thousand fifty-one," the student is to model that number on the abacus with washers. In this case, there would be three washers in the hundred-thousands place, five washers in the tens place, and one washer in the ones place. In this way, the teacher can assess the child's ability to model the number. If the child is asked to write the number, the abacus helps identify the place-value positions that have no washers but which require a zero to serve as a place holder.

Activity

MENTAL MATH

Number Tiles

Materials: Number tiles for digits 0–9

Procedure: The teacher gives a set of number tiles to each child. The following directions are given as the children form the numbers with the tiles:

1. Choose three tiles. Form a three-digit number. Read your number to others in your group. Can the child tell just from the symbols what the number is to read it? Compare the numbers and order from greatest to least.

2. Take your three-digit number and exchange the tile in the hundreds place with the tile in the ones place. Is the new number greater than or less than the former number? By how much? Compare and order these numbers in your group?

3. Choose any three tiles and form the greatest number possible.

4. Choose any three tiles and form the least number possible.

5. Form a three-digit number that is greater than 500 but less than 715. Read your number aloud to the group and check if it fits the given parameters.

6. Increase your number by 100, decrease the number by 10, increase it by 200.

7. Continue such activities for larger numbers.

8. Work at doing these activities in your head without the tiles.

Summary

To help children acquire number meanings they must represent numbers with physical materials, explore groupings and number relationships, and extend their knowledge of whole number to fractions, decimals, and integers. The instructional goal is to focus on the underlying structure of the number system.

This chapter emphasized different numeration systems. The base 10 system was developed using base 10 blocks as manipulative models. Other bases were also explored. Both proportional and non-proportional models were used in the development. The computer was used to explore large numbers and other number bases. The diagnosis and remediation section explored student difficulties with too many zeros and writing numerals.

Chapter 7 introduces the basic facts and their properties. The elementary school curriculum devotes extensive time to computation and facts. The basis of student success with the basic facts and also with algorithms (Chapter 8) is a thorough understanding of our numeration system—the emphasis of this chapter.

Exercises

Directions: Read all questions *before* answering any one exercise. Frequently the last question in one category leads to the first question in the next category.

A. **Memorization and Comprehension Exercises**
   Low-Level Thought Activities

   1. Explain the difference between a numeration system and a number system.

2. What is the answer to this problem?

$$798^0 = ?$$

3. Explain how you got your answer in A2.

4. If a child writes sixty-seven as 607, what is the child's misconception?

B. Application and Analysis Exercises
Middle-Level Thought Activities

1. Apply what you have learned about the configuration of different number bases to draw the multi-base blocks that represent these numbers:

(Given A = 10, B = 11, C = 12, etc.)
$18_{(nine)}$ $20A_{(twelve)}$ $45_{(six)}$
$21_{(three)}$ $1111111_{(two)}$

2. Analyze what the number 214 looks like in base 3 (be careful!), in base 5, in base 9, in base 15.

3. Show what the numbers in B2 would look like if represented on a chip trading mat. What do you adjust to deal with the base 2 number on the chip trading mat?

4. Using the evaluation form found in Appendix A, select commercial software for any three topics from this chapter. In the selection, include drill and practice, simulation, tutorial, and problem-solving software. For each software program selected, describe how you would integrate it in a lesson plan.

5. To reinforce math across the curriculum:
  a. List five mathematical questions/problems that come from popular storybooks for children. Show how they are related to the topics of this chapter.
  b. List five ways to use the topics of this chapter when teaching lessons in reading, science, social studies, health, music, art, physical education, or language arts (writing, English grammar, poetry, etc.).

6. Search professional journals for current articles on research findings and teaching ideas with numeration. You may wish to look at the first activity under Synthesis and Evaluation Activities below and coordinate activities.

Follow this form to report on the pertinent findings:

Journal Reviewed: _____

Publication Date: _____ Grade Level: _____

Subject Area: _____

Author(s): _____

_____

Major Finding:

Study or Teaching Procedure Outlined:

Reviewed by: _____

Some journals of note are:
Journal for Research in Mathematics Education
Arithmetic Teacher
Mathematics Teacher
School Science and Mathematics

7. Computer applications for classroom use are found in the following journals. Their programs are by no means the only ones. Search the periodical section of the library for more. You may wish to look at the first activity under Synthesis and Evaluation Activities to coordinate activities.
Teaching and Computers
Classroom Computer Learning
The Journal of Computers in Mathematics and Science Teaching
Computers in the Schools
Electronic Learning
Computer Teacher

C. Synthesis and Evaluation Exercises
High-Level Thought Activities

1. Find a concept on the Scope and Sequence Chart that you would like to teach.
  a. Search professional journals for current articles on research findings and teaching ideas for numeration. Follow the preceding journal report form. The journals noted there are still pertinent to the study of numeration. Look for computer applications.
  b. Use current articles from the professional journals to help plan the direction of the lesson. Document your sources.

c. Include at least one computer software program as a part of the lesson. Show how it will be integrated with the rest of the lesson.

d. Develop behavioral objectives. Show how you will use these to evaluate the lesson.

e. Use the research-based lesson plan format of a) review, b) development, c) controlled practice, and d) seatwork.

2. Using the chip trading mat, create regroupings or "trades" in the different bases as indicated. Use A, B, C, D, and E for numerals in bases larger than 9. Choose a bunch of each color at random and regroup from there. Work in base 4 and base 14.

3. Create a number line in base 7 using the multi-base blocks seen in the introduction as a model.

4. Prepare a lesson having students create a database on the space program. Have students include dates, distances, speeds, uniqueness (i.e., first animal in space, first artificial planet of the sun, etc.) and other information students see as relevant. Prepare problem-solving questions and include specific math objectives. Include an evaluation plan.

5. Create your own number system with numeration symbols. Do it in Logo. You can use the program BAB in Appendix D for some ideas on where to start.

6. What are the unique qualities of our base 10 numeration system, and how can they be shown as the basic building blocks of many number concepts to come?

Bibliography

Beattie, Ian. "The Number Namer: An Aid to Understanding Place Value." *Arithmetic Teacher* 33 (January 1986): 24–28.

Burnett, Peg Hampton. "A Million! How Much is That?" *Arithmetic Teacher* 29 (September 1981): 49–50.

Burns, Marilyn. *The Math Solution: Teaching for Mastery through Problem Solving.* Sausalito, Calif.: Marilyn Burns Education Associates, 1984.

Commission on Standards for School Mathematics of the National Council of Teachers of Mathematics. *Curriculum and Evaluation Standards for School Mathematics.* Working Draft. Reston, Va: The Council, 1987.

Grossnickle, Foster, E., and Leland M. Perry. "Man's Romance with Number." *School Science and Mathematics* 87 (January 1987): 7–11.

Harrison, Marilyn, and Bruce Harrison. "Developing Numeration Concepts and Skills." *Arithmetic Teacher* 33 (February 1986): 18–21.

Hofstadter, Douglas R. "Metamagical Themas: Number Numbness, or Why Innumeracy May Be Just as Dangerous as Illiteracy." *Scientific American* 246 (May 1982): 20, 24, 28, 32–34.

Jensen, Rosalie, and David R. O'Neil. "Some Aids for Teaching Place Value." *Arithmetic Teacher* 25 (November 1981): 6–9.

Jones, Philip S. "Notes on Numeration: Arithmetic on a Checkerboard, Numerals for the Blind." *School Science and Mathematics* 78 (October 1978): 481–488.

Kouba, Vicky L., Catherine A. Brown, Thomas P. Carpenter, Mary M. Lindquist, Edward A. Silver, and Jane O. Swafford. "Results of the Fourth NAEP Assessment of Mathematics: Number, Operatives, and Word Problems." *Arithmetic Teacher* 35 (April 1988):14–19.

Labinowicz, Ed. *Learning from Children: New Beginnings for Teaching Numerical Thinking, a Piagetian Approach.* Menlo Park, Calif.: Addison-Wesley, 1985.

McBride, John W., and Charles E. Lamb. "Number Sense in the Elementary School Mathematics Classroom." *School Science and Mathematics* 86 (February 1986): 100–107.

Rappaport, David. "Numeration Systems—A White Elephant." 77 *School Science and Mathematics* (January 1977): 27–30.

Schwartz, David. *How Much is a Million?* New York: Lothrop, Lee, and Shepard Books, 1985.

7

The Whole Number System

Key Question: *How can the basic number facts be taught to children in efficient and meaningful ways?*

Activities Index

Introduction

A large percentage of a child's elementary school years is spent working with the whole number system. The operations, properties, and basic facts become an important part of daily mathematics instruction. *The whole number system is the set of counting numbers and zero, where any whole number, a, precedes the whole number, b, in a definite order such that a < b and b > a* (e.g., 3 precedes 5 and 3 < 5 or 5 > 3).

This means that operations in the whole number system are predictable, with definite properties and ordered relationships.

In most textbooks, children study the whole number operations in the following order: addition, subtraction, multiplication, and division. Some mathematics educators advocate the study of addition and multiplication together and then the study of subtraction and division together.

The reader is encouraged to think of justifications for both points of view as the relationships among the operations are presented.

After students develop concepts and explore problem situations involving operations, the teacher should provide experiences to develop "operation sense" (Commission on Standards for School Mathematics, 1987). The NCTM curriculum standards defines this term as understanding properties and relationships for each operation, developing relationships among operations, and acquiring intuition about the effects of operating on a pair of numbers, such as increasing each addend by one.

An overview of the operations and properties appear in this section. The basic facts and how to teach them appear later in the Teaching Strategies.

The Operation of Addition

The operation known as addition is the process of joining things together. When any counting number (the set of whole numbers greater than zero), a, is joined together with counting number, b, there is one and only one counting number, c, that can result. For example, there is one and only one answer when 2 is added to 3. This is true for all addition of whole numbers. The operation of addition has four properties (also called principles) which can help students to learn the basic facts. They appear in Table 7.1, showing the corresponding symbols and student manipulatives for each property. The following terms are used in addition:

$$
\begin{array}{rl}
3 & \text{(addend)} \\
+\,2 & \text{(addend)} \\
\hline
5 & \text{(sum)}
\end{array}
\qquad
\left.
\begin{array}{r}
3 \\
2 \\
+\,4 \\
\hline
9
\end{array}
\right\}
\begin{array}{l}
\text{All are} \\
\text{called} \\
\text{addends} \\
\text{(sum)}
\end{array}
$$

The Operation of Subtraction

Subtraction is defined as the inverse of addition such that $a - b = c$ if and only if $c + b = a$. Note that a must be greater than or equal to b for the answer to be in the whole number system.

$$a = 8, b = 3, c = 5$$
$$a = 3, b = 5$$

$8 - 3 = 5$ works because $5 + 3 = 8$

$3 - 5$ does not fit the definition because 3 (a) is not greater than or equal to 5 (b)

The concept of subtraction is more complex than addition in that the computation can be used to find solutions to situations in three different ways. The three different meanings are defined as take away, comparison, and missing addend. All three concepts have the same subtraction problem and provide a means of calculating a difference between two groups—the minuend and the subtrahend.

$$
\begin{array}{rl}
5 & \text{(minuend)} \\
-\,2 & \text{(subtrahend)} \\
\hline
3 & \text{(difference)}
\end{array}
$$

5 cupcakes
− 2 were eaten
3 cupcakes left

The "take away" idea means that a subset of the original set is actually removed. This is probably the most easily understood and the most often applied concept of subtraction. Concrete examples are found rather easily in the experiences of children. Eating cookies, spending money, popping balloons, or birds flying away are all representations of this subtraction concept. Minus is a word used for the sign "−," not for the concept of "take away." However, many young children refer to the take away meaning as "minus."

The second meaning of subtraction is comparison. This concept is harder for children to understand and is introduced after first grade.

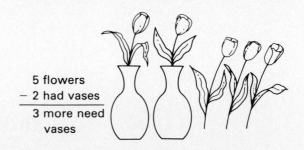

5 flowers
− 2 had vases
3 more need vases

The comparison concept means that two sets are compared to one another and the difference, either more than or less than, is found by the process of subtraction. Everyday examples of this may be experienced by children when passing papers, matching books to children in a reading group, setting the table, or getting chairs for a group. The matching of one-to-one in groups

Table 7.1 Properties/Principles of the Whole Number System for Addition and Subtraction

Definition	Symbols	With Manipulatives

Addition

Identity Property
A whole number remains itself (keeps its own identity) after an operation is performed with it. Zero is called the identity element; whatever number joins to it does not change.

$a + 0 = a$
or
$0 + a = a$

$2 + 0 = 2$
$0 + 2 = 2$

$2 + 0 = 2$

Commutative Property
When any two whole numbers are grouped together, their order may be reversed without changing the outcome.

$a + b = b + a$
$4 + 2 = 2 + 4$

$4 + 2$ = $2 + 4$

Associative Property
When any three whole numbers are grouped together, their grouping order may be reversed without changing the outcome. Any two of the three may be grouped together first and then combined with the third.

$(a + b) + c = a + (b + c)$
$(2 + 3) + 4 = 2 + (3 + 4)$
$5 + 4 = 2 + 7$
$9 = 9$

$(2 + 3) + 4 = 2 + (3 + 4)$
$5 + 4 = 2 + 7$
$9 = 9$

(continued)

Counting Principle

The counting principle states that adding 1 to any whole number gives the next higher whole number. If counting has been approached as an additive process where each succeeding number is one greater than the last, this generalization is one which children may find by intuition.

$$1 + 1$$
$$2 + 1$$
$$3 + 1$$
$$4 + 1$$
$$5 + 1$$
etc.

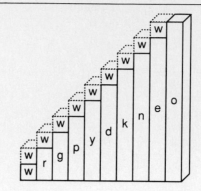

w = white
r = red
g = green
p = purple
y = yellow
d = dark green
k = black
n = brown
e = blue
o = orange

Subtraction

Identity Property

Subtraction is said to have a right identity but not a left identity. Only when zero is subtracted from a number does that number maintain its identity (remain itself).

YES	NO
$A - 0 = A$	$0 - A \neq A$

or with numerals . . .

$$2 - 0 = 2 \qquad 0 - 2 \neq 2$$

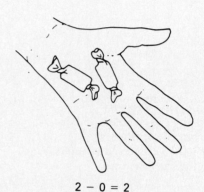

$$2 - 0 = 2$$

Commutative Property

Subtraction is non-commutative.

$$7 - 3 \neq 3 - 7$$
$$4 \neq -4$$

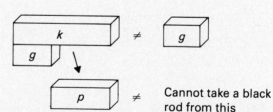

Cannot take a black rod from this

Associative Property

Subtraction is non-associative.

$$(7 - 3) - 2 \neq 7 - (3 - 2)$$
$$4 - 2 \neq 7 - 1$$
$$2 \neq 6$$

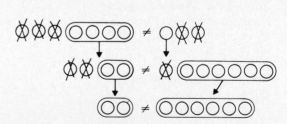

Counting Principle

The counting principle of subtraction states that subtracting 1 from a number gives the next smaller counting number.

$$7 - 1 = 6$$
$$6 - 1 = 5$$
etc.

With the number line . . .

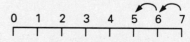

of equal size is the concept necessary for counting and leads to subtraction with differences of 0. When unequal sets of objects are matched one-to-one, if a difference greater than zero is obtained, this quantity is calculated by the subtraction process.

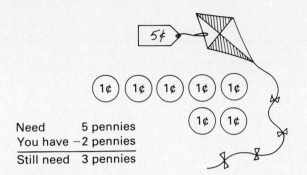

Need   5 pennies
You have   −2 pennies
Still need   3 pennies

The third concept of subtraction is the missing addend or completion. This involves the relationship of subtraction to addition. In addition, two sets are joined to form a total, and if either of these two sets is missing, then the known set may be subtracted from the total to find the missing addend. This idea of finding how many more are needed is represented in the real world. Many experiences call for children to decide the missing part when the total amount and one of the parts is known. Knowing the cost of an item and how much money you have will enable you to decide how much more money, if any, is needed to buy the item.

The properties of subtraction appear in Table 7.1. Notice that the not equal sign ($\neq$) is introduced to students when they find inequalities in some of the properties of the number system (shown in Table 7.1 under subtraction). Students can learn that there is a way of handling impossibilities, and that mathematicians have accounted for all conditions.

The Operation of Multiplication

Multiplication is the mathematical operation used to combine sets of equal size or sets containing the same number of things into a new set. The aspect of multiplication that distinguishes it from addition is that for multiplication the sets must be the same size. Multiplication can be defined as repeated addition where the addends are of equal size (2 + 2 + 2 + 2). Multiplication can also be defined in terms of

Cartesian products where all the elements of one set can be matched one-to-one with all the elements of another set whose total product is the same number (i.e., set A $\times$ set B equals the same number of elements as set B $\times$ set A).

If 2 sets A and B have the following elements:

Set A = {fur hat, top hat, cowboy hat}
Set B = {(person 1, person 2, person 3, person 4)}

Then all 3 elements of set A can be matched one-to-one with all 4 elements of set B whose total product shows the same number of elements (matches), in this case, 12.

Set *A* = Hats

Set *B* = People	Fur (a)	Top (b)	Cowboy (c)
Person *A*	Aa	Ab	Ac
Person *B*	Ba	Bb	Bc
Person *C*	Ca	Cb	Cc
Person *D*	Da	Db	Dc

Students must understand the exact wording of multiplication. For example, instead of saying "4 times 2," the teacher needs to say, "4 sets of 2." The use of the word "of" is very important because it makes the nature of multiplication (equal sets) clearer than does the word "times." This also helps children in later elementary grades when they study multiplication of rational numbers, covered in Chapter 10 of this text. The examples in Figure 7.1 show the use of this terminology.

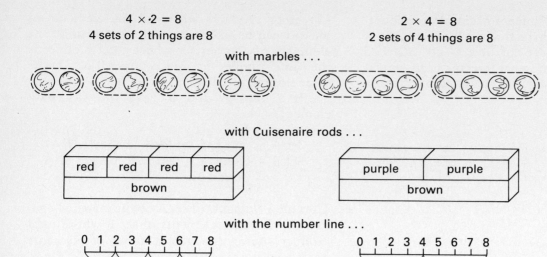

Figure 7.1

Operation of Division

Division is the math operation used to separate a set into equal parts. Division is the inverse operation of multiplication—where multiplication is the operation used to combine sets of equal size into a new set. Division can be considered the opposite of multiplication or as repeated subtraction. There are two concepts of division, both of which can be easily modeled with manipulatives: partitive division and measurement division. The difference between these two situations is noted in the following chart:

Measurement Division	Partitive Division
Know:	*Know:*
1. Number of objects in the original set	1. Number of objects in the original set
2. Number of objects to be in each new set	2. Number of new sets to be formed
Need to Find:	*Need to Find:*
3. How many new sets can be made?	3. How many objects are in each new set?

Measurement Division: involves the process of taking an original set of objects and counting out (measuring) the number of objects wanted in each new set until all objects have been distributed. Then the number of equal sets is counted. Here is an illustration of measurement division:

Pat has 10 marbles and wants to give 2 marbles to each friend. The question is how many friends will get two marbles. (Compare this situation to partitive division—how many marbles would go to each friend in a fixed group.)

Therefore, Pat gives 2 marbles to one friend, 2 to another friend, 2 to another, 2 to another, and 2 to one more friend. Then Pat counts and finds that the marbles lasted long enough to distribute to 5 friends.

Figure 7.2 shows that if there are 10 marbles, and 2 are taken away each time, this can happen 5 times. This process can also be thought of as repeated subtraction. Or, in equation form, as $10 \div 2 = 5$.

Partitive Division: involves the process of taking an original set of objects and sharing (partitioning) objects until they are all gone, or none are left for another complete round. Then the number in each new set is counted. Dealing cards to people in a game is a common situation where partitive division is used. Another example of partitive division is:

Lee knows the number of friends present, and also knows that each one will expect an equal share of the candy. It would be safer to start by giving (partitioning) one piece to each person, rather than risk having to take back some candy. The division process would go like this:

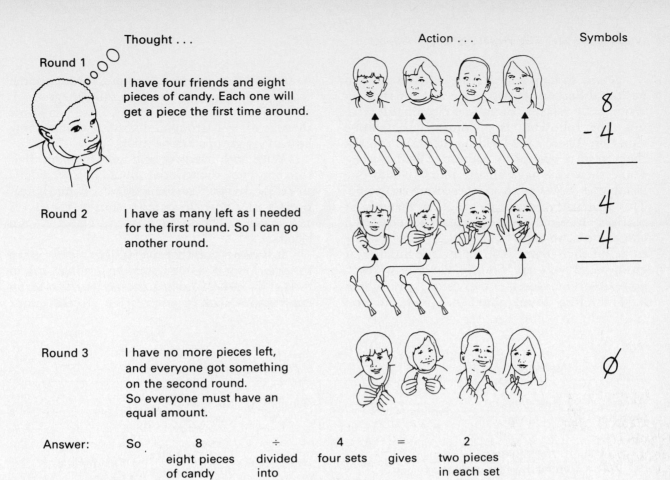

Thought . . .   Action . . .   Symbols

Round 1   I have four friends and eight
           pieces of candy. Each one will
           get a piece the first time around.
                                                                                                                    8
                                                                                                                  − 4
                                                                                                                  ─────

Round 2   I have as many left as I needed
           for the first round. So I can go
           another round.
                                                                                                                    4
                                                                                                                  − 4
                                                                                                                  ─────

Round 3   I have no more pieces left,
           and everyone got something
           on the second round.
           So everyone must have an
           equal amount.   Ø

Answer: So   8   ÷   4   =   2
                eight pieces   divided   four sets gives   two pieces
                of candy   into   in each set

Partitive Division

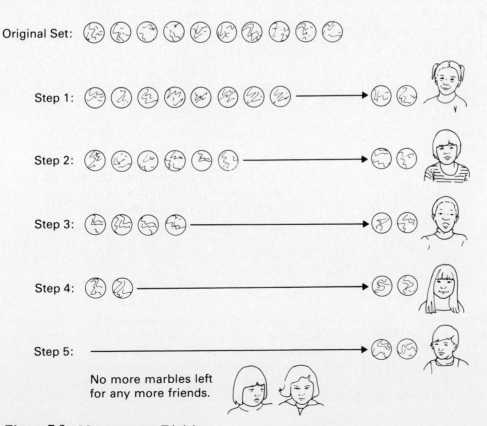

Original Set:

Step 1:

Step 2:

Step 3:

Step 4:

Step 5:

No more marbles left
for any more friends.

Figure 7.2 Measurement Division

A way to help you and the children remember the difference between the two types of division is by relating them in a meaningful fashion. The following is a common illustrative situation. There are 24 kids in the class. You ask them to form groups by counting aloud 1–2–3–4 and then you have all the fours go into one group, the threes go to another group and so on. This is partitive division. Another example of partitive division is when you have four children chosen as team captains and they select people to be on their team. You know the number of groups and you are finding the size of each group. For measurement division, you would have children count aloud 1–2–3–4 and you would call that group 1. The next four children would form group 2, the next four are group 3, and continue in such a manner. Here you know the size of each group and you are determining how many groups can be made.

With both partitive and measurement division concepts, the student should be encouraged to relate division as the inverse of multiplication. Notice how the comparisons in Figure 7.3 can be made. The comparisons in Figure 7.4 are also true.

It is important that the student clearly grasp the concept of division before beginning work in formal division. Students should be given many experiences with manipulative devices until

	Measurement Division	Multiplication
General Meaning	$\dfrac{\text{Total Objects}}{\text{Number in each set}}$ = Number of sets	$\dfrac{\text{Number of}}{\text{sets}} \times \dfrac{\text{Number in}}{\text{each set}}$ = Total objects
Example	$\dfrac{\text{Ten marbles}}{\text{Two marbles for each person}}$ = Five people	Five people $\times \dfrac{\text{Two marbles}}{\text{for each person}}$ = Ten marbles
Symbols	$\dfrac{10}{2} = 5$	$5 \times 2 = 10$
Illustration		

Figure 7.3

	Partitive Division	Multiplication
General Meaning	$\dfrac{\text{Total Objects}}{\text{Number of sets}} = \dfrac{\text{Number in}}{\text{each set}}$	$\dfrac{\text{Number of sets}}{} \times \dfrac{\text{Number in}}{\text{each set}}$ = Total Objects
Example	$\dfrac{\text{Six flowers}}{\text{Three people}} = \dfrac{\text{2 flowers for}}{\text{each person}}$	Three people $\times$ 2 flowers = Six flowers
Symbols	$\dfrac{6}{3} = 2$	$3 \times 2 = 6$
Illustration		

Figure 7.4

Table 7.2 Properties/Principles of the Whole Number System for Multiplication and Division

Definition	Symbols	With Manipulatives

Multiplication

Identity Property
The identity property of multiplication is that property which allows any counting number to maintain its identity when multiplied. The identity element is 1.

$A \times 1 = A$
$1 \times A = A$
$3 \times 1 = 3$
$1 \times 3 = 3$

3 flowers with 1 petal.
How many petals?

$3 \times 1 = 3$

Zero Property
When a counting number is multiplied by zero, the resulting product is zero.

$A \times 0 = 0$
$0 \times A = 0$
$3 \times 0 = 0$
$0 \times 3 = 0$

3 flowers with 0 petals.
How many petals?

$3 \times 0 = 0$

Commutative Property
When two factors are multiplied together, their order may be reversed without changing the outcome.

$A \times B = B \times A$
$5 \times 3 = 3 \times 5$

With the pan balance . . .

5 groups of 3   3 groups of 5

Associative Property
When three whole numbers are grouped together, their order may be reversed without changing the outcome.

$(A \times B) \times C = A \times (B \times C)$
$(2 \times 3) \times 4 = 2 \times (3 \times 4)$
$6 \times 4 = 2 \times 12$
$24 = 24$

With Cuisenaire rods . . .

$(2 \times 3) \times 4$ | g | g |
6×4 | d | d | d | d |
24 | o | o | p |

$2 \times (3 \times 4)$ | p | p | p |
2×12 | o | r | o | r |
24 | o | o | p |

(continued)

Definition	Symbols	With Manipulatives
Distributive Property Involves the two operations of addition and multiplication and states that the first factor is dispersed over the second factor.	$A \times (B + C) = (A \times B) + (A \times C)$ $3 \times (2 + 4) = (3 \times 2) + (3 \times 4)$ $3 \times 6 \quad = \quad 6 + 12$ $\quad 18 \quad = \quad \quad 18$	With Cuisenaire rods . . .

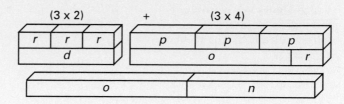

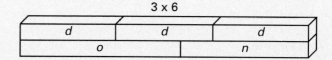

Division

Division by Zero Does NOT compute.	$A \div 0 = $ UNDEFINED	????????????

Commutativity
Division is not commutative.

$A \div B \neq B \div A$
$2 \div 4 \neq 4 \div 2$
$\quad 0.5 \neq 2$

How many sets of 4 in 2? ≠ How many sets of 2 in 4?

0.5 or $\frac{1}{2}$ ≠ 2

Associativity
Division is not associative.

$(24 \div 4) \div 2 \neq 24 \div (4 \div 2)$
$\quad 6 \div 2 \neq 24 \div 2$
$\quad \quad 3 \neq 12$

How many sets of 4 in 24? How many sets of 2 in 4?

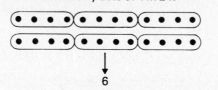

6 2

How many sets of 2 in 6? How many sets of 2 in 24?

3 12

they have a firm understanding of division as the separation of a set into equal-sized parts. Children will enjoy the experiences even more if their names are used in examples. Clay, buttons, plastic chips, or even cut-up squares of sponges can be used as manipulatives to represent the objects.

The Special Property of Division

The properties of division are presented in Table 7.2. The property of zero in division is a special case and one that can be confusing to children initially learning the division basic facts. The property of division by zero proclaims that dividing a number by zero does not compute. What does this mean? How can it be shown to children?

Division has been called the inverse of multiplication, but that does not hold true when zero enters the picture. For example, what is the answer to $8 \div 0$? When expressed in inverse fashion (which works with other division facts such as $8 \div 4 = 2$ becomes $2 \times 4 = 8$), this would become $? \times 0 = 8$. Remember that any number multiplied by zero is zero, so any number used instead of the ? would produce a product of 0. Then 8×0 would equal 8, but that is not true. The only number that could replace 8, to make the inverse possible also, is zero. Since this does not permit computation to take place meaningfully, the situation inherently becomes "does not compute."

Mathematicians declare that division by zero is "undefined." Did you ever wonder why this word was chosen? The answer is intriguing to adults with a philosophical bent. Philosophy was the parent discipline from which mathematics emerged. All mathematical issues were once philosophical ones. As the systematic treatment of relationships of thoughts, forms, and figures could be quantified symbolically, mathematics came into being. A thought that has yet to be defined quantitatively, and remains in the realm of philosophy, is the idea of division by zero. A philosopher/mathematician might reason this way:

> If I have 8 things and I divide them by nothing, I am acknowledging that nothing must be something if I can divide by it. But by definition, nothing is void of something; so how can I divide by a void?
>
> Since zero is a void, it has no unique answer by which to define it, therefore, it remains "undefined."

No wonder it is often difficult for a student to understand. Piagetian theorists would say that such abstract issues cannot be understood until the stage of formal operational thought (12 years old and beyond). Yet students must face the issue when they are learning the basic facts of division around the third grade level, when the average student is 8 or 9 years old. About all we can say is, "it won't work." But children will want to try. As with other concepts, the student should be encouraged to do some examples showing the division by zero property.

Pose the problem of $8 \div 0$ as explained above. The number sentence format of the inverse of $8 \div 0 = ?$ would be $? \times 0 = 8$. Whatever number the student puts in the box, the product is still zero. This would indicate that $0 \neq 0$; there is no answer for this problem and division by zero is undefined. But there is a way to write it. It may be represented as:

$$8 \div 0 = \text{UNDEFINED}$$

or

$$? \times 0 = 8 \neq 0 \text{ UNDEFINED}$$

Scope and Sequence

Addition	Subtraction	Multiplication	Division
Kindergarten			
Readiness with pictures.	Readiness with pictures.		
First Grade			
Sums to 10.	Differences from 10.	Readiness to skip count by 2, 5, 10.	
Family of facts.	Family of facts.		
Missing addend.	Differences from 18.		
Relationship to subtraction.			
Sums to 18.			
Commutative property.			

(continued)

Addition	Subtraction	Multiplication	Division
Second Grade			
Sums to 18.	Differences to 18.	Readiness by counting.	
Relation to subtraction.	Family of facts.	Products 0 to 5.	
Commutative property.	Adding to check.	Repeated addition.	
Associative property.	Comparative notation.		
Bridging to 10.			
Family of facts.			
Add 3 addends.			
Third Grade			
Sums to 18.	Differences from 18.	Factors through 9.	Factors to 9.
Checking by adding up.	Check by adding.	Identity property.	Inverse operation.
Commutative property.	Inverse of addition.	Commutative property.	Family of facts.
Associative property.		Associative property.	Zero property.
Fourth Grade			
Review.	Review.	Review.	Review.
Fifth Grade			
Review.	Review.	Review.	Review.
Sixth Grade			
Review.	Review.	Review.	Review.
Seventh Grade			
Review.	Review.	Distributive property.	Review.
Eighth Grade			
Review.	Review.	Review.	Review.

Teaching Strategies

Readiness to Learn Basic Facts

An understanding of the four basic operations must be developed along with a knowledge of the basic facts. Basic facts are combinations of single-digit numbers, such as $3 + 7, 4 \times 8$. Time must be given for explorations with materials. Children should use manipulatives to discover number patterns and relationships, to model particular situations, and to explore the joining, separating, and comparing of sets. After extensive informal experiences with physical objects, mathematical terms and symbols can be introduced. When children acquire operation sense, they will have a solid conceptual framework to understand the algorithms when students encounter them.

The NCTM Curriculum and Evaluation Standards (Commission on Standards for School Mathematics, 1987) present the following sequence for learning basic facts. Easy facts for all operations should begin in kindergarten. This exploration with models continues in first grade along with the introduction of strategies for easy addition and subtraction facts. In second grade strategies for harder addition and subtraction facts are presented along with strategies for easy multiplication and division facts. Timed tests can begin in second grade for the easy facts for addition and subtraction. In third grade recall of facts using timed tests continues for all operations except the hard facts for multiplication and division. All facts should be mastered in grade four.

It is important to consider a child's readiness to learn the basic facts. Understanding these prerequisites will assure greater success in getting children to have quick recall of the facts. Many first graders are "taught" addition and subtraction facts before they can understand them. Parents are often misled into believing that early recall of facts implies an understanding of the operation.

Consider what Piaget tells us about children's learning and how this relates to readiness to learn basic facts. Number conservation, knowing that numbers remain the same despite various changes in configuration, is one important factor in evaluating readiness. If a child has 5 beans and forms them into a set of 3 and a set of 2, there are still 5 beans. Until a child conserves number, counting is used to name the sum rather than the child knowing the sum remains unchanged. Refer to Chapter 5 of this text for additional details about this topic.

Piaget illustrated that number is not something innately known by children. The ability to conserve number must precede the memorization of basic facts. Teachers should provide adequate time for children to internalize the invariance of number. For a child to verify the understanding of conservation, many experiences are needed with joining, separating, and comparing sets. The child who is a nonconserver will not see the relationships between families of number facts. Many of the strategies discussed later in this chapter hold little meaning for nonconservers.

The principle of class inclusion (the ability to see that all of one group can be part of another group at the same time) also deserves careful study and evaluation before beginning a program of memorizing the basic facts. Review Chapter 5 for greater details on the bead experiments. Children younger than seven have difficulty seeing the whole as being larger than its parts. When shown a box of wooden beads with all but two beads colored brown, many children cannot see that the brown beads are a subset of the wooden beads. They are tied to the larger *visible* set—the brown beads. The logical relationship, the inclusion relation, might not be comprehended until a child is about seven years old.

The class inclusion problem can also be a class addition problem. To solve the problem, children must consider the whole set and its parts at the same time. Children need to add the parts to obtain the whole and to be able to reverse this process (reversibility). Addition is an operation relating the parts to the whole.

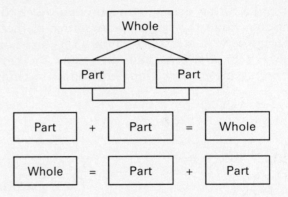

According to Piaget, until the child understands class inclusion and reversibility of thought, basic facts are learned at a meaningless level. Teaching facts using strategies described later in this chapter may be a questionable practice. Research done in the United States supports the belief that first grade children can memorize basic facts without a firm understanding of number (Van Engen and Steffe, 1964). The child who has not mastered the concept of class inclusion may not relate various number pairs with the same sum.

$$6 + 1 = 7$$
$$2 + 5 = 7$$
$$3 + 4 = 7$$
$$5 + 2 = 7$$
$$1 + 6 = 7$$

To understand addition as an operation, the child must be able to look at various pairs of addends and realize that the number can be expressed as the same sum. The child must also be able to realize that the sum (7) can become all these combinations. Number combinations can be developed with concrete materials such as lima beans painted red on one side and the child counts a given number into a small cup. The beans are spilled from the cup and the child reads the number of painted beans as the first addend and the number of white beans as the second addend. This activity is repeated many times to build the principle of invariance of number as well as to see the possible combinations for a given sum. Figure 7.5 illustrates one way to develop number invariance.

Another prerequisite for working with number facts is to develop meaning for the symbols. An excellent idea is to use physical objects to model number combinations in many ways. When children can successfully create and label sets, they are ready to attach the symbols to the actions. Invite children to create number stories using concrete materials and story boards (a picture of a forest, ocean, barn, store on which the children build their stories). A source of these manipulative materials is found in Baratta-Lorton (1978). Patterns and directions for constructing these items along with some teaching

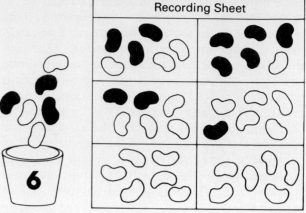

Figure 7.5

suggestions offer the teacher a rich source of help. As children model the number sentence and communicate what they are doing, the symbols have meaning.

Activity

Oral Language Development

Directions:

1. Give the child a drawing or picture of an airport landing field.

2. Have 10 small toy airplanes in two colors available to model the addends. For example, a child may place 2 pink airplanes and add 5 green airplanes and tell that there are 7 airplanes in all on the landing field.

3. Emphasize the use of complete sentences that reinforce verb usage and mathematical actions. With rich language interaction, the meaning of the symbols and the operation deepens.

4. When the child is ready to write the symbols, the concrete experiences can be continued until the child indicates a readiness to discontinue them.

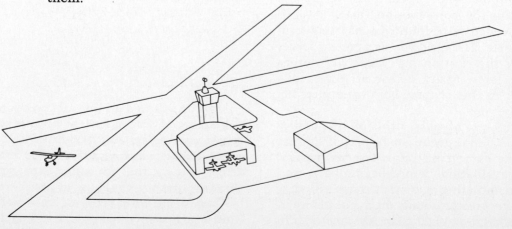

Understanding the concept does not imply that facts will be mastered. Learning facts by rote does not mean understanding the operation. This sequence suggests that a solid mathematics program incorporates all three components. Meaningful problem-solving situations should accompany each step in the learning sequence.

Thinking Strategies for the Basic Facts of Addition

The beginning step in developing basic facts is to build an understanding of the operation. Manipulatives are the important key. Children should have opportunities to represent the number sequence with physical models. The suggested models to use will be discussed with each operation.

Teaching basic facts should center around the structure of mathematics. Simply going through the number combinations is not enough. Research has shown that certain thinking strategies help children learn the basic facts (Carpenter and Moser, 1984; Fuson, 1986; Rathmell, 1978; Steinberg, 1985; and Thornton, 1978). Children should be encouraged to develop other thinking strategies or associations that may hold particular meaning for them. The NCTM curriculum standards (Commission on Standards for School Mathematics, 1987) state that teaching children strategies and encouraging them to develop their own strategies leads to learning basic facts in less time as well as helping students to reason mathematically. Teachers must help children build understanding by getting them to organize what they know and to disregard inefficient strategies. When children resort to using their fingers to get answers, it should simply tell us that they are not ready for the symbolic level. More time should be spent at the concrete and connecting level.

+	0	1	2	3	4	5	6	7	8	9
0	0	1	2	3	4	5	6	7	8	9
1	1	2	3	4	5	6	7	8	9	10
2	2	3	4	5	6	7	8	9	10	11
3	3	4	5	6	7	8	9	10	11	12
4	4	5	6	7	8	9	10	11	12	13
5	5	6	7	8	9	10	11	12	13	14
6	6	7	8	9	10	11	12	13	14	15
7	7	8	9	10	11	12	13	14	15	16
8	8	9	10	11	12	13	14	15	16	17
9	9	10	11	12	13	14	15	16	17	18

The 100 Addition Facts

Figure 7.6

The basic 100 addition facts (Fig. 7.6) are those problems that are composed of two single-digit addends. Thus the facts include $0 + 0 = 0$ through $9 + 9 = 18$. They serve as the basis for all further addition work and need to be developed from a concrete basis of understanding to an automatic response level if they are going to be useful in computation.

For many children the two basic forms, horizontal $2 + 3 = 5$ and vertical,

$$\begin{array}{r} 2 \\ +3 \\ \hline 5 \end{array}$$

provide no real stumbling block to learning the addition facts. In teaching basic facts, both forms need to be used and generally the horizontal form is introduced first. Children may be given help recognizing that the combinations are the same by actually writing both forms of the same fact. This should be done at the same time children are performing actions with multiple embodiments (various models).

Activity

COMPUTERS

Logo and BASIC

Directions:

1. Look at the Logo and BASIC programs first shown in Chapter 2 under the name ADD.

2. Try a few problems and see how a teacher might use such programs to reinforce basic facts.

Activity

CALCULATORS

Using the Calculator with Basic Facts

Directions:

1. Think of how the calculator could be used to help children learn the basic facts. How different is the calculator program from the two computer programs shown above?

2. Try some basic facts and see the difference.

3. See the calculator's ability to store a constant by trying this activity. To practice adding on 3: push 4, push + key, push 3, push = key, new number, push = key, new number, push = key, etc. Do you see how it is storing the number each time? Try it with other numbers. (If this activity does not work on your calculator, experiment with your calculator to determine how it handles the constant function.)

Understanding the mathematical principles associated with each operation is important to enhance learning. An important property of addition is the *commutative principle*. Figure 7.7 shows that 45 addition facts are learned readily if the principle of commutativity is applied. The facts that are related by the commutative principle are shaded in the figure.

To internalize the principle of commutativity, students need experiences with concrete materials. Figure 7.8 shows several effective manipulative devices to use. Explorations with a computer and a calculator may also be meaningful. Students need to realize the power of this principle in reducing the number of facts to learn.

+	0	1	2	3	4	5	6	7	8	9
0		1	2	3	4	5	6	7	8	9
1	1		3	4	5	6	7	8	9	10
2	2	3		5	6	7	8	9	10	11
3	3	4	5		7	8	9	10	11	12
4	4	5	6	7		9	10	11	12	13
5	5	6	7	8	9		11	12	13	14
6	6	7	8	9	10	11		13	14	15
7	7	8	9	10	11	12	13		15	16
8	8	9	10	11	12	13	14	15		17
9	9	10	11	12	13	14	15	16	17	

Figure 7.7

Connecting Cubes

Painted Beans

Colored Counters

Dominoes

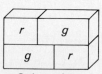

Cuisenaire Rods

Figure 7.8

Activity

COMPUTERS

CALCULATORS

Commutativity with BASIC

1. This BASIC program is saved under the name ADDEND.

2. Type LOAD ADDEND and follow the directions after the word RUN.

3. Compare it to the calculator exploration on the right. What are the differences and similarities?

Computer Exploration

Type the following program:

```
NEW
10 INPUT "ADDEND=";A
20 INPUT "ADDEND=";B
30 PRINT A;"+";B;"=";A+B
40 PRINT B;"+";A;"=";B+A
50 END
RUN
```

You will see

```
ADDEND=
ADDEND=
```

Type   6 [Press Return]
        2 [Press Return]

Type RUN to begin.

Calculator Exploration

Test the commutative property. Notice the different actions required when using the calculator instead of the computer.

Is $6 + 2 = 2 + 6$?
Is $4 + 3 = 3 + 4$?
etc.

Enter two more basic fact numbers at a time and see what happens.

Practice alone is not sufficient to facilitate memorizing the basic facts. Students need to increase their memorization abilities by using an organized list of the facts. This approach places the facts into categories according to the structure of mathematics and their order of difficulty. It also capitalizes on the natural thinking strategies invented by children for learning the facts. Table 7.3, the table of fact strategies, assumes that the commutative principle has been developed and continues to be emphasized. As you read the table, consider the definition and notice the systematic way the facts are related to each other. Also think about how organizing the facts in this manner reduces the number of facts to be learned and encourages retention.

The fact strategies should be shown and developed with concrete materials to ensure understanding. Many of these devices can be made inexpensively and quickly and are well worth the time. Activities with manipulatives are suggested for teaching the following fact strategies:

adding one; counting on; near doubles; and bridging to ten.

• *Adding One:* Cuisenaire rods are a good manipulative to show this principle. Have children build a staircase with the ten different colored rods. Take a white rod and "walk" it up the staircase. With each step, the children see it is one more. This procedure is shown in Figure 7.9 and in Table 7.1 under the counting principle.

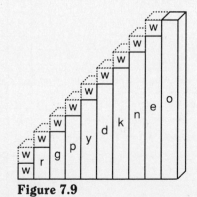

w	= white
r	= red
g	= green
p	= purple
y	= yellow
d	= dark green
k	= black
n	= brown
e	= blue
o	= orange

Figure 7.9

Table 7.3   Thinking Strategies for the Basic Facts of Addition*

Definition	Facts to be Learned

Identity Element:
Adding 0 to any number does not change the number.

19 basic facts are learned.

cuisenaire rods

+	0	1	2	3	4	5	6	7	8	9
0	0	1	2	3	4	5	6	7	8	9
1	1									
2	2									
3	3									
4	4									
5	5									
6	6									
7	7									
8	8									
9	9									

Adding One:
Depends on understanding number seriation—adding one to any number is the same as naming the next counting number. See assessment in Chapter 5.

17 facts are learned.

cuisenaire rods

+	0	1	2	3	4	5	6	7	8	9
0										
1			2	3	4	5	6	7	8	9	10
2		3								
3		4								
4		5								
5		6								
6		7								
7		8								
8		9								
9		10								

Counting On:
Most effective when addend is 1, 2, or 3 more than any given number.

Involves two skills . . .
1. knowing which number is greater and that counting begins here.
2. knowing counting sequence beginning at any one-digit number.

Additional 28 facts learned.

+	0	1	2	3	4	5	6	7	8	9
0										
1										
2			4	5	6	7	8	9	10	11
3			5	6	7	8	9	10	11	12
4			6	7						
5			7	8						
6			8	9						
7			9	10						
8			10	11						
9			11	12						

Definition	Facts to be Learned

Doubles:
Both addends are the same number.

Children learn easily, possibly due to children's games and poems.

6 facts are learned.

+	0	1	2	3	4	5	6	7	8	9
0										
1		2								
2			4							
3				6						
4					8					
5						10				
6							12			
7								14		
8									16	
9										18

Near Doubles:
Use when addends are consecutive numbers.

If children know 8 + 8 = 16, then prompt them to reason that
8 + 9 is one more or 17
and . . .
8 + 7 is one less or 15.

10 additional facts are learned.

+	0	1	2	3	4	5	6	7	8	9
0										
1										
2				5						
3			5		7					
4				7		9				
5					9		11			
6						11		13		
7							13		15	
8								15		17
9									17	

Bridging to Ten:
Use when one addend is close to ten.

More difficult for children to understand because . . .
1. mastery of sums to 10 must be developed first.
2. ability to separate a number into two parts is required.
3. must mentally keep track of changes in both addends—
   . . . increasing one addend by one or more.
   . . . decreasing other addend by same amount.

14 additional facts are learned.

+	0	1	2	3	4	5	6	7	8	9
0										
1										
2										11
3									11	12
4								11	12	13
5									13	14
6										15
7					11					16
8				11	12	13				17
9			11	12	13	14	15	16	17	

(continued)

Definition	Facts to be Learned

Sharing:
When addends differ by two, middle number is found and then doubled.

This can also be considered as a "doubles + two" strategy.

Last 6 facts are learned.

+	0	1	2	3	4	5	6	7	8	9
0										
1										
2										
3										
4							10			
5								12		
6					10				14	
7						12				
8							14			
9										

*NOTE: The blocks with numbers in the main body of the table denote new facts to be learned. The shaded part denotes facts already learned by a previous strategy. A shaded block with a number included indicates an overlap between a previously learned strategy and a new strategy.

• *Counting On:* Much time could be spent in quick practice sessions with counting one, two, or three more than a given number. Such sessions can be done easily when children are waiting in line to wash hands, to leave for another class, or when children are lining up for recess, music, or whatever. An effective aid that can be quickly made is shown in Figure 7.10. The first addend is written to the left with the second addend being plus two or plus three. Attached as flip cards are the counting numbers the child would say to reach the sum. If the child needs this prompt, the cards can be flipped over for help. If the child wants to check the answer, it appears on the back.

A simple game is to have a deck of cards of numbers which serve as the first addend. The child rolls a die that has two sides marked with 1, two sides marked with 2, and two sides marked with 3. The die indicates the number to

be added to the card drawn from the deck. If the correct sum is given, the child can move on the gameboard the number of spaces indicated on the die (Fig. 7.11).

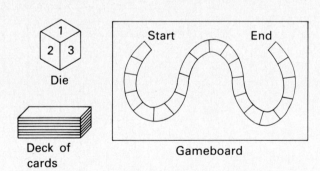

Figure 7.11

• *Near Doubles:* Since this strategy reinforces something a child already knows and feels comfortable with, the doubles, a teaching/practice device for the near doubles is shown in Figure 7.12. We must always look for ways to build on prior knowledge to form new associations.

Flip cards in up position . . .

. . . With cards down   . . . Back view

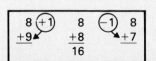

Figure 7.10

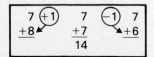

Figure 7.12

• *Bridging to 10:* Wirtz (1974) developed some materials called "ten trays" that build understanding of this strategy in a tactile, visual manner. The next activity shows teaching techniques to develop this concept. The trays may be made from egg cartons. The sides of plastic fruit baskets, plastic canvas, or chicken wire may be used to project on the overhead for whole class discussion. Ten trays consist of a 2 × 5 array where one side must be filled to 5 before the other side is started. Counters are placed in the squares starting at the upper left and proceeding across until that row is filled.

Activity

MANIPULATIVES

MENTAL MATH

Bridging to 10

Directions:

A set of dot cards for the 10 trays. Counters and egg cartons.

Teacher flashes a card to child for a second. The child puts that number of counters in the egg carton trays and says the total amount. Teacher continues in similar fashion for other numbers.

Teacher can also flash a dot card to the child for a second and the child tells how may more dots are needed to form 10.

Counters and egg cartons to make two 10 trays.

Teacher (or child) shows addition fact with 7, 8, or 9 as one addend. The other addend is large enough to form a sum greater than 10. A 10 is made with the first addend by adding counters from the second 10 tray.

The answer appears as 10 + remaining number.

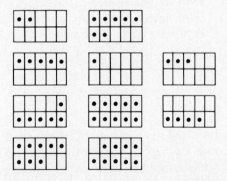

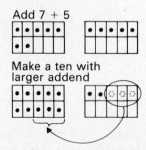

Another manipulative device to use for bridging to 10 can be constructed easily and inexpensively and is shown in Figure 7.13. It is composed of 18 plastic beads on a pipe cleaner. Ten beads are one color and 8 beads are another color. When combinations are formed, the two colors indicate the two-digit number that results as the sum. The device allows for all facts

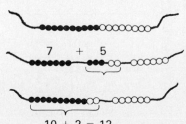

Figure 7.13   10 + 2 = 12

through 9 + 9 (sum 18 or the total number of plastic beads) and visually indicates when a bridging to 10 occurs.

Children often regard learning the 100 addition facts as an insurmountable task. When the facts are broken into groups, the task becomes manageable and reasonable to attain. When thinking strategies and fact strategies are used rather than random memorization, the facts are learned faster and retained longer. Strong evidence suggests that using current knowledge about how children learn basic facts produces good achievement and retention (Commission on Standards for School Mathematics, 1987).

Thinking Strategies for the Basic Facts of Subtraction

There are 100 basic subtraction facts that are formed as the inverse of addition. There appears to be two different philosophies of when subtraction facts should be introduced to children: at the same time as the related addition facts or in groups (facts to 6, facts to 10, facts to 18) after the similar groups of addition facts have been introduced. Whichever approach is supported, the child should have many experiences with concrete objects to build understanding. Children may also have experiences with computer programs such as the following activity.

Activity

COMPUTERS

CALCULATORS

Subtraction Basic Facts with BASIC

Directions:

1. This BASIC program is saved under the name MINUEND.

2. Type   LOAD MINUEND   and run the program. Compare it to the calculator exploration on the right. What are the differences and similarities?

Computer Exploration

```
NEW
10 INPUT "MINUEND=";A
20 INPUT "SUBTRAHEND=";B
30 PRINT A;"-";B;"=";
40 INPUT C
50 IF C=A-B THEN 70
60 PRINT A;"-";B;" DOES NOT=";C
70 END
RUN
```

Calculator Exploration

Explore the subtraction facts on your calculator. How would you explain a negative answer?

How does the computer program handle a negative answer? Can you tell *before* running the program? Try some examples with the minuend *smaller* than the subtrahend and see what happens. Remember to type RUN to start the program before each new set of numbers.

Activity

COMPUTERS

Logo and Subtraction

Directions:

1. Study the Logo program in Chapter 2 called ADD.

2. What changes need to be made so the program will work with subtraction?

3. CLUE . . .

      If "1 + RANDOM 10 . . . creates random numbers from 1 to 10

      If "2 + RANDOM 10 . . . creates random numbers from 2 to 11

      If "3 + RANDOM 10 . . . creates random numbers from 3 to 12

What can you make "A and "B so that . . .

      . . . A will generate random numbers between 9 and 18?

      . . . B will generate random numbers between 0 and 9?

4. What other lines in the program will you need to change so children will know the program is performing subtraction?

5. To see how well you planned, load the Logo program saved under the name SUBTRACT.

6. Type   `ED "SUBTRACT`   to see the program procedure.
Type together the CONTROL and G keys to leave the editor mode.
Do a few examples with the program by typing   `SUBTRACT`

For many children, subtraction is a more difficult operation to understand and the facts are memorized more slowly. In a study by Brush (1978), children as young as four years old understood informal notions of subtraction. The difficulty comes when formal instruction is given and the various types of subtraction (take-away, missing addend, comparison) confuse them. The take-away concept is often over-emphasized in the classroom to the detriment of the other two. Many teachers read the symbol for subtraction (−) as "take away" rather than minus which may interfere with the children's understanding of the compare and equalize problems as subtraction situations. This practice fixes the take-away concept firmly in children's minds so when comparison subtraction is involved, children are confused because nothing is removed from the set.

Another problem with subtraction is that students' textbooks rely heavily on pictures to teach the concept of subtraction. The pictures depicting the operation of subtraction need to be *action* pictures that show first the total beginning set, then the subset to be removed, and finally, the resulting set (Fig. 7.14). Such is seldom the case, which creates confusion for children. The result is Figure 7.15 which is representative of the pictures shown in first grade students' texts.

Typical Approach of Most Textbooks

3 − 1 = _____

Figure 7.15

Best Approach for Textbooks

3   − 1   =   2

Figure 7.14

Interviews with young children reveal many misconceptions and faulty perceptions about the message conveyed by the pictures. This situation forces children to move to the symbolic level too quickly with remediation often an unfortunate outcome. Research by Campbell (1984) indicates that the child's ability to interpret pictures is developmental.

Mastery of the subtraction facts should follow development of a numerical understanding of subtraction. Learning the basic facts will become more than symbol manipulation when efficient strategies for memorization are taught.

Many of the thinking strategies to help learn subtraction facts are closely related to the addition facts. Because of this, the authors support teaching the two operations together. In the words of Mary Baratta-Lorton (1976), "By experiencing both operations simultaneously, the children are prevented from getting locked into one concept . . . Mathematics stops being many separate concepts and begins to reveal itself as an interrelated pattern which forms a whole." (p. 210) Fact strategies and the structure of the number system can be used to teach subtraction facts as shown in Table 7.4.

Table 7.4 Thinking Strategies for the Basic Facts of Subtraction*

Definition	Facts to be Learned

Subtract Zero:
Easily learned—subtracting 0 from any number does not change the number.

−	0	1	2	3	4	5	6	7	8	9
0	0	1	2	3	4	5	6	7	8	9
1										
2										
3										
4										
5										
6										
7										
8										
9										

Subtract the Whole—Almost the Whole:
Subtracting the number from itself results in 0 left.

Almost the whole means to subtract a number that is one less than the number:
$$8 - 7 = 1$$
18 facts learned.

−	0	1	2	3	4	5	6	7	8	9
0	0	1								
1	1	2								
2	2	3								
3	3	4								
4	4	5								
5	5	6								
6	6	7								
7	7	8								
8	8	9								
9	9	10								

Definition	Facts to be Learned

Counting Back:
Most effective when the number subtracted is 1, 2, or 3. Related to "counting on."

24 facts learned.

−	0	1	2	3	4	5	6	7	8	9
0										
1		2	3	4	5	6	7	8	9	10
2			3	4	5	6	7	8	9	10	11
3				4	5	6	7	8	9	10	11	12
4										
5										
6										
7										
8										
9										

Counting On:
Use when 3 or less is to be subtracted. Helpful to associate with counting on for addition.

Missing addend approach is emphasized: $5 + ? = 8$.

12 additional facts learned.

−	0	1	2	3	4	5	6	7	8	9
0										
1										
2										
3										
4		5	6	7						
5		6	7	8						
6		7	8	9						
7		8	9	10						
8		9	10	11						
9		10	11	12						

Fact Families:
Emphasizes inverse operation of addition/subtraction and interrelates them.

3	4	7

$3 + 4 = 7; 4 + 3 = 7$
$7 - 4 = 3; 7 - 3 = 4$

Last 36 facts learned.

−	0	1	2	3	4	5	6	7	8	9
0										
1										
2										
3										
4					8	9	10	11	12	13
5					9	10	11	12	13	14
6					10	11	12	13	14	15
7					11	12	13	14	15	16
8					12	13	14	15	16	17
9					13	14	15	16	17	18

NOTE: The blocks with numbers in the main body of the table denote new facts to be learned. The shaded part denotes facts already learned by a previous strategy. A shaded block with a number included indicates an overlap between a previously learned strategy and a new strategy.

The fact strategies for subtraction are: subtract zero; subtract the whole — almost the whole; counting back; counting on (counting up); and fact families. Keep in mind that it is critical to have children model these concepts with objects before the ideas are taught and before children are expected to use these strategies in meaningful ways. Practice with concrete manipulatives will help develop understanding and memory of these fact strategies. Many teachers and parents consider only flash cards when some other devices are more effective. Since missing addends is a related concept for several strategies, several teaching aids can be made to reinforce this concept. Flip cards, illustrated in Figure 7.16, enhance learning the several strategies called subtract the whole — almost the whole, counting back, and counting on. Several studies (Carpenter and Moser, 1984; Fuson, 1984; and Steinberg, 1985) have shown that counting back (counting down) presents children with considerable difficulty and research by Fuson (1986) in-

dicates that counting up is an effective approach for first grade children.

Analyzing the relationships between the facts of addition and subtraction helps the recall of fact families. Facts are organized into "families" to help children use a known fact to recall an unknown fact.

Triominoes are an effective device that can be made easily and inexpensively to develop fact families (Fig. 7.17). The triominoes can be color coded in clusters of facts such as: facts to 6, facts to 10, facts to 14, facts to 18. The teacher selects the facts on which a child needs practice and gives that triomino to the child, who then writes the four related equations.

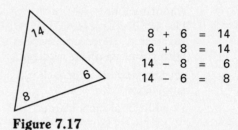

$$8 + 6 = 14$$
$$6 + 8 = 14$$
$$14 - 8 = 6$$
$$14 - 6 = 8$$

Figure 7.17

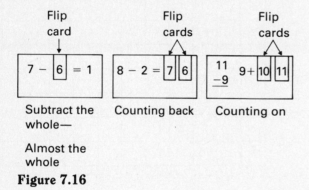

Flip card	Flip cards	Flip cards
7 − 6 = 1	8 − 2 = 7 6	11 / −9 9 + 10 11
Subtract the whole—	Counting back	Counting on
Almost the whole		

Figure 7.16

When errors are noted, the teacher should ask the child to model the equation with concrete objects. This procedure will become self-correcting and a valuable learning experience. Triominoes can also be used as a practice device. If the top corner is covered, you are asking for a sum. If one of the side corners is covered, a missing addend is needed.

Activity

MANIPULATIVES

Subtract the Whole/Almost the Whole

Materials: Four sets of 0–9 number cards (40 cards in all)

Procedure: Shuffle the cards. Put them in a pile between two players. On alternating turns, a player draws a card from the pile. If the player can form an equation with two numbers that will give an answer with a difference of one, the player says, "UNO." The player says the number sentence (i.e., 4 take away 3 is 1) and lays down the two cards, 4 and 3. Play continues until all cards have been drawn. The player with the most equations wins.

Thinking Strategies for the Basic Facts of Multiplication

Readiness activities for multiplication should begin with talking about groups of the same size and exploring counting patterns. Activities that encourage children to visualize groups along with the related languages should precede any formal use of the multiplication table or the formal word "times."

Children are eager to begin the multiplication chapter, usually included in the second grade text. Multiplication means the world of "big kids" and young children are eager to quote products of large numbers to teacher and others.

Multiplication should begin as ideas about repeated addition—equal groups. Children see multiplication as counting groups of objects rather than single objects. Activities using materials as shown in Figure 7.18 should begin in the early grades to show multiplication as repeated addition. Learning experiences should include children describing real-life situations that suggest multiplication.

Remember when language becomes important, these groups should be expressed as all showing three groups of two. In the resulting number sentence, 3×2, the first factor defines the number of groups and the second factor defines the size of each group. Have the child build rows, stacks, or groups according to your directions and allow them to discover relationships such as the commutative property. Involve children in physical groupings also: give three books to two children, have four children hold up five fingers, put children's desks or chairs into equal groups, group five children and decide how many eyes or toes.

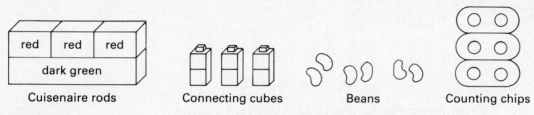

Cuisenaire rods   Connecting cubes   Beans   Counting chips

Figure 7.18 $3 \times 2 = 6$

Activity

Multiplication as Repeated Addition

The first factor names the number of groups and the second factor names the size of each group.

Check this on your calculator:

$48 + 48 + 48 + 48 = ?$

Now make this into a multiplication equation.

number of groups   *size of each group*

4        $\times$ 48   = ?

Is the answer the same?

Which way requires fewer steps?

Try some others:

$5 + 5 + 5 + 5 + 5 = ?$
$5 \times 5 = ?$
$9 + 9 + 9 = ?$
$3 \times 9 = ?$

(continued)

Write the related multiplication equation and check on your calculator:

1. $3 + 3 + 3 + 3 = ?$
2. $8 + 8 + 8 + 8 + 8 = ?$
3. $1 + 1 + 1 = ?$
4. $6 + 6 + 6 + 6 = ?$
5. $4 + 4 + 4 + 4 + 23 + 23 = ?$
6. $7 + 7 + 7 = ?$

Try this on your calculator $8 + = = =$. What happens?

When formal drill on the basic facts is needed, the following thinking strategies should be considered. There are 100 basic multiplication facts composed from all single-digit factors.

Thus the facts include $0 \times 0 = 0$ through $9 \times 9 = 81$. The resulting table is shown in Figure 7.19.

Activity

COMPUTERS

CALCULATORS

Exploring Basic Facts for Multiplication

Directions: This program is saved on the BASIC disk under the name MULTIPLICATION. Type RUN to start the program before each new set of numbers. Answer some correctly and incorrectly.

Computer Exploration

```
10 LET A = INT (RND(1)*10)
20 LET B = INT (RND(1)*10)
30 PRINT A;"X";B;"=";
40 INPUT C
50 IF C = A * B THEN 70
60 PRINT A;"X";B;"="; A * B
70 END
```

Calculator Exploration

Practice multiplying by 6.
Push 6, push X key, push 5,
push = key, push new number,
push = key, push new number,
push = key, etc.
Try with other numbers.

Type NEW to go to this program, then . . .
Type   LOAD MULTIPLICAND   and   RUN

Computer Exploration

```
10 INPUT "MULTIPLICAND=";A
20 INPUT "MULTIPLIER=";B
30 PRINT A;"X";B;"=";A*B
40 PRINT B;"X";A;"=";B*A
50 END
RUN
```

Calculator Exploration

Try these:
$9 \times 8 =$
$8 \times 9 =$
$5 \times 7 =$
$7 \times 5 =$
Try others

Activity

COMPUTERS

PROBLEM SOLVING

Multiplication with Logo

Directions:

1. Study the Logo program in Chapter 2 named ADD.

2. Think what you would need to change for the program to work with multiplication.

3. You have studied the changes made in ADD to produce SUBTRACT earlier in this chapter.

4. Use the edit instructions in Appendix B and try creating the new program. If you have your own copy of the disk, give your new program a different name from the others on the Logo disk and save it to use in your own teaching.

The mathematical principle called the *commutative principle* is an important property of multiplication (as it is for addition) that can be applied to learn many multiplication facts. This is one of the most valuable strategies because by using it, the number of facts left to be learned is 45. Figure 7.20 indicates the facts that remain to be learned when the commutative principle is used. Students need to feel confident that changing the order of the factors will not alter the answer or the product. Practice with manipulatives helps develop understanding the language of multiplication. Language development is important for correct interpretation of the concept.

Most children develop some strategies for recalling the multiplication facts. Many of these strategies involve the properties of the number system and counting patterns. The fact strategies are shown in Table 7.5 along with diagrams to identify which facts are solved with which strategy. Remember that easier strategies should be taught first. The main fact strategies for multiplication are: the identity element; zero property; skip counting; multiples of 9; and doubles.

x	0	1	2	3	4	5	6	7	8	9
0	0	0	0	0	0	0	0	0	0	0
1	0	1	2	3	4	5	6	7	8	9
2	0	2	4	6	8	10	12	14	16	18
3	0	3	6	9	12	15	18	21	24	27
4	0	4	8	12	16	20	24	28	32	36
5	0	5	10	15	20	25	30	35	40	45
6	0	6	12	18	24	30	36	42	48	54
7	0	7	14	21	28	35	42	49	56	63
8	0	8	16	24	32	40	48	56	64	72
9	0	9	18	27	36	45	54	63	72	81

The 100 Multiplication Facts

Figure 7.19

x	0	1	2	3	4	5	6	7	8	9
0		0	0	0	0	0	0	0	0	0
1	0		2	3	4	5	6	7	8	9
2	0	2		6	8	10	12	14	16	18
3	0	3	6		12	15	18	21	24	27
4	0	4	8	12		20	24	28	32	36
5	0	5	10	15	20		30	35	40	45
6	0	6	12	18	24	30		42	48	54
7	0	7	14	21	28	35	42		56	63
8	0	8	16	24	32	40	48	56		72
9	0	9	18	27	36	45	54	63	72	

Figure 7.20

Table 7.5   Thinking Strategies for Basic Facts of Multiplication*

| Definition | Facts to be Learned |

Identity Element:
Multiplying any number by 1 results in the same number.

17 basic facts are learned.

x	0	1	2	3	4	5	6	7	8	9
0										
1		1	2	3	4	5	6	7	8	9
2		2								
3		3								
4		4								
5		5								
6		6								
7		7								
8		8								
9		9								

Zero Property:
Multiplying any number by 0 results in 0 for an answer.
19 basic facts are learned.

x	0	1	2	3	4	5	6	7	8	9
0	0	0	0	0	0	0	0	0	0	0
1	0									
2	0									
3	0									
4	0									
5	0									
6	0									
7	0									
8	0									
9	0									

Skip Counting by 2, 5, 3:
Relate "twos" to doubles in addition;
"fives" to counting nickels or telling time.

39 additional facts are learned.

x	0	1	2	3	4	5	6	7	8	9
0										
1			2			5				
2		2	4	6	8	10	12	14	16	18
3			6	9	12	15	18	21	24	27
4			8	12		20				
5		5	10	15	20	25	30	35	40	45
6			12	18		30				
7			14	21		35				
8			16	24		40				
9			18	27		45				

Definition	Facts to be Learned

Multiples of 9:
Many interesting patterns and relationships can be used. Finger multiplication.

9 more facts are learned.

x	0	1	2	3	4	5	6	7	8	9
0										
1										9
2										18
3										27
4										36
5										45
6										54
7										63
8										72
9		9	18	27	36	45	54	63	72	81

Doubles:
The number multiplied by itself— sometimes easier facts to remember.

x	0	1	2	3	4	5	6	7	8	9
0										
1										
2			4							
3				9						
4					16					
5						25				
6							36			
7								49		
8									64	
9										81

NOTE: The blocks with numbers in the main body of the table denote new facts to be learned. The shaded part denotes facts already learned by a previous strategy. A shaded block with a number included denotes an overlap between a previously learned strategy and a new strategy.

Children should have many experiences with objects, calculators, or arrays to help develop the concept for multiplication and the strategies. There are several simple computer programs that show the effect of multiplying by 1 (identity element) and multiplying by 0 (zero property). Some children like to run these programs to prove that *any* number when multiplied by 1 remains unchanged or when multiplied by 0 results in 0 for an answer. Multiplying by 0 confuses some children. They think that if you start with "something" and you multiply it by zero, you still have the "something" with which you began.

Activity

Exploring Multiplying by 1 or 0

Directions:

1. Type the program after you have booted the BASIC disk. This helps you get acquainted with typing in BASIC.

2. Type the exact way it is printed here. If you make a typing mistake, just type the whole line over underneath the program.

3. Type   LIST   to see the whole program.

4. Type   RUN   to use the program with each new example.

Computer Exploration

```
NEW
10 INPUT "ANY NUMBER=";A
20 LET B = 1
30 PRINT A;"X";B;"=";A*B
40 END
```

Calculator Exploration

Explore multiplying any number by 1. What is the result? Try 0 × 1.

Computer Exploration

```
NEW
10 INPUT "ANY NUMBER=";A
20 LET B = 0
30 PRINT A;"X";B;"=";A*B
40 END
```

Calculator Exploration

Use your calculator to multiply any number. What is the result? Try 4 × 5 × 89 × 0 × 2.

Skip counting is another major strategy that shows patterns and the structure of the number system. Skip counting by two begins in most first grade texts. Using examples in the child's world builds a visual frame of reference for these facts: 2 wheels on each bike, 2 socks in each pair, 2 rows in a six-pack of soda, 2 rows in an egg carton, 2 eyes on each child.

Skip counting with the "fives" facts is familiar because of money (counting nickels) and time (counting five minute intervals). The counting pattern for fives is presented in most second grade books and in many first grade books. If we look at the multiples of five on a hundreds chart, a pattern quickly emerges. These activities present some ways to develop an appreciation and curiosity in children about the interesting patterns of the multiples and counting numbers.

Activity

Finding Multiples

Multiples of 5

Materials: Hundreds chart, chalkboard, paper.

Procedure:

1. List the multiples of 5.

2. What digits appear in the ones column?

1	2	3	4	5	6	7	8	9	10
11	12	13	14	15	16	17	18	19	20
21	22	23	24	25	26	27	28	29	30
31	32	33	34	35	36	37	38	39	40
41	42	43	44	45	46	47	48	49	50
51	52	53	54	55	56	57	58	59	60
61	62	63	64	65	66	67	68	69	70
71	72	73	74	75	76	77	78	79	80
81	82	83	84	85	86	87	88	89	90
91	92	93	94	95	96	97	98	99	100

Hundreds Chart

3. What pattern is found for the tens digits?
   Extend the chart and see if this pattern continues.

4. Add the digits together $(25 = 2 + 5 = 7)$.

5. What kind of numbers are the "5s" facts?

Multiples of 9

Procedure: The teacher displays a larger number line or writes one on the chalkboard. Students underline or circle all multiples of 9. Make separate column listing of these numbers. Encourage students to find the many patterns such as:

1. The tens digit in the product is 1 less than the number being multiplied by 9.

$$9 \times 4 = 36 \quad \text{(3 is 1 less than 4)}$$
$$9 \times 6 = 54 \quad \text{(5 is 1 less than 6)}$$

2. The sum of the digits in the products is equal to 9.

$$2 \times 9 = 18, 1 + 8 = 9$$
$$3 \times 9 = 27, 2 + 7 = 9$$
$$7 \times 9 = 63, 6 + 3 = 9$$

3. The digits in the ones column decrease by 1 beginning with 9 and the digits in the tens column increase by 1 beginning with 1.

$$9$$
$$18$$
$$27$$
$$36$$
$$45$$

4. The numbers that are one more and one less than a multiple of consecutive numbers can be added together and form another multiple of 9.

$$\dots 17, 18, 19 \dots$$
$$17 + 19 = 36$$

(continued)

Multiples of 3

Procedure:

1. Circle the multiples of 3. What patterns do you notice?

2. Say the multiples of 3 to 30. How many multiples are in each row (decade) of the chart?

3. List the multiples of 3:   3   6   9
                              12   15   18
                              21   24   27

4. Add the digits of the two-digit numbers. What do you see?

5. Starting with 3, tell whether the number is odd or even. What pattern is there?

6. Many other patterns can be found for the multiples of 3.
   How many can you discover?

The multiples of nine is another strategy that emphasizes number patterns and relationships. Finger multiplication is a fun way to determine facts of nine. Children enjoy the "magic" of this activity as they internalize some number combinations. Use both hands with fingers spread apart. Label the fingers consecutively from 1 to 10, as indicated in Figure 7.21. Bend the "multiplier finger," i.e., 4 × 9, bend the fourth finger. To read the product, count the fingers to the left of the bent finger as the tens digit (3) and the fingers to the right of the bent finger as the ones digit (6). Study the examples in Figure 7.21.

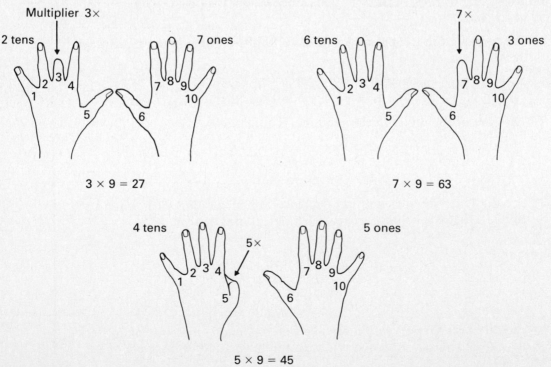

3 × 9 = 27

7 × 9 = 63

5 × 9 = 45

Figure 7.21

If we look at the table of facts, Figure 7.22, we see which facts have been covered to this point. The remaining facts can be learned by a variety of strategies. The doubles (4×4, 6×6) are often easy facts to learn and recall. Repeated addition might be a strategy to use when one factor is less than five. The child simply changes the multiplication problem to an addition problem and solves, $3 \times 6 = 6 + 6 + 6 = 18$. This interpretation of multiplication is useful if the student has had experiences with sets, arrays, and skip counting activities.

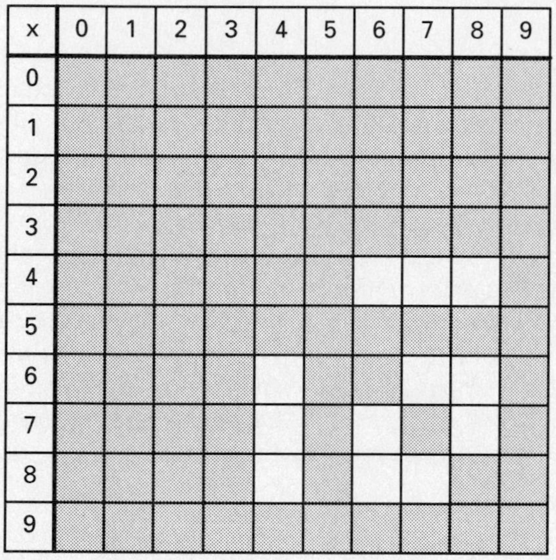

x	0	1	2	3	4	5	6	7	8	9
0										
1										
2										
3										
4										
5										
6										
7										
8										
9										

Figure 7.22

Two other strategies may help in learning any of the troublesome facts that are left. One of these is "Add Another Set." This strategy uses the distributive law of multiplication. The child uses an easier known fact to find the answer to a more difficult fact. For example, if a child knows 5×8 (5 eights) is 40, then 6×8 would be 6 eights. One more set of eight must be added to 40 to make 48.

Think: I know $5 \times 8 = 5$ groups of 8

$8 + 8 + 8 + 8 + 8 = 40$

So, $6 \times 8 = 6$ groups of 8 so add 8 more to 40

The focus of the last strategy is for multiples of 4, 6, and 8, where the child doubles a known fact to answer more difficult facts. For example, $6 \times 7 = ?$

Think: I know $3 \times 7 = 21$. since 3 is half of 6, I need to double the answer.

This will give me 6 groups of 21.

$$\text{Or} \quad \begin{array}{r} 3 \times 7 = 21 \\ + \; 3 \times 7 = 21 \\ \hline 6 \times 7 = 42 \end{array}$$

Figure 7.23 indicates the facts that can easily be solved by this strategy. Other facts can be obtained with this strategy, but, a regrouping is required and the process becomes more difficult to do quickly and mentally.

x	0	1	2	3	4	5	6	7	8	9
0										
1										
2										
3					12		18		24	
4				12	16	20	24	28		
5					20					
6				18	24			42	48	
7					28		42			
8				24		40	48		64	
9										

Figure 7.23

For both of the last two strategies, students need to be encouraged to split one factor apart and figure the answer to build confidence that the strategy works. Arrays and Cuisenaire rods work well to help develop this concept. Another way to visually show this concept is graph paper. Cut the paper into a specific array, then apply the distributive law to fold it into two parts. Students can write the product and equation for each part on the back (Fig. 7.24). Group work with these concrete materials helps children check for wrong interpretations and helps to motivate them to remain on task.

When we look at these strategies, almost all the facts have been covered. The two remaining facts, 8×7 and 3×7, can be included in "add another set." However, regrouping is required. Not all of the strategies presented can be used efficiently with all of the basic facts. However,

Fold →

2 x 6 = 12

2 x 6 = 12

$4 \times 6 = (2 \times 6) + (2 \times 6)$
$4 \times 6 = 12 + 12$
$4 \times 6 = 24$

5 groups of 5

3 groups of 5

8 groups of 5

$8 \times 5 = (5 \times 5) + (3 \times 5)$
$8 \times 5 = 25 + 15$
$8 \times 5 = 40$

Figure 7.24

instructional time spent on developing thinking strategies will help mastery occur and develop confidence in one's ability to find the answer when quick recall cannot be done.

Thinking Strategies for the Basic Facts of Division

"Since most division problems involve remainders, it is unnecessary and unwise to spend an excessive amount of time on memorizing divi-

sion facts. . . . This increases the importance of mastering multiplication facts and being able to use multiplication in this type of division problem. An argument could even be made to learn division strategies, but not recall the division facts" (Commission on Standards for School Mathematics, 1987).

There are 90 basic division facts since division by zero is undefined. There is no solution for $4 \div 0$ because the check by multiplying does not work, $4 = ? \times 0$. The most useful strategy for learning division facts is to think of the prob-

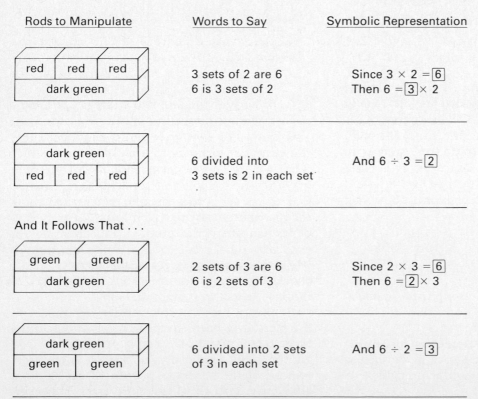

Rods to Manipulate	Words to Say	Symbolic Representation
red | red | red / dark green	3 sets of 2 are 6 / 6 is 3 sets of 2	Since $3 \times 2 = \boxed{6}$ / Then $6 = \boxed{3} \times 2$
dark green / red | red | red	6 divided into 3 sets is 2 in each set	And $6 \div 3 = \boxed{2}$

And It Follows That . . .

green | green / dark green	2 sets of 3 are 6 / 6 is 2 sets of 3	Since $2 \times 3 = \boxed{6}$ / Then $6 = \boxed{2} \times 3$
dark green / green | green	6 divided into 2 sets of 3 in each set	And $6 \div 2 = \boxed{3}$

Figure 7.25

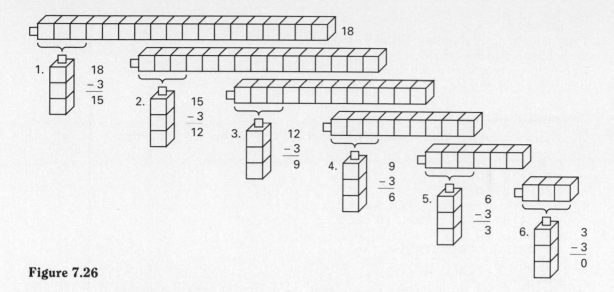

Figure 7.26

lem as a missing factor and recall a related multiplication fact (28 ÷ 4 = ?; think ? × 4 = 28). This makes learning fact families for multiplication and division more important as this approach emphasizes looking for the missing factor.

$$6 \times 3 = 18$$
$$3 \times 6 = 18$$
$$18 \div 3 = 6$$
$$18 \div 6 = 3$$

Triominoes are a device to help children practice fact families. Color coding keeps them in sets that allow for easier management. Facts can be clustered into categories, as described in the multiplication section, according to the strategies involved. When a corner of the card is covered, it means looking for the missing factor.

Figure 7.25 shows the relationship between the two operations of multiplication and divi-

sion, using Cuisenaire rods as the manipulative. Since there are no specific fact strategies for the division facts, the interrelatedness of the two operations must be emphasized and shown with concrete materials.

Division can also be viewed as repeated subtraction. An array or sets of concrete objects will help build this concept. Connecting cubes in long trains can be made into stacks or groups of equal size. As each set is split from the train, it can be recorded as a series of repeated subtractions (Fig. 7.26).

Since division is the inverse of multiplication, multiplication facts form the foundation for the strategies to learn division facts. The physical materials suggested in beginning multiplication work can be used in developing division concepts. The calculator and the computer can be used to explore the properties of division.

Activity

PROBLEM
SOLVING

Properties of Division in BASIC

Directions: You can type these directly in the computer after you have booted the BASIC disk.

Special clue: The division sign in BASIC is made with a /. When you see the binking cursor

COMPUTERS

Type   ? 8/0   (if Apple BASIC) or type   PRINT 8/0

Press return and see what happens.
Do again with other numbers.
As a teacher, encourage children to make up their own problems using the same format and writing on paper what they discovered from the pattern.

(continued)

Division by 1

Type   ?8/1   (if Apple BASIC) or type   PRINT 8/1
Press return and see what happens.
Follow the procedure from the preceding example. Children need to understand that division by one *does not change* the number divided.

Commutative Principle

Write a new program that students will be able to use. Type the following:

```
NEW
10 INPUT C
20 INPUT D
30 IF C/D = D/C THEN 60
40 PRINT "NOT COMMUTATIVE"
50 GOTO 70
60 PRINT "COMMUTATIVE"
70 END
RUN
```

Directions:

1. When the blinking cursor appears, type in a number and press [Return]. Do this twice (once for input C and again for input D).

2. Type RUN before each new set of numbers that you wish to try.

Associative Principle

Does the associative principle work for division? Explore with the computer.
Adapt the preceding program to work with associativity.
Type in your ideas and see if they work.

Clue: A / (B / C) = (A / B) / C

Did the combinations turn out to be equal?

Remember: Type RUN before trying each new set of numbers.

Try other division combinations in the same format.

Is there any case when division will be associative? Try possibilities. What is your conclusion?

How well did you program in BASIC?
You can check your program with the one saved on the disk under the name
ASSOCIATIVITY IN DIV
Type   LOAD ASSOCIATIVITY IN DIV
Type   LIST   to see the program.

Activity

CALCULATORS

Division by 1

Will you get the same answers on the calculator as you did on the computer?

    4 ÷ 1 =   Try these: 641 ÷ 1 =   28 ÷ 1 =   31 ÷ 1 =

Don't be too sure without trying. Write an explanation of what you discovered.

Even though children experience division problems in everyday situations, division is one of the most difficult operations to understand. Manipulation of materials and verbalizing division in problem-solving situations are keys to enhance understanding.

Activity

CALCULATORS

Division as Repeated Subtraction

Enter 32 in the calculator. Subtract 4 repeatedly until you have 0. Keep track of the number of times you can subtract 4. Write the related division equation for this experience.

$$32 - 4 - 4 - 4 - 4 - 4 - 4 - 4 - 4 = 0$$

Try with other numbers

$$56 - 7 - 7 \ldots$$
$$18 - 2 - 2 \ldots$$
$$30 - 5 - 5 \ldots$$

Write a division equation for each example. What does each number in the equation represent?

Activity

PROBLEM SOLVING

Exploring the Basic Operations

Directions: There are two ways to think about these subtraction problems. Study the problems and think about what equation you would use to solve it.

1. Jim has 24 pencils. He can group them in several ways to have an even number in each group. Name five ways he can group the 24 pencils.

2. Barbara baked 21 cookies. If 13 cookies are eaten, how many cookies are left? How many more cookies did Barbara bake than were eaten?

Mastering the Facts

If children are to be successful mastering the basic facts, the operations must be developed through multiple embodiments. Children should have opportunities for informal activities with materials to model situations involving the different operations. When drill and practice are introduced, short, frequent sessions are recommended. Facts should be leveled in terms of strategies and difficulty. Not all 100 addition facts should be studied and tested in a timed setting at the outset.

Beginning speed tests with the easiest facts assures success and a feeling of confidence. Once a satisfactory score is earned, the next level is tackled. Drill these selected new facts, then test these facts in addition to the previously learned facts. In this way, the tests gradually build until the total set of 100 addition facts is included.

Teachers should make a "big deal" out of passing each test and keep individual charts and/or class charts. This helps monitor progress and motivate children to continue mastering more facts. Encourage parents to reinforce drill at home by keeping them informed of the facts on which their child is currently working. To tell parents to "work with your child on basic facts," is almost hopeless in terms of definite results. Typically, the parents will purchase a set of flash cards and drill will be intense for only a few weeks. This approach is inappropriate because the parents are trying to practice all the facts rather than taking them in small groups that relate to thinking strategies. Both parties (parent and child) become frustrated and soon the practice ends. Praise and positive reinforcement should be given for learning each new group of facts rather than waiting until the entire set of facts has been mastered.

One of the authors has studied the performance of thousands of elementary students on mastering basic facts. She has noted a definite pattern of performance levels for the 4 basic operations which is:

- *Addition Facts*—easiest to learn and master
- *Multiplication Facts*—next easiest to learn and master
- *Division Facts*—next to master
- *Subtraction Facts*—hardest to learn and last operation to master

Perhaps this tells us that addition and subtraction facts should be taught together more often and mastered together. Frequently addition facts are learned first after much drill and practice, and limited time remains in the school year to master subtraction facts. More research is needed in this area and on the effectiveness of the integrated approach.

If mastery of basic facts is begun in grades 3 and above, the same general procedures should be followed with emphasis on thinking strategies to help recall. It is also valuable to assess which facts are not mastered, then develop strategies to learn those facts. Two easy procedures are suggested.

Give a timed fact test for each operation with time guidelines of three seconds per problem (five minutes for 100 facts and four and a half minutes for 90 facts). Ask students to work the problems row by row answering only those facts which they know automatically without any extra thought or counting. This test should be repeated on several occasions to ensure a higher validity. Isolate the "problem facts" into clusters and work on these with many different embodiments and strategies.

Another procedure, which takes more time but seems to have more validity, is to show flash cards of facts to an individual child. In a timed sequence of no more than three seconds, show each fact to the child. The student must answer as quickly as possible. Correct cards are put in one pile and missed cards are put in another pile. This procedure can be repeated until you feel more assured that all the facts in the "correct pile" are known at an automatic level. Now drill and practice can begin on groups using strategies for these unknown facts as described earlier.

A clever drill device is to use *glow-in-the-dark* paint to write the troublesome facts. The facts are made into a mobile. The numbers should be large enough to be seen in bed when the lights are off. This approach helps when children find it hard to sleep. The novelty is enjoyed and the facts are learned quickly. (Fig. 7.27)

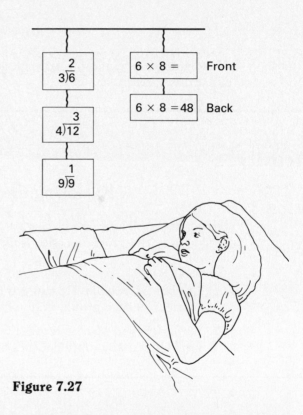

Figure 7.27

Another example of a teacher-made drill device is a canister learning machine. This aid can be made with two canisters or tubes of graduated sizes that fit snugly one inside the other. Follow these directions to make the canister.

1. Choose two canisters that fit snugly, one inside the other. The inner canister should be longer than the outer canister to allow the machine to turn (Fig. 7.28). Cut four holes in the outer canister.

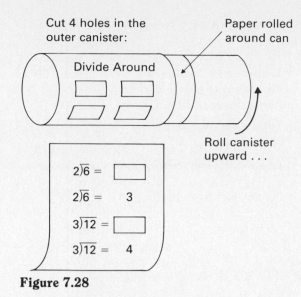

Cut 4 holes in the outer canister:

Paper rolled around can

Divide Around

Roll canister upward . . .

2)6̄ = ☐

2)6̄ = 3

3)12̄ = ☐

3)12̄ = 4

Figure 7.28

2. Three windows are covered with clear plastic. Only the lower right window is left open for the students to write answers. The paper is rolled shelf paper.
3. Cut the holes in the canister first. Then slip the paper inside and make the problems in the form seen at the left of Figure 7.28.

4. The student answers the problem by writing in the window at the lower right. Then the student moves the paper upward, and the correct answer appears at the lower right. The answer written by the student cannot be changed because it is now under the plastic. One more turn and a whole new problem appears.

> *Note:* Students should be encouraged to make new paper strips for the machine and give the strips to other students to answer. Students have to plan the problems they will use (problem solving), and it promotes cooperative learning among peers.

Remember that drill activities should be varied and interesting. Many simple inexpensive devices can be made to practice the facts. Another effective way to provide drill is with teacher-made (or commercial) games such as board games, dice games, and card games. A number of teacher-controlled microcomputer software programs are available that provide instant reinforcement in a timed setting. Some of these are in the format of game challenges.

Activity

MATH
ACROSS THE
CURRICULUM

Basic Facts and Music

Directions: The following ideas present opportunities to blend mathematics with music in a variety of ways:

Records

1. There are many excellent records on the market with "catchy" tunes. These can be purchased and used in the classroom to learn the basic facts. A catalog of teacher materials will show what is available. The records generally take one fact table ("Let's learn the facts for sixes") at a time rather than focusing on fact strategies.

2. Some excellent records are:

"Melody Math"

The Hap Palmer Series

Multiplication Rock (1973) by Bob Dorough

Pat, Clap, and Snap Rhythm Chant

1. A chant can also be used. The tempo to the chant may start slowly and increase the speed quickly.

2. Choose a student who is talented in rhythm activities to lead the group.

(continued)

3. The chant may be adapted to all of the operations. A sample using multiplication is presented below:

Movement with Hands	Words to Be Spoken
1. Pat palms on upper part of legs two times	"5" (with the first pat) "times" (with the second pat)
2. Clap hands together two times.	"3" (with first clap) "is" (with second clap)
3. Snap fingers of right hand once.	"fif" (with first snap)
4. Snap fingers of left hand once.	"teen" (with second snap)

Tapes

Consider taping students when they are doing the rhythm chant. The tape has two distinct advantages:

1. The tape itself can be used as the leader for other rounds, giving the teacher time to observe carefully which children are having trouble remembering the facts quickly.

2. It also serves as a motivation for a new class just beginning the basic facts to hear another group doing extremely well on tape. It gives the new class a goal toward which to work.

There are many activities from which a teacher may choose to help students learn the basic facts. The teacher who uses many forms of drill—calculators, games, computers, video and audio tapes—will find better attitudes and performance by students. It is the teacher who must determine which activities and strategies are suited to the needs and interests of the students.

Diagnosis and Remediation

Simultaneous and Successive Thought Processing

Students preferring to use successive thought processing (the parts-to-whole thought strategy) may understand addition and multiplication (combining parts to make the whole) better than they may understand subtraction and division (moving from the whole to the detailed parts). It is just the opposite for students preferring simultaneous processing. Both groups can learn the basic facts more efficiently and more thoroughly by applying the following teaching techniques.

- *Subtraction and Division Basic Facts: Help for the Successive Processor.* When a successive learner is faced with subtraction and division basic facts, it may be best to look on the combinations as "missing addends" for addition and "missing factors" for multiplication. The teacher should encourage the children to ask questions like those in Figure 7.29.

- *Addition and Multiplication Basic Facts: Help for the Simultaneous Processor.* Simultaneous learners may be helped by using the idea of "subtracting as much as they can and adding as little as they can" when working with addition and multiplication. Subtracting from 10 or multiples of 10 will eliminate the largest amounts and students are left with as little as possible to add or multiply. Study the examples in Figure 7.30.

Original Problem: 17 − 8 = ?

What adds with 8 to make 17?
8 + □ − 17

Original Problem: 36 ÷ 4 = ?

What multiplies with 4 to make 36?
4 × □ = 36

Figure 7.29

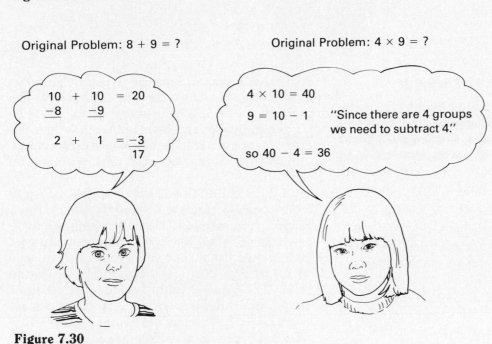

Original Problem: 8 + 9 = ?

$$10 + 10 = 20$$
$$\underline{-8} \quad \underline{-9}$$
$$2 + 1 = \underline{-3}$$
$$17$$

Original Problem: 4 × 9 = ?

4 × 10 = 40
9 = 10 − 1   "Since there are 4 groups
we need to subtract 4."
so 40 − 4 = 36

Figure 7.30

Basic Fact Tables

Basic fact tables for each operation were illustrated in this chapter. Most elementary textbooks present the table in its incomplete form and ask the children to fill it in, a task which requires successive processing. The table is completed square by square in a systematic way, so number relationships develop as each number combination is answered. The activity in the textbook may need to be modified for students preferring simultaneous processing approaches.

They need to view the completed table before relationship patterns can be seen clearly. They become more confused by moving from row to row and column to column filling in blank squares.

It is effective to pair both learners together, allowing the successive processor to complete the table while the simultaneous processor watches for the overall patterns. Both learners can share what they discovered about the tables when the exercise is completed, another cooperative learning venture (Fig. 7.31).

Figure 7.31

Correcting Common Errors

Some children will still have difficulty learning basic facts. An excellent way to check understanding of the operations is to give a child physical objects and ask the child to model a number sentence. A common error is to show sets for each number but fail to perform the operation. For example:

Teacher: Displays the equation 3 + 4 = 7. "Show me what this number sentence means using these counters."

Child: The child models a set of 3 counters under the numeral "3," a set of 4 counters under the numeral "4," and a set of 7 counters under the numeral "7."

Problem: The child is merely modeling sets for the numbers rather than showing addition as the union of two discrete sets.

Remediation: Give the child many opportunities to represent the addition process in a problem-solving situation where the action of the operation is demonstrated.

Many children experience difficulties with subtraction facts over 10 once the regrouping algorithm for subtraction has been learned. They apply that procedure for the two-digit numbers of the subtraction facts. When children have this problem, the teacher must show them that the regrouping yields the same number as the original number (i.e., 13). Generally this error will diminish with more time given to the subtraction algorithm.

Learning disabled children should be given facts in small chunks and speed tests in a similar fashion. Too many facts are disturbing and overwhelming to the LD child. One-minute quizzes over leveled facts work much better for children with special needs. It is suggested by Thornton et. al. (1983) that a child who can write 50 digits per minute should be able to correctly write answers to 30 basic facts per minute. This 5 to 3 ratio may prove a useful guide in designing speed tests.

If students are having trouble going from the concrete level to the symbolic level, they may profit from the use of an individual number line placed at the top of their desks. Students can make their own from a sample. When such number lines are laminated, they have been known to last a whole school year under rugged use. If a student is growing over-dependent on the number line as a crutch, it can be covered with a book and used only at the end of a session as a check. This makes an excellent connecting level technique.

Failure to understand and apply the commutative law is another area of difficulty. Here the child can work 9 + 2 by counting on, but does not readily change the order of the addends to help obtain the answer to $2 + 9$. The triominoes and fact families are techniques to emphasize this concept.

Children often employ faulty reasoning when working with the basic facts. Some of the more common problems are described below.

- **Faulty Reasoning 1**

Can Answer Correctly	*Cannot Answer Correctly*
$\begin{array}{r}9\\+3\\\hline\end{array}$ or $\begin{array}{r}7\\\times 4\\\hline\end{array}$	$\begin{array}{r}3\\+9\\\hline\end{array}$ or $\begin{array}{r}4\\\times 7\\\hline\end{array}$
$9 + 3 =$	$3 + 9 =$
$7 \times 4 =$	$4 \times 7 =$

If the pattern above is seen in a variety of basic facts, the children are likely to be "position counters"—they take whatever numeral appears in the top position or in the left position and add the numeral in the next position to it. Their count becomes confused when the larger numeral follows in the second position because there are more chances for mistakes in counting.

A teaching strategy to help remediate this problem is to develop a better understanding of the commutative principle. Even if commutativity is too difficult for early primary children to understand, they can be taught the counting-on strategy as outlined earlier in this chapter. A modified approach can be seen with multiplication. The child sees 7×4 as 7 taken 4 times (or 4 groups of 7) and calculates the basic fact something like this:

7	and again	7	and	14
$\begin{array}{r}7\\+7\\\hline 14\end{array}$		$\begin{array}{r}7\\+7\\\hline 14\end{array}$		$\begin{array}{r}14\\+14\\\hline 28\end{array}$

When taught that the order of the numbers can be changed and the product is not affected, children can do the multiplication more quickly.

- **Faulty Reasoning 2**

Children answer problems in the following way:

In subtraction	*In division*
$\begin{array}{r}4\\-1\\\hline 1\end{array}$ $\begin{array}{r}9\\-4\\\hline 4\end{array}$	$3\overline{)6}$ $6\overline{)24}$

Children who answer basic facts in this manner, often do not know the basic fact in question. To obtain a simple solution, they reason, "Sometimes the answer is the same number that is taken away (the subtrahend). Since I cannot think of an answer, I'll use the one that is already there" (in the subtrahend).

A teaching strategy is to review the fact strategy for doubles in addition. They can learn to reason that if $1 + 1 = 2$ and $2 - 1 = 1$, then $1 + 1$ cannot equal 4 and $4 - 1$ cannot equal 1. The subtrahend and the difference can only be alike in doubles. What is commonly seen in subtraction and division can also happen with addition and multiplication, but it is usually less common. Examples would be:

$\begin{array}{r}2\\+8\\\hline 8\end{array}$	$\begin{array}{r}7\\\times 5\\\hline 5\end{array}$

- **Faulty Reasoning 3**

Children answer problems in the following way:

$\begin{array}{r}9\\-1\\\hline 7\end{array}$	$\begin{array}{r}10\\-8\\\hline 1\end{array}$

Children may miss the correct answer by one number. As stated in the teaching strategies section of this chapter, counting-back strategies are often found to be helpful in such a situation. Sometimes mistakes like the ones above occur because some children count rapidly and lose track of the number being removed. Sometimes children have been taught the counting-back strategy but when they try to apply it mentally, they reason that $9 - 1 = 7$ by saying, "When I count back it is 9, 8, 7. If I take one away it will look like this:

$$9 \quad 8 \quad 7$$

Since I have 9, the one to be taken away is the 8 and that means 7 is the next number that is left."

Another reasoning pattern is to think that $10 - 8 = 1$ by saying, "I'm at 10 and I need to know how many 8 is away from 10. It is 10, 9, 8. There is one number in the middle so 8 is one away from 10."

Baroody (1984) noted a number of problems using the counting down procedure. Not only must children be able to count backwards with ease, but attention must be given simultaneously to the keeping-track process which is done by counting forward. Baroody states " . . . it is important to determine what *level of automaticity* must be achieved with backward counting to engage efficiently in the demanding simultaneous processing required by the algorithm." (p. 210)

To remediate this error, a number line can help rectify these misconceptions ONLY IF the teacher emphasizes that the spaces between numerals are important in the counting. Many commercially-prepared number lines appear like the one shown in Figure 7.32. Without augmentation, many children will use this aid incorrectly, following the misguided reasoning outlined above.

Figure 7.33 shows the same number line with the curved segment between numerals to emphasize the spaces in counting. Some clever elementary teachers refer to the spaces as "hops" up or down the number line. A teacher may say that $10 - 1 = 9$ because a person starts at the numeral 10 and hops back one space and lands on 9.

Some children are just poor counters. Therefore, using any counting strategy for remediation will be unsuccessful. Such children should be encouraged to use derived fact strategies, known as DFS. The DFS have been explained earlier in this chapter. Research shows that the following ones are helpful to children needing help (Steinberg, 1985).

1. doubles
2. doubles $+ 1$ and $- 1$
3. doubles $+ 2$ and $- 2$
4. bridging to 10, such as:

   $6 + 8$ is $6 + 4 = 10$ with 4 more left to add to 10 so $10 + 4 = 14$

   or

   $12 - 7$ is $10 - 7$ which is 3 with 2 left so $3 + 2 = 5$

• **Faulty Reasoning 4**

This faulty reasoning pattern belongs to the teacher rather than the children. Some teachers believe that students must move quickly from the use of concrete manipulatives to the symbolic stage where the children do worksheet after worksheet without the use of manipulatives or pictures. Children frequently give such teachers clues to their learning difficulties by their responses to basic fact problems on textbook pages or on worksheets. Figure 7.34 shows an actual worksheet of a primary child.

The child makes very few mistakes but it is evident that the child needs some form of concrete or semi-concrete aid to arrive at the answers. The tally marks are needed in almost every problem. The child seems to understand the property/principle of zero in addition with assurance; no tally marks are needed to check out the answers.

What should be done about this situation? More work with fact strategies and concrete materials is necessary. To demand that a child not use the tally marks would only force the child to use materials, eye blinks, fingers out of the teacher's view, or just guess the answer. Tally marks can help bridge the gap between concrete materials and symbolic numerals. This child needs to be encouraged to place the tally marks on a separate page where the marks can be separated for each problem, and prevent the mislearning that occurs when the child thinks the answer counted was correct.

Some children will become overdependent on the tally marks. They will need to be "weaned" from them gradually, starting with

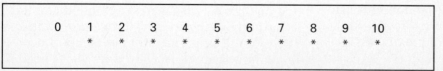

Commercially-prepared number line

Figure 7.32

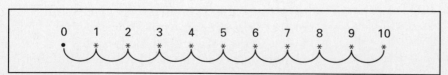

Number line changed by teacher to help children's misconceptions

Figure 7.33

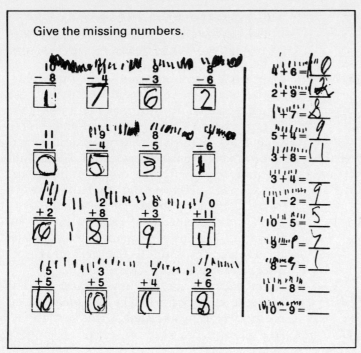

Figure 7.34

the basic facts involving adding one or taking one away. These facts can be reasoned more easily. The next step is to adopt a systematized way to introduce the teaching strategies discussed earlier in the chapter.

Teacher Assessment of Student Work: Steps in Analytical Thinking

Research (McMillen, 1986) indicates that people can profit from writing down their thought processes in mathematics. It can bring clarity to processes that would otherwise be easily confused. An example of such a thought process is presented here as it relates to the assessment of a child's problems in learning basic addition and subtraction facts. Excellent teaching involves the analysis of what a child understands or does not understand, and what misconceptions must be remediated.

A model outlining steps in analytical thinking is presented here so the teacher can see a way to approach the task of diagnosing a child's learning difficulties from answers a child has given to a textbook assignment. Exercises at the end of the chapter will ask the teacher to analyze the work of two other students. The analytic process applied here will be of help at the end of

the chapter as well. Figure 7.35 shows the work of a first grade student who has several misconceptions about the basic facts of subtraction.

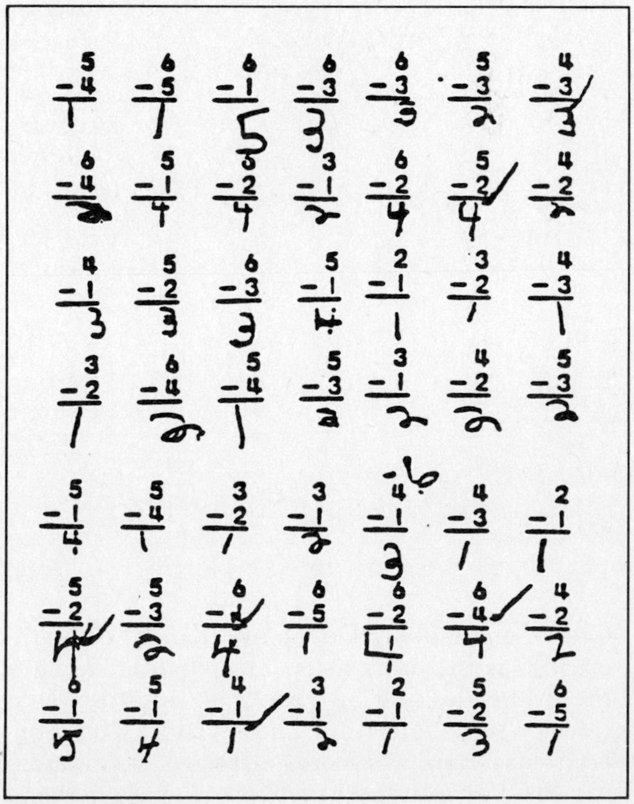

Figure 7.35

214 CHAPTER SEVEN   THE WHOLE NUMBER SYSTEM

Steps in analytical thinking:
1. Find the basic facts answered incorrectly (marked with a check).
2. Determine if the same basic facts as those missed appear more than once on the worksheet. If so, see if the student missed the problem and/or answered the problem in the same way as those marked incorrect.
   a. If the same mistake is made consistently, look for patterns like those mentioned in the learning styles section of the chapter and in this section (diagnosing learning difficulties).
   b. If the student sometimes answers correctly, make a chart such as the one below to analyze the answers:

   Notice: This student is correct more often than not. Now a teacher must decide if the incorrect answers were the result of sloppy work habits or a real misconception.

   c. If sloppiness is the cause, answers tend to deteriorate the further down the worksheet the student goes. Is this the case here? No.
   d. If a real misconception is the cause, further scrutiny is needed. Often the best teachers can do is to make an intelligent guess based on the pattern of answers. It appears that the student is doing each problem separately without seeing its connection with the other problems on the page. Hence, the student answers correctly sometimes but not others.
3. Look for the pattern of reasoning used by the child. The most frequent pattern seen is the faulty reasoning 2. The next most frequent pattern is faulty reasoning 3.
4. Refer back to the teaching strategies that help the child who makes the mistakes in faulty reasoning as listed in this section.

Basic Fact	Answered	Analysis
5 − 2 = 4	5 − 2	2 correct / 2 incorrect—missed by one each time, same answer each time
6 − 1 = 4	6 − 1	2 correct / 1 incorrect—missed by one
6 − 4 = 4	6 − 4	2 correct / 1 incorrect—missed by two
4 − 1 = 1	4 − 1	2 correct / 1 incorrect—missed by two
4 − 3 = 3	4 − 3	2 correct / 1 incorrect—missed by two

Summary

One of the most talked about element of mathematics is the basic facts. This chapter explains the properties and operations of addition, subtraction, multiplication, and division. The properties are emphasized for effective understanding of the number system. The Scope and Sequence Chart provides details of where each topic is included in the curriculum; however that sequence is being challenged by a number of mathematics educators who support the NCTM Curriculum and Evaluation Standards. Teaching strategies for the properties are carefully introduced including common mental arithmetic ideas (strategies) for each property. Manipulatives are included to provide adequate concrete experiences. Thinking strategies and problem solving are emphasized throughout the chapter.

The diagnosis and remediation includes common faulty reasoning when working with basic facts. The next chapter, algorithms, shows how the ideas presented here are extended to multidigit numbers.

Exercises

Directions: Read all questions *before* answering any one exercise. Frequently the last question in one category leads to the first question in the next category.

A. Memorization and Comprehension Exercises
Low-Level Thought Activities

1. The chapter asserts that the counting-back strategy is not effective when counting back more than ___ because children lose track of the counting.

2. Give the definition of the counting-on strategy. Is the following a good example of the counting-on strategy?

$$5 + 4$$

3. Skip counting works well only with ___ factors.

4. Sketch an example of the property/principle of commutativity and associativity using Cuisenaire rods.

5. Think of another concrete material (other than Cuisenaire rods) that could show the property/principle of commutativity and associativity. Sketch those examples and explain how the principles would be taught.

6. Pick two Cuisenaire rods and indicate how you could use them to show addition, subtraction, multiplication, and division of the basic facts. Sketch your examples.

7. Explain, in your own words, the difference between partitive division and measurement division. Give an example of each one.

8. Explain, in your own words, why division by zero is considered "undefined" by mathematicians.

B. Application and Analysis Exercises
Middle-Level Thought Activities

1. Think of a set of triominoes to teach the division facts. Sketch some examples and explain how they could be used.

2. Using the evaluation form found in Appendix A, select commercial software for any three topics from this chapter. In the selection, include drill and practice, simulation, tutorial, and problem-solving software. For each software program selected, describe how you would integrate it in a lesson plan.

3. To reinforce math across the curriculum:
   a. List five mathematical questions/problems that come from popular storybooks for children. Show how they are related to the topics of this chapter.
   b. List five ways to use the topics of this chapter when teaching lessons in reading, science, social studies, health, music, art, physical education, or language arts (writing, English grammar, poetry).

4. Apply the use of an EPR (discussed in Chapter 2) to teach the concept of commutativity of a basic fact in addition and one in multiplication. Show three different ways to make an EPR.

5. Pick a grade level and analyze how the basic fact strategies are presented in the textbook for that grade level. Find three word problems that could be solved by more than one strategy. Show the solutions using different strategies. Label each strategy as you use it.

6. Search professional journals for current articles on research findings and teaching ideas with the basic facts of the whole number system. You may wish to look at the first activity under Synthesis and Evaluation Activities below and coordinate activities.

Follow the form below to report on the pertinent findings:

```
┌─────────────────────────────────────────────┐
│                                               │
│  Journal reviewed: _____  │
│                                               │
│  Publication Date: _____  Grade Level: ___ │
│                                               │
│  Subject Area: _____ │
│                                               │
│  Author(s): _____ │
│                                               │
│             _____ │
│                                               │
│  Major Finding:                               │
│                                               │
│                                               │
│  Study or Teaching Procedure Outlined:        │
│                                               │
│                                               │
│  Reviewed by: _____ │
│                                               │
└─────────────────────────────────────────────┘
```

Some journals of note are listed below:
Journal for Research in Mathematics Education
Arithmetic Teacher
Mathematics Teacher
School Science and Mathematics

7. Computer applications for classroom use are found in the following journals. Their programs are by no means the only ones. Search the periodical section of the library for more. You may wish to look at the first activity under Synthesis and Evaluation Exercises to coordinate activities.
Teaching and Computers
Classroom Computer Learning
Electronic Learning
Computers in the Schools
The Journal of Computers in Mathematics and Science Teaching
The Computing Teacher

C. Synthesis and Evaluation Exercises
High-Level Thought Activities

1. Find a concept on the Scope and Sequence Chart that you would like to teach. Using the information in 6 and 7 above, create (synthesize) a model lesson plan for teaching the basic facts. Evaluate how all the components can work together. Include the following components in the plan:
   a. Use the current articles from the professional journals to help plan the direction of the lesson. Document your sources.
   b. Include at least one computer software program as a part of the lesson. Show how it will be integrated with the rest of the lesson.
   c. Develop behavioral objectives. Show how you will use these to evaluate the lesson.
   d. Use the research-based lesson plan format of i) review, ii) development, iii) controlled practice, and iv) seatwork.

2. The multiplication table in Figure 7.36 is the work of an actual student. Follow the analytical thinking steps outlined in the section on diagnosing learning difficulties. Here are some questions to help you start the analytic process:
   a. Does this student understand the multiplication property/principle of zero? What makes you answer yes or no? Justify your answer.
   b. Tally marks are used in some of the answers. In what section of the table do these appear? What does that tell the teacher about the set of problems that are not understood? It will be helpful to review the section on teaching strate-

Figure 7.36

gies as it relates to the multiplication table.

c. At what point in the table does the student relate one answer (product) to the next answer (product) without regard for the multiplier and the multiplicand?

d. What pattern begins to emerge after the basic fact of 4 × 4?

e. What parts of the table does the student know well?

3. Plan a program of remediation to help the above student learn the basic multiplication facts with which he/she is having trouble.

a. What would be a realistic beginning point of instruction? Justify your answer.

b. In what order would you teach the remaining deficient skills? Justify your choice of sequence.

c. Are concrete manipulatives needed? At what point? Which ones discussed in this chapter would you recommend? Justify your answers.

d. At what point would you "wean" the student from concrete manipulatives to the pictorial or semi-concrete models?

e. At what point would you encourage the student to use the symbolic representation without other help?

4. The worksheet in Figure 7.37 is another example of a student's work. Analyze what the student knows and what the student does not know.

a. Plan a program of remediation for the student. The questions asked in 3 may be helpful here as well. Your plan should include:
   i) analysis of the error pattern
   ii) an evaluation of the thought processing required in the examples and the thought processing displayed by the student
   iii) the sequential development of concrete, pictorial, and symbolic models to reteach the concept.

5. Prepare a cooperative learning center activity using the microcomputer with spreadsheet software. Have students prepare a spreadsheet and record the high and low temperatures each day. The data is usually available in the local newspaper. An option is to record the temperature at several fixed times during each school day.

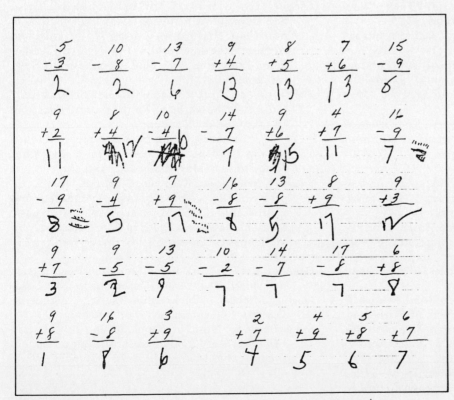

Figure 7.37

The following table is an example:

Day	High Temp	Low Temp
1	75	43
2	80	55
3	72	62
4	80	60
5	77	47
6	82	55
7	84	70
8	88	45
9	75	55
10	77	45
11	80	65
12	85	60
13	88	62
14	89	70
15	87	52

Include in the lesson, problem-solving activities for national temperatures and previous year comparisons, and any related activities where the spreadsheet is a valuable learning tool to understand basic math concepts.

6. Read the NCTM curriculum standards regarding the issue of teaching, testing, and emphasis of basic facts. Take a position for or against this recommendation and write a position paper stating your stand and defending your position.

7. Suppose you are a fifth-grade teacher and you find many students do not know their basic facts. What approach would you take? Explain in detail what materials, teaching strategies, and other resources you would use to rectify the situation.

8. How can the basic number facts be taught to children in efficient and meaningful ways?

Bibliography

Baratta-Lorton, Mary. *Mathematics Their Way.* Menlo Park, Ca.: Addison-Wesley, 1976.

————. *Workjobs II.* Menlo Park, Calif.: Addison-Wesley, 1978.

Baroody, Arthur J. "Children's Difficulties in Subtractions: Some Causes and Questions." *Journal for Research in Mathematics Education* 15 (May 1984): 203–213.

Brush, Lorelei R. "Preschool Children's Knowledge of Addition and Subtraction." *Journal for Research in Mathematics Education* 9 (January 1978): 44–54.

Campbell, Patricia F. "Using a Problem-Solving Approach in the Primary Grades." *Arithmetic Teacher* 32 (December 1984): 11–14.

Carpenter, Thomas P., and James M. Moser. "The Acquisition of Addition and Subtraction Concepts in Grades One through Three." *Journal for Research in Mathematics Education* 15 (May 1984): 179–202.

Commission on Standards for School Mathematics of the National Council of Teachers of Mathematics. *Curriculum and Evaluation Standards for School Mathematics.* Working draft. Reston, Va.: The Council, 1987.

Dorough, Bob. *Multiplication Rock.* Xerox Films. Middletown, Conn.: Xerox Educational Publishing, 1973.

Dossey, John A., Ina V. S. Mullis, Mary M. Lindquist, and Donald L. Chambers. *The Mathematics Report Card, Are We Measuring Up? Trends and Achievement Based on the 1986 National Assessment.* Princeton, N.J.: Educational Testing Service, June 1988.

Fuson, Karen C. "More Complexities in Subtraction." *Journal for Research in Mathematics Education.* 15 (May 1984): 214–225.

————. "Teaching Children to Subtract by Counting Up." *Journal for Research in Mathematics Education* 17 (May 1986): 172–189.

Fuson, Karen C., and Kathleen T. Brinko. "The Comparative Effectiveness of Microcomputers and Flash Cards in the Drill and Practice of Basic Mathematics Facts." *Journal for Research in Mathematics Education* 16 (May 1985): 225–232.

Greene, Gary. "Math-Facts Memory Made Easy." *Arithmetic Teacher* 33 (December 1985): 21–25.

Hagen, Michael. "Sticks and Bones." *Arithmetic Teacher* 33 (September 1985): 44–45.

Jones, Philip S. "Notes on Numeration: Arithmetic on a Checkerboard, Numerals for the Blind." *School Science and Mathematics* 78 (October 1978): 481–488.

Kouba, Vicky L., Catherine A. Brown, Thomas P. Carpenter, Mary M. Lindquist, Edward A. Silver, and Jane O. Swafford. "Results of the Fourth NAEP Assessment of Mathematics: Number, Operations, and Word Problems." *Arithmetic Teacher* 35 (April 1988): 14–19.

Lessen, Elliott I., and Carla L. Cumblad. "Alternatives for Teaching Multiplication Facts." *Arithmetic Teacher* 31 (January 1984): 46–48.

McMillen, L. "Science and Math Professors are Assigning Writing Drills to Focus Students' Thinking." *The Chronicle of Higher Education* (22 January 1986): 19–21.

Miller, Geoffery. "Graphs Can Motivate Children to Master Basic Facts." *Arithmetic Teacher* 31 (October 1983): 38–39.

Rappaport, David. "Numeration Systems—A White Elephant." *School Science and Mathematics* 77 (January 1977): 27–30.

Rathmell, Edward C. "Using Thinking Strategies to Learn Basic Facts." In *Developing Computational Skills* 1978 Yearbook of the National Council of Teachers of Mathematics. Reston, Va.: National Council of Teachers of Mathematics, 1978. 13–38.

Russell, R. L., and Herbert P. Ginsburg. *Cognitive Analysis of Children's Mathematics Difficulties.* Rochester, N.Y.: University of Rochester, 1981.

Smith, Lyle R. "Mathematics on the Balance Beam." *School Science and Mathematics* 85 (October 1985): 494–497.

Steinberg, Rugh M. "Instruction on Derived Facts Strategies in Addition and Subtraction." *Journal for Research in Mathematics Education* 16 (November 1985): 337–355.

Thornton, Carol A. "Emphasizing Thinking Strategies in Basic Fact Instruction." *Journal for Research in Mathematics Education* 9 (May 1978): 214–227.

Thornton, Carol A., Benny F. Tucker, John A. Dossey, and Edna F. Basik. *Teaching Mathematics to Children with Special Needs.* Menlo Park, Calif.: Addison-Wesley Publishing, 1983.

Van Engen, H., and Leslie P. Steffe. *First Grade Children's Concept of Addition of Natural Numbers.* Madison: University of Wisconsin, Research and Development Center for Learning and Re-Education, 1964.

Wirtz, Robert. *Individualized Computation.* Washington, D.C.: Curriculum Development Associates, 1974.

Yannone, Denise Stamp. "Mathletes, One and All." *Arithmetic Teacher* 32 (May 1985): 13.

8

Algorithms

Key Question: *What are some ways to help students remember the essential steps required to answer multi-digit problems in mathematics and understand the reasoning behind the answers they obtain?*

Activities Index

Introduction

The basic facts and place value concepts are applied in the computational skills required to work the algorithms. *An algorithm is a set of rules for solving a problem, a step-by-step sequence, a method that continually repeats some basic process.* This chapter covers the algorithms for the addition, subtraction, multiplication, and division of multi-digit numbers. These numbers, containing two or more place values, go beyond the basic facts, which are combina-

tions of single-digit numbers. Algorithms are efficient ways of incorporating the basic facts into larger, multi-digit numbers.

Algorithms also apply to operations with integers. *Integers are the set of numbers both positive or negative and zero; the entire class is shown as 0, ± 1, ± 2, ± 3, ± 4 . . . and on.* Positive integer is another name for the counting or natural numbers that children begin to study when entering school. Integers are generally included in most sixth grade textbooks and are covered by many teachers in an informal way during the sixth grade year. The formal study of integers usually begins in seventh and eighth grade, but children are exposed to situations involving negative and positive numbers long before formal study begins. The thermometer is one common use of integers. As children use the calculator, they are bound to produce a negative number occasionally. Teachers may find themselves explaining the concept of integers long before they had planned. The teaching strategies section will present techniques for working with both positive and negative integers.

The Development of Algorithmic Models

Elementary textbooks devote a great amount of time to the development of algorithms. Usually a textbook will present only one algorithm for each operation. It is important to remember that there is no one right way to solve a problem. There are many algorithms that have been developed over time for each operation. Figures 8.1–8.4 are examples of some of the interesting algorithms found in elementary school textbooks dating from 1848 until 1930.

Manipulative materials to teach the algorithms may be proportional or non-proportional models. Teachers must consider carefully the attributes of the models before determining which ones to use and in which sequence.

Proportional vs. Non-proportional Models

Some algorithmic models are proportional in the sense that the concrete and pictorial models are based on the *structure* of the numeration system

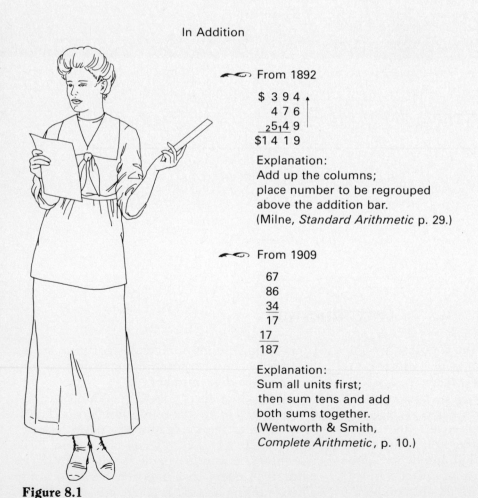

In Addition

From 1892

$ 3 9 4
  4 7 6
 2 5 4 9
$1 4 1 9

Explanation:
Add up the columns;
place number to be regrouped
above the addition bar.
(Milne, *Standard Arithmetic* p. 29.)

From 1909

67
86
34
17
——
17
——
187

Explanation:
Sum all units first;
then sum tens and add
both sums together.
(Wentworth & Smith,
Complete Arithmetic, p. 10.)

Figure 8.1

seen in Chapter 6. Every original unit in a problem is clearly seen as the unit is regrouped to a larger and larger place value.

Some algorithmic models are non-proportional in the sense that the concrete and pictorial models do not account for every original unit as it is regrouped to larger place values. The inclusion of every original unit into the next larger place value is assumed but not clearly visible.

Examples of proportional and non-proportional models were shown in Chapter 6. They will be presented throughout this chapter to show how such models work with the regrouping of multi-digit numbers.

IN SUBTRACTION

The following is a copy of the actual page in a 1930s text encouraging students to use whichever of the three methods would work best for them. Study the thought process involved in each one.

---

10   HIGHER ARITHMETIC

Result + subtrahend = minuend

There are three distinct methods used in subtraction. They are illustrated in the following exercises.

FIRST METHOD:

6027
3574
——
2453

Think, "4 from 7 is 3. 7 from 12 is 5. 6 from 10 is 4. 4 from 6 is 2."

SECOND METHOD:

6027
3574
——
2453

Think, "4 from 7 is 3. 7 from 12 is 5. 5 from 9 is 4. 3 from 5 is 2."

THIRD METHOD:

6027
3574
——
2453

Think, "4 and 3 make 7, write 3. 7 and 5 make 12, write 5 and carry 1. 6 and 4 make 10, write 4 and carry 1. 4 and 2 make 6, write 2."

In the following exercises use the method with which you are familiar. In this book the third method, known as the additive method, will be used.* If you wish to change from one of the other methods to the additive method, or from any of the three methods to either of the other two, be sure to practice using the new method for several days in order to insure accuracy and speed.

*See Stone's pamphlet *How We Subtract*, published by Benj. H. Sanborn & Co.

Figure 8.2

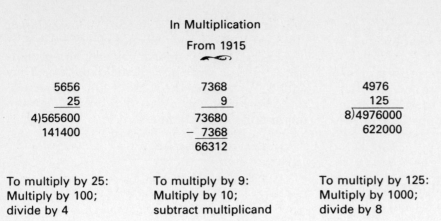

In Multiplication

From 1915

5656	7368	4976
25	9	125
4)565600	73680	8)4976000
141400	− 7368	622000
	66312	

To multiply by 25:   To multiply by 9:   To multiply by 125:
Multiply by 100;   Multiply by 10;   Multiply by 1000;
divide by 4   subtract multiplicand   divide by 8

(Wentworth & Smith, *Essentials of Arithmetic*, p. 11.)

Figure 8.3

In Division

From 1848

dividend 8467|8   divisor
          8 |1058 quotient
          046
           40
          067
           64
            3

Note: All the parts are there;
it's the arrangement that's different.

([French Text]
*New Treatment of Arithmetic for the
Decimal and Metric System*, p. 55.)

Figure 8.4

From 1877

Divide 1592 by 35, that is 7 × 5

7)1592
 5)227 . 3 first remainder
    45 . 2 second remainder

Explanation given in 1877 text:
1592 has 227 sevens and 3 ones & 277
sevens has 45 thirty-fives and 2 sevens
& 2 sevens = 14 ones; so the true
remainder is 14 + 3 = 17.

(Davies & Peck, *United Course:
Complete Arithmetic*, p. 58.)

Representational vs. Non-representational Models

Another way to show algorithms is through the use of representational or non-representational models. *Representational models show the regrouping strategy that occurs between powers of ten represented by the place value of numerals.* Both proportional and non-proportional models of algorithms can be representational models. Algorithm 1 in Figure 8.5 shows the representational regrouping for subtraction of multidigit numbers. *Non-representational algorithms are based on the properties of the number sys-* *tem and NOT on the structure of the numeration system.* (See Chapter 6 for the difference between numeration and the number system.) The non-representational model is shown as Algorithm 2 in Figure 8.5.

Algorithm # 1	Algorithm # 2
Representational	Non-representational
325	325
−187	−187
138	138

Figure 8.5

Non-representational strategies cannot be easily seen as a regrouping from one power of ten to another. Algorithm 2 works because problems such as:

$$\begin{array}{r} 11 \\ -\ 8 \\ \hline 3 \end{array} \qquad \begin{array}{r} 12 \\ -\ 9 \\ \hline 3 \end{array}$$

yield the same answer. Adding one to the subtrahend is the same as taking one away from the minuend. It is based on the property of counting numbers (as seen in Chapter 7) that adding one or taking one away yields a unique, predictable number in the counting sequence.

You will see many different algorithms as you proceed through the chapter, some representational and some non-representational.

Analyze the algorithms with the knowledge that there is no one right algorithm. There are still more to be created. One of the higher-thought activities at the end of this chapter asks you to design your own algorithm for operations on multi-digit numbers. There is intellectual prowess in the idea that any algorithm can be seen through progressively more abstract models, arriving at a correct answer no matter what model is used.

A special attempt has been made to choose algorithms that are quite different from one another. Some elementary and middle school students may find one algorithmic model more understandable than another. It is hoped that the teacher will see the variety of options as a way of meeting the diversified needs of students.

Scope and Sequence

First Grade
- Add two-digit numbers without regrouping.
- Subtract two-digit numbers without regrouping.
- Readiness for regrouping in addition and subtraction.

Second Grade
- Add two-digit numbers with or without regrouping.
- Subtract two-digit numbers with or without regrouping.
- Add three two-digit numbers with or without regrouping.
- Add and subtract with money notation.

Third Grade
- Add any two numbers or more to 4 digits.
- Subtract any two numbers to 4 digits.
- Add and subtract with money notation.

Fourth Grade
- Add any two or more numbers to 5 digits.
- Subtract any two numbers to 5 digits.
- Estimate sums and differences.
- Add and subtract with money notation.
- Multiply any number to 4 digits by one-digit number.
- Divide any number to 4 digits by any one-digit number.
- Multiply any number to 3 digits by two-digit number.
- Divide any number to 3 digits by two-digit number.

Fifth Grade
- Add and subtract any numbers to 6 digits.
- Estimate sums and differences.
- Multiply any number to 4 digits by any number to 3 digits.
- Estimate products.
- Divide any number to 4 digits by two-digit numbers.
- Find averages.

Sixth Grade
- Mastery expected on all multi-digit operations.

Seventh-Eighth Grades
- Mastery expected on all multi-digit operations.
- Operations with integers.
- Order of operations.
- Evaluate algebraic expressions with variables.

Teaching Strategies

The student's first experience with algorithms occurs in textbooks around the second grade level (see Scope and Sequence Chart) when the algorithms for regrouping for addition and subtraction are introduced. This does not mean that students cannot be exposed to the concept of regrouping prior to second grade if concrete materials are used and problem-solving situations require it. In fact, many children devise their own algorithms to solve problems with large numbers and without formal instruction in the process; many interesting and innovative algorithms are invented by children.

The teacher's editions of textbooks offer ideas to teach the algorithms using careful sequential development. Techniques often include ideas for using concrete materials like base 10 blocks or bundles of sticks. It is important to allow time for children to become completely familiar with the algorithm at a concrete level before proceeding to the textbook's pictures (semi-concrete or pictorial level). If ample opportunities have been included for experiences with the trading or regrouping process during the numeration or place value section, students should have sufficient understanding for developing the addition and subtraction algorithms.

Oral interviews with children indicate that many of them find it difficult to relate mathematical concepts to real experiences. They memorize a set of rules at a verbal level and have difficulty transferring ideas to new situations. A child who can work $42 - 18$ correctly may not be able to illustrate that concept with counters, money, or concrete materials. Operations may be introduced too quickly without enough time spent on modeling, manipulating, and dramatizing the procedure. When this happens, children do not have an "operation sense."

Teachers must guard against teaching the algorithms in a rote, meaningless way followed by pages of computations, a strategy that is easy to do and often seems expedient. The algorithm becomes a mechanical procedure without any association to manipulation of materials. An example of this rigidity is apparent in the following example: A child was asked why the regrouped digit was placed above the tens place. Could the digit be placed below the line in the tens place and still represent the same thing? (See Fig. 8.6) The child emphatically replied that the method was wrong and would never work

because the teacher and textbook had routinely and rotely taught only one exact procedure. Therefore, the steps of the algorithm should be connected with manipulation of materials and verbal explanations followed by many opportunities for practice by each student.

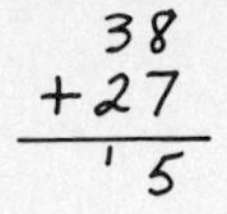

$$
\begin{array}{r}
38 \\
+27 \\
\hline
^1 5
\end{array}
$$

Figure 8.6

Results of the fourth mathematics assessment of the National Assessment of Educational Progress (Dossey et al., 1988) did not measure for complex computations involving numbers with more than three digits. The assessment placed less emphasis on whole number computation and did not include items on long division or multiplication involving regrouping. Third-grade students performed well on addition of two-digit numbers (84 percent) and subtraction (70 percent), however around half of the students performed well on three-digit subtraction. Seventh-grade students made large gains on three-digit subtraction (around 85 percent). As noted earlier, limited items on multiplication and division restrict much comparison (Kouba et al., 1988).

A large percentage of teaching time is spent on the algorithms for whole numbers. When we consider the impact and availability of calculators now, plus the increased use expected in the future, can we remain content to devote so much instructional time to the algorithms? Among many educators and in several national reports such as the NCTM curriculum and evaluation standards, there is a cry for a de-emphasis on tedious computational algorithms. This does not imply that teaching the algorithms should be discontinued but as Beattie (1986) states, " . . . it does mean that the reason for teaching algorithms should change from obtaining correct answers through the rote manipulation of symbols to understanding the meaning of the operation and the rationale behind each step in the algorithm. And the way in which algorithms are taught should change from a mechanical procedure to methods that will facilitate *understanding*. This understanding will become more, not less, important in the calculator and computer age." (p. 23) The NCTM curriculum standards (Commission on Standards for School Mathematics, 1987) propose a change in the typical grade level placement and sequence of the algorithms. They call for a change in the expectations for paper-and-pencil computational proficiency. Students are expected to demonstrate computational proficiency involving two- and three-digit numbers by the end of the third grade. More complex computations are to be done with a calculator. In grades 5–8 students will be expected to multiply two-digit numbers by two-digit numbers and to divide two- or three-digit numbers by one-digit numbers. The standards also assume that a calculator will be available at all times for students' use. The emphasis is a shift from computational proficiency with multi-digit numbers to many other types of computational methods that include not only paper-and-pencil but mental arithmetic, estimation, calculators, and computers.

Teaching the Addition Algorithm

The addition algorithm (addition with regrouping) should begin with proportional place-value materials such as connecting cubes, bean sticks, pocket charts, or base 10 blocks (Fig. 8.7). Whatever the model, it should include some place-value boards to help children focus on a logical organizational procedure. This allows greater transfer from the concrete to the symbolic level.

Proportional Models

Labinowicz (1980) has sequenced a variety of instructional materials based on their relative positions in abstraction. His sequence begins with the most concrete—lima beans in small cups. The materials are in one-to-one correspondence with the number represented. Next, he lists connecting cubes where the interlocking nature allows quick groups of ten to be formed. These materials are still in one-to-one correspondence with the number. Beansticks or base 10 blocks are next. The groupings are fixed but the one-to-one correspondence is still present. The next level is base 10 blocks that are unscored into specific units (sometimes called Dienes blocks). With this material, the proportion of ten-to-one is shown but the materials do not allow the child to decompose and the one-to-one correspondence is lost (Fig. 8.7).

Non-proportional materials (Fig. 8.8) can also be used as instructional devices for the algorithms, but only after proportional materials have been introduced. It is important to have a sound developmental sequence because of the range in levels of representation between the various models. Labinowicz labels an abacus or chip trading as even more abstract models. Each

Proportional Models

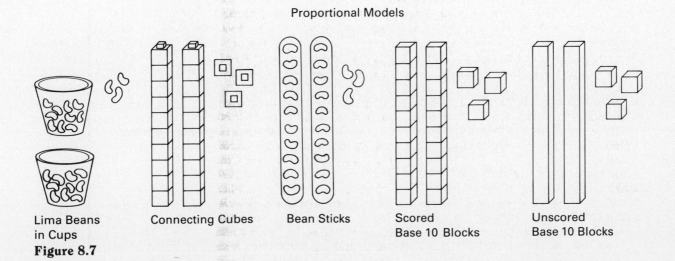

Lima Beans in Cups Connecting Cubes Bean Sticks Scored Base 10 Blocks Unscored Base 10 Blocks

Figure 8.7

Non-proportional Models

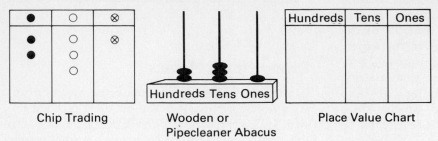

Chip Trading   Wooden or   Place Value Chart
                     Pipecleaner Abacus

Figure 8.8

is a non-proportional model that is representative in nature and should be presented last in the instructional sequence.

Children are introduced to addition with two-digit numbers without regrouping in most first grade textbooks. They are taught to start adding with the ones place, then the tens place, but it is difficult to monitor whether this procedure is being followed. Since the same answer results working from left to right or from right to left, children may be using either procedure without applying any place-value aspects to the process. For this reason, the authors support teaching addition with regrouping and without regrouping together so that the mechanical aspect is replaced with thoughtful attention to the process. Kamii and Joseph (1988) and Madell (1985) found that when first- and second-grade children are allowed to work double-column addition, they invented their own natural way that usually proceeds from left to right. This method means that place value is not separated from addition instruction. Kamii advocates encouraging children to invent all kinds of ways to work the addition problems. When word problems and situational problems accompany the presentation, children are less likely to ask, "Do I have to regroup on this one?" Manipulative materials are the key to understanding and their use will result in fewer remediation problems.

Activity

MANIPULATIVES

Teaching Addition with Regrouping Using Proportional Models

Materials: Base 10 blocks
Place-value mats
Student chalkboards—optional

Procedure: The situational problem: You have collected 26 baseball cards. Your friend, Brett, will sell you 17 baseball cards for $4.50. If you decide to buy the cards, how many will you have then?

Step 1: Place 2 tens and 6 ones on the mat in the appropriate columns.

Step 2: Place 1 ten and 7 ones on the mat.

Step 3: How many ones are there? (13) Can you make a trade of ones for tens?

Step 4: Trade 10 ones for 1 ten and place the regrouped ten in the tens place.

Step 5: Combine the tens. How many tens are there? How many in all?

Step 6: Record the steps and answer on your chalkboard. (Do this step only when the children are ready for the connecting level.)

Materials: Connecting cubes
               Place-value mats

Procedure:

Step 1: Lay out 2 sticks of tens and 6 individual cubes. Place on mat in appropriate columns.

Step 2: Lay out 1 stick of ten and 7 individual cubes. Place on mat in appropriate columns.

Step 3: How many ones in all? Count the ones (13). Are there enough to make another ten?

Step 4: Join the 10 individual cubes into a stick of ten. Place on mat in tens place.

Step 5: How many in all?

Materials: Pocket chart
               Tags and rubber bands

Procedure: Steps are the same except that the child places bundles of tens and individual tags. When a group of ten is made from the 13 ones, the 10 tags are bundled together with a rubber band. This is the regrouped ten.

Materials: Bean sticks
               Place-value mats

Procedure: The steps are the same as for the base 10 blocks. A trade is made of 10 loose beans for a ten's stick.

Although the foundation for the algorithm begins with two-digit numbers, children need to continue to have concrete experiences with regrouping for hundreds and thousands. At this point, many teachers claim to not have sufficient time or materials to provide direct manipulation by children with place-value models. Time spent on these activities is well worth it in terms of long-range results. If a child is to truly internalize and "own" the concept, time and concrete experiences are required. Figure 8.9 shows the teaching strategy for multi-digit addition.

Situational Problem: Monday night 258 people watched the school play. Tuesday night 379 people watched the school play. How many people watched the play?

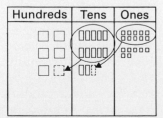

Place 258 blocks on the mat in place-value positions: 2 hundred blocks (FLATS), 5 tens (LONGS), and 8 units. Below it, place 379 blocks in place-value positions: 3 hundreds blocks (FLATS), 7 tens (LONGS), and 9 units. Combine the units (17) and regroup 10 ones for 1 ten. In the tens place, combine the LONGS (13) and trade 10 tens for 1 hundred. Combine the FLATS in the hundreds place and get 6.

Figure 8.9

Students can practice multi-digit addition with the calculator. Since the calculator will be used by most adults to do similar problems, it helps prepare students for the adult world. Notice that the calculator worksheet has been prepared to induce interest as they work with it.

Activity

CALCULATORS

Addition Practice

Directions: Find the sums. Add horizontally and vertically. Use the calculator.

1)

876	+		=	1371
+		+		+
709	+		=	1341
=		=		=
	+		=	

4)

964	+		=	1177
+		+		+
587	+		=	923
=		=		=
	+		=	

2)

892	+		=	1292
+		+		+
641	+		=	998
=		=		=
	+		=	

5)

108	+		=	760
+		+		+
493	+		=	610
=		=		=
	+		=	

3)

666	+		=	1069
+		+		+
275	+		=	1094
=		=		=
	+		=	

6)

563	+		=	811
+		+		+
719	+		=	1178
=		=		=
	+		=	

Teaching the Subtraction Algorithm

Much has been written about the subtraction algorithm and alternative algorithms to use. The most common algorithm is the decomposition method, which is traditionally the one presented in student textbooks. It can be related easily to place-value models and emphasizes the inverse relationship of addition and subtraction. There are several situations which involve different interpretations of subtraction. This chapter will address teaching strategies for understanding the operation of subtraction using the decomposition algorithm. Although different physical situations model the subtraction operation, the standard algorithm is identical for each one.

The Decomposition Algorithm:

$$\begin{array}{r} 42 \\ -17 \\ \hline 25 \end{array}$$

In ones column, we can't take 7 from 2. Regroup 4 tens as 3 tens and add 10 ones to 2 (12 ones). Now 7 from 12 is 5 and 1 ten from 3 tens is 2 tens.

The first kind of interpretation for subtraction is the take-away approach. It is the easiest for a child to understand, the most common one that a child encounters in real-life situations, and is simple to represent physically. Modeling operations and algorithms with objects will allow better understanding and will prepare children to interpret the textbook pictures more clearly. The following is an example of the teaching strategies for the take-away interpretation of subtraction.

Activity

MANIPULATIVES

Teaching Take-Away Subtraction with Regrouping

Materials: Base 10 blocks
Place-value mats

Procedure: The situational problem: Jenny had 52 cents. She bought a pencil for 27 cents. How much money does she have left?

Step 1: Show 5 tens and 2 ones in the correct place-value positions.

Step 2: Start in the ones place. Can you take 7 away from 2? You must regroup (trade) a ten.

Step 3. Regroup 1 ten for 10 ones. This leaves 4 tens and 12 ones.

Step 4. Now take away 7 ones from 12 ones. How many are left? (5)

Step 5. Take away 2 tens from 4 tens. How many are left? (2)

Answer: 25.

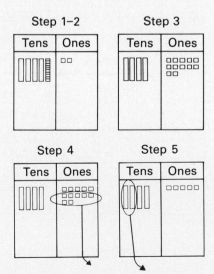

The procedure is the same for bean sticks (trade a ten stick for 10 loose beans), connecting cubes (break apart a ten stick into 10 individual cubes), and pocket charts (unbundle a ten into 10 tags).

The second interpretation of subtraction is the comparison idea. Here a comparison is made between the two sets or numbers. Usually children do not encounter this interpretation in their textbooks until second grade. The questions for comparison involve ideas of how much more, how many more, how much less, how much older, and so on where the two sets must be compared. This is significantly more difficult for children to master. The algorithm is the same; it is a question of interpretating problem-solving situations and knowing that the subtraction operation is to be used. Many children confuse the word "more" with the idea of addition. Also in the comparison interpretation, the modeling involves making both sets and determining the difference by one-to-one correspondence. The teaching strategy using manipulatives follows.

Activity

MANIPULATIVES

Teaching Comparison Subtraction with Regrouping

Materials: Base 10 blocks
Place-value mats

Procedure: The situational problem: Ramona has 52 cents. Phillip has 27 cents. How many more cents does Ramona have than Phillip?

Step 1. Show 52 cents as 5 tens and 2 ones. Show 27 cents as 2 tens and 7 ones in the correct place-value positions.

Step 2. Compare the ones. You see that 7 is greater than 2. Regroup or trade a ten for 10 ones. Add these ones to the 2 ones making 12 ones. This leaves 4 tens.

Step 3. Compare 12 ones to 7 ones. Count the difference as 5 ones.

Step 4. Compare the tens: 4 tens to 2 tens—the difference is 2 tens. Answer: 25.

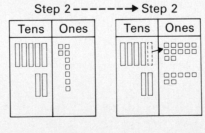
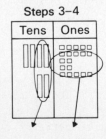

The third interpretation for subtraction is the missing addend concept. Here the problem involves knowing what you start with or have now and knowing how many you need in all. The idea is finding how many more are needed. Writing the problem as an addition problem with a missing addend may help visualize the procedure to use to solve the problem.

Activity

MANIPULATIVES

Teaching Missing Addend for Subtraction with Regrouping

Materials: Number balance with tags

Procedure: The situational problem: Together Sue and Tracy have 35 cents. Sue has 19 cents. How many cents does Tracy have?

Step 1. Put 35 on the right side of the number balance.

Step 2. Put 19 on the left side of the number balance.

Step 3. How much more should be added to the left side so both will balance or be equal?

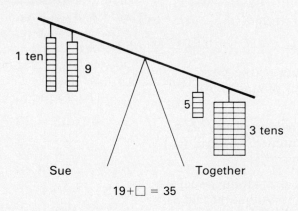

Many of the difficulties that children have with word problems requiring addition or subtraction result from the various interpretations of the operations that are possible. Explanations of addition as "putting together" and subtraction as "taking away," place limits on the operations that are inaccurate when we consider the many other situations calling for these two operations. Thompson and Hendrickson (1986) describe the many other contexts of problem-solving situations that children encounter for addition and subtraction and offer some instructional procedures. They identify three categories of problems that suggest addition and subtraction: change, combine, and compare. There are six kinds of change problems, two types of combine, and six kinds of compare problems. Each problem requires careful instructional procedures and an opportunity to encounter the various problem-solving situations.

Students sometimes invent unique ways to work problems. They create their own algorithm. Teachers must be sensitive to this possibility when they give the rule, "Always show your work." True, it might help decide where the error in thinking or computation occurs, but it could curtail the child's mental computations. Children are creative in adapting the techniques we teach them. The student may not be able to explain the thinking, and the teacher may need to watch the child working the problem while asking questions in a reassuring, encouraging manner, using the elaborating technique discussed in Chapter 2. Without such acceptance, Hamic (1986) warns that some students will give up their efforts and do mathematics the "proper" way, while others may become discouraged by teachers who insist on only one way. The following method of regrouping was invented by a first grader.

$$\begin{array}{r} 36 \\ -19 \\ \hline \end{array} \quad \text{(absolute difference between 9 and 6 is 3)}$$

$$30 - 10 = 20 - 3 = 17$$

Teaching the Multiplication Algorithm

Before considering teaching strategies for the multiplication algorithm, a quick review of the language and mathematical concepts of multiplication is recommended. Multiplication also has several interpretations: repeated addition, a rectangular array or row-by-column, and a combination-type. It is important not to limit the children's exposure to only one approach or they will also be limited in their abilities to decide when a problem-solving situation calls for multiplication. Hendrickson (1986) lists the many different situations that call for multiplication or division with some teaching suggestions.

Interpretation of multiplication:

$4 \times 5 = 4$ groups of $5 = 5 + 5 + 5 + 5$

To change from equation to working form:

$$\begin{array}{r} 5 \\ \text{4 groups of 5} = \times 4 \\ \hline 20 \end{array}$$

Notice: The vertical position places the 4 as the second number and still is read as 4 groups of 5.

Multiplication without regrouping usually appears in student's textbooks around the end of the third grade. This concept is easily taught and understood by children who have a reasonable proficiency with the basic facts and a firm understanding of place value. It can be modeled with arrays or base 10 blocks as shown in Figure 8.10.

Expanded notation and the distributive property help teach the multiplication algorithm with regrouping by showing how the partial products can be obtained and integrated into the shortened form of the algorithm. In the example, 4×23, the problem can be rewritten as $(4 \times 20) + (4 \times 3)$. The teaching strategies use base 10 blocks, connecting cubes, and graph paper for arrays. Remember that all problems should be introduced in problem-solving situations and solid concept development with manipulatives should be established with smaller numbers since physical manipulations are difficult for larger numbers.

Partial Products	Expanded Notation

$$\begin{array}{r} 4 \times 23 = 23 \\ \times 4 \\ \hline 12 \\ +80 \\ \hline 92 \end{array} \qquad \begin{array}{r} 20 + \quad 3 \\ \times 4 \quad \times 4 \\ \hline 80 + \quad 12 = 92 \end{array}$$

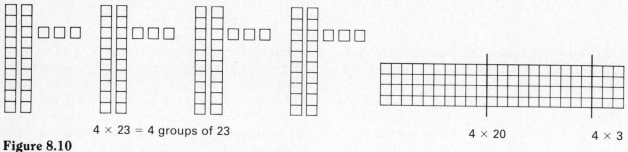

$4 \times 23 = 4$ groups of 23

4×20 4×3

Figure 8.10

Activity

PROBLEM
SOLVING

Multiplying with Manipulatives

Situational Problem: P. J. was given 4 books of tickets to sell. Each book contained 23 tickets. How many tickets had to be sold?

With Base 10 Blocks or Cubes

1. Form 4 groups of 23 (2 ten sticks and 3 ones). Place on place-value mat.
2. Count ones = 12
3. Trade 10 ones for 1 ten (base 10 blocks). Snap 10 ones together to make 1 ten (connecting cubes).
4. Count tens = 9 tens
5. Answer: 92

Place Value Mat with Blocks

Tens	Ones

Base 10 Blocks

Connecting Cubes

Tens	Ones

With Graph Paper

1. Cut graph paper to show 4 rows of 20 and 4 rows of 3.
2. Count ones (4 × 3 = 12) and make into 1 ten and 2 ones.
3. Count tens: 4 groups of 2 tens = 8 plus 1 regrouped ten = 9 tens.
4. Answer: 92

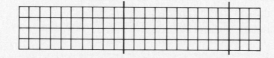

An instructional sequence for multiplication should begin with multiplying by 10 and 100. This skill is important for estimation activities and testing the reasonableness of answers. Children should see the pattern of the zeros and develop some general rules about what happens. These ideas can be tested using a calculator. The next step is to multiply numbers by multiples of 10 and 100 such as 20, 30, 40, 400, 500. Beware of using the term "adding zeros" as this is erroneous thinking and conflicts with the zero property of addition. A more appropriate term to use is "annexing zeros."

$$4 \times 2 = 8$$
$$40 \times 2 = 80$$
$$400 \times 2 = 800 \qquad \text{What pattern do you see?}$$

Even though this skill is considered easy by children, it is important that they understand the relationship between the factors and the product so mental estimation can be used later with success and confidence. Expanded computational forms using place-value names is another valuable activity that reinforces the concept: $4 \times 70 = 4 \times 7$ tens $= 28$ tens $= 280$.

Two-digit multipliers come next in the instructional sequence for multiplication. The first step is to have a two-digit number multiplied by a multiple of 10 such as 45×80. The student should be encouraged to round the first factor to a multiple of ten, estimate the product, then compare the actual computed answer to the estimate. When given experiences with expanded forms and the distributive property of multiplication, many children learn to mentally compute these problems with partial products. For example:

$$45 \times 80 = (40 + 5) \times 80 = (40 \times 80)$$
$$+ (5 \times 80) = 3200 + 400 = 3600.$$

Rounded estimate:

$$50 \times 80 = 4000, \text{ answer is reasonable.}$$

Students need to be encouraged to use the commutative property as well as the distributive property for multiplication when necessary. If they are not confident about this property, they will perform as shown where they "put the larger number on top":

$$
\begin{array}{r}
375 \times 5000 = 5000 \\
\times 375 \\
\hline
25000 \\
350000 \\
1500000 \\
\hline
1875000
\end{array}
$$

rather than:
$$
\begin{array}{r}
375 \\
\times \quad 5000 \\
\hline
1875000
\end{array}
\quad \text{or} \quad
\begin{array}{r}
375 \\
\times 5000 \\
\hline
1875000
\end{array}
$$

The following is an example of how to show the algorithm for a two-digit number multiplied by a two-digit number using squared paper or base 10 blocks. The algorithm is known as the copy method because a "copy" of the total problem is shown in an area (geometric) approach.

Activity

MANIPULATIVES

Copy Method

The copy method can be used with graph paper and with the base 10 blocks. The same problem will be shown using both manipulatives.

The problem is $16 \times 24 =$ ____?____

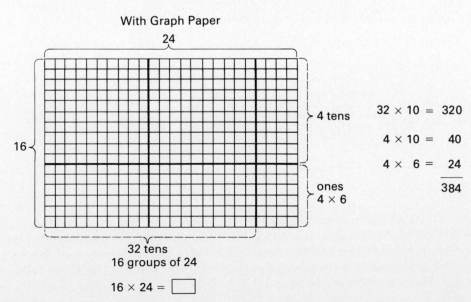

With Graph Paper

24

16

4 tens

ones
4×6

32 tens
16 groups of 24

$$16 \times 24 = \boxed{}$$

$$32 \times 10 = 320$$
$$4 \times 10 = 40$$
$$4 \times 6 = 24$$
$$\overline{384}$$

Explanation of How It Works with the Squared Paper

1. Determine how many ones. (24)
   Regroup as 2 tens and 4 ones.

2. Determine how many tens.
   4 + 12 + 2 regrouped tens as 18 tens
   Regroup as 1 hundred and 8 tens

3. Determine how many hundreds.
   2 + 1 regrouped = 3 hundreds.
   Total = 3 hundreds + 8 tens + 4 ones = 384

With Base 10 Blocks

Learning Mat looks like this at the beginning of each problem.

All the base 10 blocks in the answer must fit in here.

Learning Mat with full representation of all the base 10 blocks filled in.

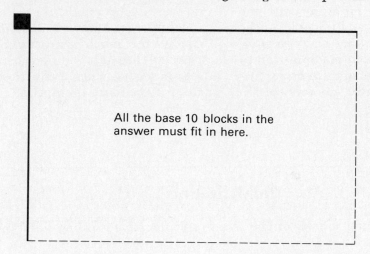

With symbols . . .

2 × 100 = 200
16 × 10 = 160
24 × 1 = 24
384

Explanation of How It Works with Base 10 Blocks

1. Place the 24 on top of the dark line with two sets of ten and 4 individual ones. (The LONGS are not scored to show each individual unit but the numeral, 10, is shown to remind the reader that this represents ten, sometimes called a LONG in math texts).

(continued)

2. Place the 16 on the left side of the dark line, using one set of ten and six ones. Let the darkened square at the upper left be a reminder that the blocks of one factor cannot overlap the blocks of the other factor. Both must stop when they touch the darkened square.

3. Fill in the rectangle to the right of the 16 and below the 24. All the base 10 blocks must fit in that configuration to represent the 16 groups of 24. Fill in the answer, starting with the largest representations first (FLATS), progressing to tens (LONGS) and then the ones (UNITS) at the bottom right-hand corner of the rectangle.

Once children become proficient with the solution of these problems and understand the process for regrouping, they are ready to work with larger numbers in which manipulation of materials becomes too bothersome to be productive. Estimation of reasonable answers continues to be important and the calculator should be used for larger numbers. Remember to evaluate whether too much instructional time is being spent on becoming proficient in multiplying large numbers, when estimation and the calculator may be more appropriate.

Activity

ESTIMATION

CALCULATORS

Multiplication

The following activity can be used in a cooperative learning situation:

Estimation — Multiplication
(a game for two or more players)

How to play:

1) Choose a number as your target. Circle it.

93	556	5444
	491	649
1010	237	
212	303	57

2) Choose another number and enter it in your calculator, then enter operation.

3) Quick! In five seconds or less, enter a number that you think will produce the target number when you push the equal sign.

4) Each player tries reaching the same target number. The one who gets the closest wins one point. The first player to win 10 points wins the game.

Example:

23 × _____ = (93)

33 × 3 = 99 ◄——— almost!

23 × 4 = 92 ◄——— closest

32 × 2 = 64

Display:

Enter 23					2	3
Enter X					2	3
Estimate						4
=					9	2

Teaching the Division Algorithm

The division algorithm requires many prerequisite skills that create difficulties for children because of the numerous opportunities for error. The algorithm requires a knowledge of subtraction and multiplication algorithms, estimation skills, and an understanding of place value. Many teachers regard division of multi-digit numbers as the hardest concept to teach in elementary school mathematics. A disproportionate amount of time is spent mastering this algorithm, when we consider how often adults reach for a calculator to work long division problems. Why do teachers, textbooks, competency, and standardized tests continue to emphasize long division? We cannot afford to spend so much instructional time on a skill that may not be necessary in the twenty-first century. This does not mean we recommend that teachers stop teaching the division algorithm. Instead, the focus of instructional time should be on understanding the division algorithm using one-digit divisors with three-digit or four-digit dividends. Any work with larger numbers should involve calculators and estimation. The NCTM curriculum standards (Commission on Standards for School Mathematics, 1987) state that educators must foster a solid understanding of simple calculations and not waste valuable class time teaching tedious calculations using paper-and-pencil algorithms.

In developing the algorithm, instruction should include concrete models such as base 10 blocks, connecting cubes, bean sticks, pocket charts, and/or counters and cups. Any variety of material that is available can be used. In choosing the manipulative materials to model the algorithm, remember that the actions on the materials, the language used to describe the actions, and the corresponding steps of the algorithm must be in agreement. This is especially important in division where two different interpretations or approaches to the algorithm are possible. These result from two different situations—measurement division, which is related to successive subtraction, and partition division, which is related to equal distribution (see Chapter 7 for additional explanations about the differences). Weiland (1985) comments that the distributive algorithm is well matched to what children do on their own. Interviews with children indicate children have difficulties in reconciling measurement and partitive division. The suggestion is to give children opportunities to create their own arguments and explanations about the division algorithm so that instruction can be matched to the way they think.

Before the division algorithm requiring regrouping is introduced, students are exposed to division with remainders and to division of two-digit numbers where no regrouping is required. Division is the only operation where we work from left to right, dividing up the largest places first. Dividing by tens and multiples or powers of 10 is a skill needed for estimation as well as a way to see the effect on the size of the quotient. It is advisable to relate multiplying and dividing by multiples or powers of 10 to note relationships and patterns.

$$4 \times 6 = 24 \qquad 4 \times 60 = 240$$
$$24 \div 6 = 4 \qquad 240 \div 60 = 4$$
$$40 \times 600 = 2400 \qquad 40 \times 60 = 2400$$
$$2400 \div 600 = 40 \qquad 2400 \div 60 = 40$$

Compare these two interpretations for division. Notice that the algorithm is the same but the physical manipulation and the mathematical language to explain the steps are different.

The Problem: $72 \div 6$

Partitive Division	Measurement Division
Situational Problem: 72 people. Need 6 equal teams. How many on a team?	*Situational Problem:* 72 people. Want 6 people on a team. How many teams?
1. Distribute the 7 tens evenly to the 6 teams.	1. How many groups of 6 can you make from 7 tens? One.
2. Record 1 in tens column. One ten is left.	2. Record 1 in tens column. One ten is left.
3. Break the ten into 10 ones. Add to 2; this is 12 ones.	3. Break the ten into 10 ones. Add to 2; this is 12 ones.
4. Give 2 ones to each team. Record 2 in ones column.	4. How many groups of 6 can you make from 12 ones? 2
5. So each team will have 12 players and no players are left.	5. So 12 teams of 6 players.

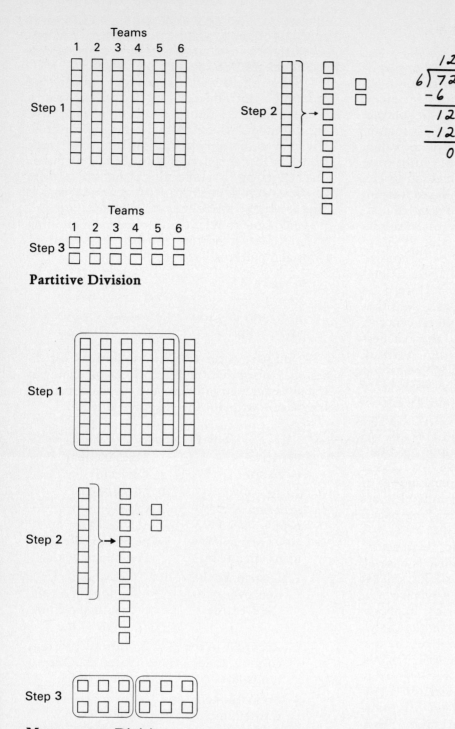

Partitive Division

Measurement Division

The procedure for problems with larger numbers is basically the same. It is important to select appropriate materials, some proportional and some non-proportional, that correctly model the action children need to perform with them. After a period of time working with the manipulatives, show children how to record the action symbolically to provide a connecting level between the concrete and symbolic levels. Once a problem-solving situation is involved, we must determine if the problem calls for partition or measurement division before the action is modeled with manipulatives.

The Problem: 144 ÷ 6

1. How many groups of 6 can you make from hundreds? You can't get any. (See below.)
2. Exchange 1 hundred for 10 tens.
3. Add 4 tens; this makes 14 tens.
4. How many groups of 6 can you make from 14 tens? 2 with 2 tens left. Record.
5. Exchange 2 tens for ones—this makes 20 ones, plus 4 = 24 ones.
6. How many groups of 6 can you make from 24 ones? 4. Record. So there are 24 groups of 6 in 144.

Whether partition or measurement interpretation is easier for students to use in learning the division algorithm has not been determined.

There is no doubt that the language of the algorithm should match the meaning of the model.

A frequent error in division is the failure to record a zero as a place holder in the quotient. If a sequence of steps is followed such as, "divide–multiply–subtract–bring down," it may offer a needed structure for the child's thinking. However, the most successful way is to show the operation concretely (Fig. 8.11). When the number is smaller than the divisor and another number is to be "brought down," this division step must be recorded with a zero in the quotient. Students can use play money in denominations of thousands, hundreds, tens, and ones to add interest and variety to long division problems. Some other ideas are presented in Cheek and Olson (1986).

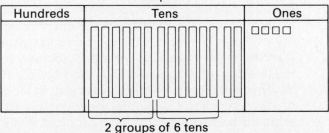

Symbolic

$$6\overline{)144} \atop \begin{array}{r} 24 \\ -12 \\ \hline 24 \\ -24 \\ \hline 0 \end{array}$$

Symbolic

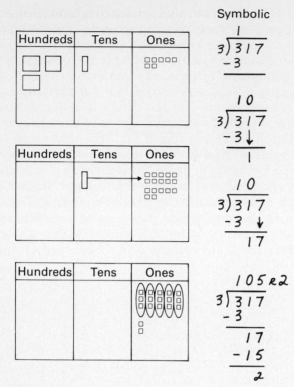

Figure 8.11

The Problem: 317 ÷ 3

1. How many groups of 3 can be made from the hundreds? (1) Record answer.
2. How many groups of 3 can be made from the tens? (0) Record.
3. Exchange 1 ten for 10 ones and add to 7 ones = 17 ones.
4. How many groups of 3 can be made from the ones? (5) Record. Two ones are left as the remainder.

Estimating the range of a quotient will help students with the problem of using zeros appropriately in the quotient. A quick check of an estimated quotient can be made by rounding the divisor and checking the product obtained with the dividend. If there is a large discrepancy, the student can try annexing a zero to the quotient and doing the procedure again.

	estimate quotient	check and compare	try again
21)422	2	20	20
	20)400	× 2	×20
		40 not 400	400
			matches dividend better

Remind students that any time a digit is brought down in the division algorithm, the "divide–multiply–subtract–bring down" cycle starts again. The key to avoiding the omission of the zero in the quotient is to follow this procedure:

- Estimate the number of digits in the quotient.
- Mark those places in the quotient to remember how many digits are needed.
- Keep dividing until there are no more digits to bring down.
- Compare the remainder to be sure it is less than the divisor.
- Check the quotient by multiplying the quotient by the divisor (add remainder if needed).

The Problem: 32)6543

1. Estimate number of digits in quotient.

$$100 \times 32 = 3200$$
$$1000 \times 32 = 32000$$

Need 6543 so quotient has three digits.
2. Mark three places in quotient.

- - -
32)6543
3. Begin "divide–multiply–subtract–bring down" cycle.

An area that causes difficulties for children is understanding how to interpret the remainders when word problems are involved. Part of the problem is that they need to reflect on the question in the word problem and label the quotient appropriately. This does not happen easily. Children work the problem and consider themselves finished with it. This is when cooperative learning experiences are valuable because students are given opportunities to verbalize the problems and to internalize meaning. Evidence of this difficulty is shown in the following situation with fourth graders. The problem was for them to decide how many six-packs of soda should be purchased for 45 people if each person was to have one can of soda. The division posed no difficulty with an answer of 7 remainder 3. The question (how many six-packs) was answered by the students: "We need 7 of them and will have 3 left over." Three what left over—cans? people? They claimed to have 3 cans of soda left over. They did not understand the need to increase the quotient. (See Fig. 8.12).

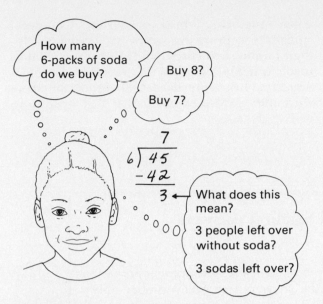

Division takes a long time to master and to understand. Time spent with manipulative materials is greatly beneficial. It is more important to spend instructional time on understanding problem-solving situations and interpreting remainders rather than mastering long multi-digit division.

Figure 8.12

Activity

CALCULATORS

Division Patterns with the Calculator

Pathways
Find the path to the END answer by dividing. Use your calculator as needed.
No diagonal lines allowed.

Example: START

1000	2	1	10
2	4	5	5
2	5	5	5
10	2	2	1
			1

END

1) START

800	4	5	1
2	2	2	2
4	6	3	2
5	2	4	5
			1

END

2) START

19683	6	9	27
5	4	3	3
18	3	5	6
6	5	3	0.9
			9

END

Combining the Operations

The following activities can be used when studying one operation alone or for review of the operations after all of them have been presented. Understanding the base 10 system can be enhanced as students apply what they have learned when exploring these activities. Computer spreadsheet programs work well with multi-digit numbers. A change in the operation can be made quickly and can be reported instantaneously on the spreadsheet. Students can create their own problems using the four operations and a calculator.

Activity

CALCULATORS

Find the Largest and Smallest Answers

Directions:

1. Using the digits 3, 4, 5, 6, 7, 8, and 9, create the greatest number possible and the least number possible in the answer. Use each number only once. Follow the forms below; fill in the spaces where the question marks are located. Use the calculator for computations.

$$
\begin{array}{r} ???? \\ +??? \\ \hline \end{array}
\qquad
\begin{array}{r} ?? \\ ?? \\ +??? \\ \hline \end{array}
\qquad
\begin{array}{r} ???? \\ -??? \\ \hline \end{array}
$$

$$
\begin{array}{r} ???? \\ \times??? \\ \hline \end{array}
\qquad
??\overline{)?????}
$$

2. Change the configurations and see what happens. Develop three more for each operation.

3. Use 0, 1, 2, with other combinations of digits. How do the problems differ? Which one is the easiest to recognize as the smallest? (This is a question that is challenging to a first or second grader, but is quickly answered by an upper grade student.)

Activity

COMPUTERS

Problem Solving with the Spreadsheet

Directions:

1. Construct a spreadsheet using the following values in the respective cells.

	A	B	C	D
1	Number	3	30	300
2	49	+A2+B1	+A2+C1	+A2+D1
3	490	+A3+B1	+A3+C1	+A3+D1
4	4900	?	?	?
5	49000	?	?	?

2. Fill in the rest of the cells by yourself. Move the arrow → and see the values that appear on the spreadsheet.

3. Remember that a "+" typed before the beginning letter designates that the next variables are to be calculated.

4. Predict the values that should appear in column E if E = 3000. Write down your predictions. Calculate to find out the appropriate value formulas.

5. Choose different numbers but keep the magnitude (B, C, etc.) the same. What patterns do you see?

6. Keep the same numbers but change the magnitude. Predict what will happen in each case.

NOW CHANGE THE VALUE CELLS TO FORMULAS FOR . . .
Multiplication   (remember to use * for multiplication)
Subtraction
Division   (remember to use / for division)

TRY THE ALTERNATE APPROACH . . .

1. The following values were derived in a spreadsheet program like the one above. What values would the numbers and the magnitudes have?

	A	B	C	D
1	Number	?	?	?
2	?	43	88	538
3	?	385	430	880
4	?	3805	3850	4300
5	?	38005	38050	38500

Activity

CALCULATORS

Creating Your Own Problems

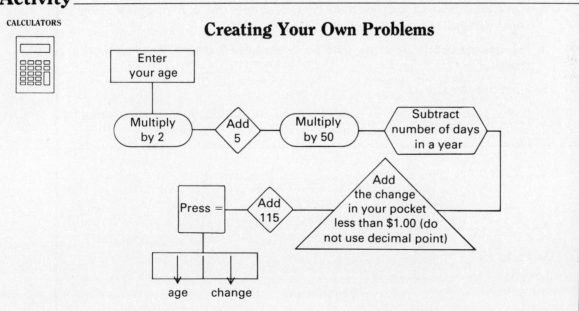

Teaching Integers

Many textbooks introduce integers with a number line to help students see the relationship of positive and negative numbers to each other.

Moving to the right of zero results in positive numbers and moving to the left of zero results in negative numbers. Some children's card games have a situation where the player might lose enough points to be "in the hole." This is an opportune time to discuss negative numbers. Another situation that involves the use and interpretation of negative numbers is with temperatures that are below zero. It is important to encourage the correct way to read and write these numbers when they appear.

Teaching students the four basic operations with integers is a difficult task. Many students learn to work problems with integers by memorizing rules such as "negative times a negative is a positive." The rule may not make sense, but it produces correct answers. Physical models are seldom used as teaching devices, except perhaps a number line. This model has limitations and only partial explanations are possible.

When students are introduced to addition and subtraction of integers, one simple, effective approach is to use checks (money received) and bills (money spent). Davis (1967) introduced young children to integers through the "Post-man Stories" from the Madison Project. In this setting, checks represent positive numbers and bills represent negative numbers. The action of "bringing" refers to addition and the action of "taking away" refers to subtraction. When the postman brings (+) a check, this indicates adding a positive number. When the postman takes away (−) a check, the action is subtracting a positive number. In the case of bringing (+) a bill, the action would be adding a negative number. If the postman takes away a bill, the effect would be to subtract a negative number. The concept emphasizes that when a check is received, we become richer but when a bill is received, we become poorer. Likewise, if a check must be given up (subtracted), we become poorer. Read these examples and try to visualize the possible events that could occur in The Post-man stories.

Activity

CALCULATORS

MATH
ACROSS THE
CURRICULUM

Economics

Directions:

1. Have students act out each story and/or tell the end result of each story—Would they be richer or poorer?

2. Have them check the result with a calculator to see how both can work together.

3. Here are the stories:

   If the postman brings(+) us a check(+) for $3 and brings(+) another check(+) for $5, we would have a total of $8 or in symbols, $^{+}3 + {}^{+}5 = {}^{+}8$.

= Bank account

Balance Record	
Bills (we owe) −	Checks (we receive) +

If the postman brings(+) us a check(+) for $6 and brings(+) a bill(−) for $2, we would have a total of $4 or in symbols: $^+6 + {}^-2 = {}^+4$.

If the postman brings(+) a bill(−) for $5 and brings(+) another bill(−) for $2, we would have a total of −$7 or in symbols: $^-5 + {}^-2 = {}^-7$.

If the postman brings(+) a check(+) for $3 and brings(+) a bill(−) for $4, we would need $1 to pay the bill, or in symbols: $^+3 + {}^-4 = {}^-1$.

If the postman brings(+) a check(+) for $9 and takes away(−) a check(+) for $2, we would be poorer in the end since the postman *takes away* a check. With symbols: $^+9 - {}^+2 = {}^+7$.

4. *For older students (middle school age):*
Create similar stories using the bank account model and buy stereos, hit records, cars, or items that young teenagers are looking forward to owning in the near future (if they don't have them already). The stories become relevant and examples using checks and bills are realistic to the modern-day desire for all the luxuries.

Students need to understand that the effect of the postman taking away a check is less money. Some teachers might want to relate this event to turning a paycheck over immediately to someone to whom you owe money so you are not richer as the end result. To help clarify this situation, Davis suggests including a time factor in the imaginary story about the postman. The use of time prevents misconceptions about the mathematical equation that is appropriate for each situation.

If the postman takes away(−) a check(+) for $5 and takes away(−) another check(+) for $2, the result would be $-{}^+5 - {}^+2 = {}^-7$. As a result of the postman's visit, we have less money because two checks were taken away from us.

If the postman takes away(−) a bill(−) for $5 and takes away(−) a bill(−) for $2, we would be richer in the end since bills were taken away. In symbols: $-{}^-5 - {}^-2 = {}^+7$. Again the time factor should be emphasized. We had to give the postman bills which makes us richer in the long run since we had counted on paying those bills and now we do not have to pay them. One method to help children visualize these events is to use play money. When a bill comes for $7, we set money aside to pay that amount and put the bill with it to remember it. If the postman makes an error (and errors could occur in the same amounts as bills), when we give the postman the bill, we can see the effect of making us richer.

The real challenge comes when trying to teach multiplication and division of integers with understanding. For multiplication, inter-

pretation of the factors is a critical feature of The Postman stories. The *second* factor is the money (bill or check) and the *first* factor tells how many times the postman brings the item or takes it away. Study these examples:

$^+2 \times {}^+3 = {}^+6$ — The postman brings two checks for $3 each.

$^+2 \times {}^-5 = {}^-10$ — The postman brings two bills for $5 each.

$^-3 \times {}^+4 = {}^-12$ — The postman takes away three checks for $4 each. Remember—this means we are poorer when we have checks taken away.

$^-2 \times {}^-6 = {}^+12$ — The postman takes away two bills for $6 each. This is good. We are richer when bills are taken away.

Many educators suggest using physical models for teaching and representing integers. Chang (1985) presents several models to teach addition and subtraction of integers: number line, play money (a technique that is similar to The Postman stories) and color-block technique that uses negative and positive blocks. The article outlines the teaching strategies for all three approaches in a clear, concise manner. The positive-negative charge model has been suggested by several authors. Battista (1983) extends the model to the four operations and illustrates how the model can show properties of the system of integers. Some of the more difficult concepts are covered in the following activity.

Activity

MANIPULATIVES

PROBLEM SOLVING

Get a Charge Out of Integers

Materials: Optional—use colored blocks or buttons to represent the two charges (negative and positive)
Large mouth jar (such as peanut butter)

Procedure:

Subtraction $^+5 - {}^+2 = {}^+3$

Start with the jar containing 5 positive charges.
Take out 2 charges, which leaves 3 positive charges.

Multiplication $^+3 \times {}^-4 = {}^-12$

The sign of the first factor indicates whether the multiplication is repeated addition or repeated subtraction. The second factor tells what kind of charges are to be put in (repeated addition) or taken out (repeated subtraction) of the jar. In the problem, we add 4 negative charges 3 times.

$$^+3 \times {}^-4 = {}^+12$$

We must remove 4 negative charges 3 times from the jar which would have the effect of making the jar more positive.

Division $^-24 \div {}^-6 = {}^+4$

Count how many times $^-6$ must be added to the jar to get $^-24$. The quotient is positive because we must repeatedly add the divisor to the jar to get the dividend.

Students in the middle grades need concrete materials and games to reinforce concepts as much as children in lower grades. The middle grade teacher may be more reluctant to use this technique, but it has been proven effective for all grades. The game "Add-Lo" (Bledsoe and Bledsoe, 1986) provides practice in adding integers, creating expressions using sums of integers, and taking the absolute value of an integer. Teaching operations with integers is an opportune time to challenge students to apply their prior experiences in new situations.

Diagnosis and Remediation

Although there are a variety of errors associated with whole number algorithms, a great many of them have to do with working with zeros and applying place value principles to the algorithm. Some errors are caused from difficulties with computation, such as not knowing the basic facts. Specific errors for each operation will be covered, along with some remediation techniques when applicable. Generally, it can be said that most remediation involves additional experiences with concrete materials when reteaching the steps of the algorithm.

Unfortunately, most elementary mathematics methods classes do not have adequate time to cover procedures in diagnostic and prescriptive teaching. This means that most teachers have had little, if any, preparation in this important dimension of teaching. How does a teacher diagnose learning difficulties related to mathematics? Some tests are available for diagnosing the child's achievement level or cognitive abilities, but these are generally paper-and-pencil tests which measure the symbolic level of understanding. Diagnosing at the concrete level is more time consuming and difficult as it requires more interpretative skill by the teacher. Oral interviews and teacher probing are valuable assessment techniques. Teachers should observe students working individually and collectively as they communicate about mathematics. In this way, the teacher can discover the strategies used by the child to arrive at answers along with any misconceptions the child has about the algorithm and the structure of the number system. The NCTM curriculum and evaluation standards (Commission on Standards for School Mathematics, 1987) state that assessment must include dialogue between the teacher and students in order to assess thinking.

Once the knowledge level of the learner has been determined, appropriate materials and instructional methods must be selected. Unfortunately, far too many teachers review the concept at the abstract or pictorial level, then give additional problems for the child to work. The overreliance of teachers on drill sheets as a means of correcting the deficiency is a major concern of mathematics educators interested in the diagnostic/remediation process.

The teacher should plan teaching strategies to use as many senses as possible in developing mathematical concepts. Approximately 80 percent of our sensory intake is visual, 11 percent is auditory, and 2 percent is tactile (Capps and Hatfield, 1979). Manipulative materials may provide the visual mode, but if the child is allowed to work with them, tactile as well as visual skills can be employed. The steps of the algorithm should be verbalized by both the teacher and the child as they exchange roles while explaining the procedure. Whenever possible, instruction should be based on concrete experiences that employ as many sensory modes as possible. This should be the situation for initial instruction, but it is of paramount importance during remediation.

Another important aspect of diagnostic/prescriptive teaching is selecting appropriate materials. There are many manipulative devices the teacher may select including both proportional and non-proportional models. Which device is most appropriate depends upon several aspects of the learner: maturity level, previous experiences, materials used for initial instruction, visual or perceptual problems, fine motor skills, learner's intact skills, distractability problems, and any other special needs of the child. Preparing an instructional hierarchy that takes these issues into account is a skill that takes years to acquire. Determining the proper instructional sequence is a most difficult task. For these reasons, the teacher must carefully read each section on diagnosis and remediation to become an effective teacher.

Correcting Common Errors

Many errors with addition and subtraction result from difficulties with regrouping. Children may not know when to regroup and will regroup when it is not necessary or will fail to regroup. As mentioned earlier, the authors advise that problems with regrouping should be introduced at the same time as problems without regrouping.

Beginners often benefit from a learning mat (Fig. 8.13) that tells them when and where to regroup. The mat can be a file folder, a shoe box, etc. The important thing is that the child sees 9 holes cut so just 9 ones can fit into the slots. The tenth one is covered in black to remind the stu-

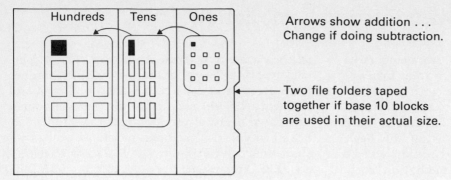

Figure 8.13

Arrows show addition . . .
Change if doing subtraction.

Two file folders taped
together if base 10 blocks
are used in their actual size.

dents that when all the holes are filled and this is the only one left, it is a sign that it is time to regroup. The same thing is done for tens and hundreds. If the learning mat is plasticized, the arrows can be drawn in crayon so that they can be changed for addition or subtraction. See Figure 8.13 for the construction of the mats with base 10 blocks or toothpicks.

Common errors for the algorithms of the four basic operations are presented here. Study the examples and decide what error the child is making. Then work the problems given using the child's error pattern. Compare answers with those given below each example. Decide what remediation procedures you would use in each situation. In all cases, remediation involves additional experiences with place-value models so the algorithm is understood rather than just a series of memorized routines. The procedures in this chapter should be followed as a guide for remediation and will not be discussed in detail in this section. Some additional teaching strategies will be mentioned when appropriate.

Addition

Consider these problems:

```
  37      297
+46     +658
───     ────
 713    81415
```

Try these using the error:

```
  48      306
+29     +495
```

(617)   (7911)

Error pattern: Adds by column without regrouping.

Remediation techniques: Estimate the answer and compare difference between the estimated and obtained sums. Use a closed abacus where regrouping is forced to occur because only 9 chips or beads will fit in each column. Use

graph paper with the rule that only one digit can be written in each section to focus attention on the need to regroup. Review addition algorithm.

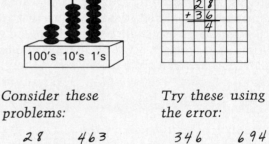

Consider these problems:

```
  28      463
+37     +358
───     ────
 515    7112
```

Try these using the error:

```
  346     694
+ 39     +137
```

(3715)   (7212)

Error pattern: Adds from left to right.

Remediation techniques: Use a form that is placed over the problem to reveal only the ones column, then the tens, and continues in a right-to-left sequence. Also you could write "A" over the ones column, "B" over the tens column, and so on to show the sequence. Color coding works well with some handicapped children who are visually oriented.

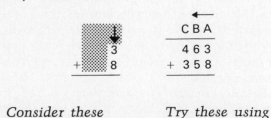

Consider these problems:

```
  27      377
+35     + 94
───     ────
 52      361
```

Try these using the error:

```
  85      219
+67     +576
```

(142)   (785)

Error pattern: Fails to add the regrouped digit.

Remediation techniques: Estimating the sum will work in some cases. May need to provide a "box" at the top of the columns to the left of the ones with a large plus sign to help remind the student of this step.

$$\boxplus\ \ 27$$
$$+\ 35$$

Subtraction

Consider these problems:

```
 300      56
-198     -29
-----    ----
 298      33
```

Try these using the error:

```
 512      309
-258     - 98
-----    -----
(346)    (391)
```

Error pattern: Takes the smaller number from the larger without regard to position.

Remediation techniques: Estimate the answer. Have the student check using addition. In many cases, the answer is larger than the original number and makes no sense. Concrete materials must be used with a take-away model (rather than the comparison model) to show that the first number, or minuend, is modeled with the blocks or chips and not the second number, or subtrahend.

Consider these problems:

```
 315      746
-138     -159
-----    -----
 172      551
```

Try these using the error:

```
 452      315
-189     - 96
-----    -----
(221)    (214)
```

Error pattern: Regroups all columns as ten and fails to add the previous digit.

Remediation techniques: Use concrete proportional models to review the subtraction algorithm.

Consider these problems:

```
 2'0'6     7'1'5
-  38      -288
-----     -----
   78       437
```

Try these using the error:

```
 465      285
- 97     -189
-----    -----
(378)    (96)
```

Error pattern: Regroups in the column farthest to the left and adds ten to each column rather than working column by column starting with the ones.

Remediation techniques: Concrete materials to reteach the steps of the algorithm. Could also use a form that allows only the column on the left to be seen at one time (like in an addition example mentioned earlier).

Consider these problems:

```
 3'6       5'2
-14       -36
-----     -----
112        16
```

Try these using the error:

```
 27        98
-19       -26
-----     -----
 (8)      (62)
```

Error pattern: Regroups all columns when regrouping is unnecessary.

Remediation techniques: This error occurs in second and third grades when the child is learning the algorithm and problems with mixed operations are included. The child focuses on the regrouping process and perseveres. Give mixed problems where the child is asked only if regrouping is needed. Also estimation will help alert the child to a problem in the answer.

Do you trade?   Circle yes or no.

```
26 Yes    39 Yes    41 Yes
+12 No   +24 No    +39 No

52 Yes    90 Yes
+27 No   +26 No
```

Multiplication

Consider these problems:

```
  37       241
x  5      x 34
-----     -----
1535      8164
         61230
         ------
         69,394
```

Try these using the error:

```
  23        186
x 47       x 42
-----      ------
(9541)   (453,852)
```

Error pattern: Fails to regroup.

Remediation techniques: Estimation will help draw attention to the large difference in products. Use graph paper with the rule that only one digit per column can be recorded. Review the multiplication algorithm with concrete materials.

Consider these problems:

$$\begin{array}{r} \overset{6\,8}{} \\ \times\ 5 \\ \hline 500 \end{array}\qquad \begin{array}{r} 68 \\ \times\,26 \\ \hline 608 \\ 1460 \\ \hline 2068 \end{array}$$

Try these using the error:

$$\begin{array}{r} \overset{2\,9}{} \\ \times\ 3 \end{array}\qquad \begin{array}{r} 52 \\ \times\,87 \end{array}$$

$$(127)\qquad(5284)$$

Error pattern: Adds carried digit before multiplying the next column.

Remediation techniques: This error could be a carry-over of the drill with regrouping in the addition algorithm. Estimation may help clue the student about the difference in the two products—estimated and obtained. Review of the algorithm with concrete materials may reinforce sequence of steps.

Consider these problems:

$$\begin{array}{r} 83 \\ \times\,27 \\ \hline 581 \\ 166 \\ \hline 747 \end{array}\qquad \begin{array}{r} 423 \\ \times\,254 \\ \hline 1692 \\ 2115 \\ 846 \\ \hline 4653 \end{array}$$

Try these using the error:

$$\begin{array}{r} 73 \\ \times\,24 \end{array}\qquad \begin{array}{r} 302 \\ \times\,58 \end{array}$$

$$(438)\qquad(3926)$$

Error pattern: Fails to annex a zero as a place holder when multiplying tens and hundreds.

Remediation techniques: Worksheets with these zeros already recorded may help. Another strategy is to use a form that exposes only the column multiplied and column labels. Outlining the correct placement of steps with colored lines on graph paper helps draw attention to the se-

quence of the algorithm. Using concrete models is encouraged along with estimation.

Consider these problems:

$$\begin{array}{r} 56 \\ \times\,38 \\ \hline 198 \end{array}\qquad \begin{array}{r} 584 \\ \times\,73 \\ \hline 4072 \end{array}$$

Try these using the error:

$$\begin{array}{r} 75 \\ \times\,29 \end{array}\qquad \begin{array}{r} 185 \\ \times\,37 \end{array}$$

$$(185)\qquad(575)$$

Error pattern: Multiplies column by column like the addition and subtraction algorithm.

Remediation techniques: Use a form to help show the sequence and use graph paper outlined with the steps. Graph paper can be color coded according to the multiplier to give visual clues. Also the distributive law will indicate the partial products and help in understanding how to complete the steps. Estimation will clue the error.

Consider these problems:

$$\begin{array}{r} 265 \\ \times\,238 \\ \hline 2120 \\ 7950 \\ 5300 \\ \hline 63{,}170 \end{array}\qquad \begin{array}{r} 468 \\ \times\,83 \\ \hline 1404 \\ 17440 \\ \hline 31{,}844 \end{array}$$

Try these using the error:

$$\begin{array}{r} 327 \\ \times\,286 \end{array}\qquad \begin{array}{r} 705 \\ \times\,356 \end{array}$$

$$(886{,}622)\ (60{,}630)$$

Error pattern: Does not keep columns straight and errors occur in the adding.

Remediation techniques: Use graph paper to help keep columns in proper alignment. Can also turn notebook paper sideways which will

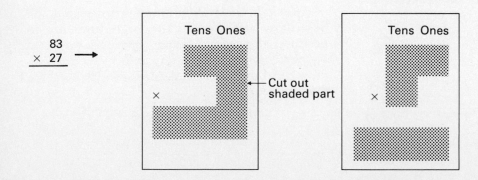

cause the blue horizontal lines to become vertical lines to help with alignment.

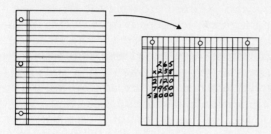

Division

Consider these problems:

$$9 \overline{)4560} \quad 56\,R\,6$$
$$-45$$
$$\overline{60}$$
$$-54$$
$$\overline{6}$$

$$6 \overline{)4818} \quad 830$$
$$-48$$
$$\overline{18}$$
$$-18$$
$$\overline{0}$$

Try these using the error:

$$8 \overline{)5840} \qquad 3 \overline{)752}$$

$$(73) \qquad (25\,R\,2)$$

Error pattern: Does not record zero as a place holder in the quotient when the division step cannot be done (the number is too small to be divided).

Remediation techniques: Estimation will offer the clue that there is an error. Concrete models are the best device to use as the child will see from where the zero comes. Using a flow chart of the sequence of steps may also help—"divide, multiply, subtract (and compare), bring down." Also, estimating the size or range of the quotient, then marking that in the quotient with lines (two for a two-digit quotient or three for a three-digit quotient), will help draw attention to the error when it occurs. However, some children will misuse this device and will simply record a zero in the ones place to satisfy the needs of the size of the quotient.

$$4 \overline{)352} \quad \text{Think} \to 4 \overline{)320}^{\;80 \leftarrow 2\text{ digits}} \quad 4 \overline{)352}^{\;\underline{\quad}\underline{\quad}}$$

Consider these problems:

$$6 \overline{)516} \quad 68$$
$$-48$$
$$\overline{36}$$
$$-36$$
$$\overline{0}$$

$$3 \overline{)273} \quad 19$$
$$-27$$
$$\overline{3}$$
$$-3$$
$$\overline{0}$$

Try these using the error:

$$8 \overline{)584} \qquad 4 \overline{)568}$$

$$(37) \qquad (241)$$

Error pattern: Records the problem (or perhaps even works the problem) from right to left as in the other three operations.

Remediation techniques: Use play money to model the division process and visualize the algorithm. Could also use a cover form to reveal numbers in the dividend beginning with the left. Stress where the quotient will go. Estimate the quotient to determine if it is reasonable.

Consider these problems:

$$4 \overline{)387} \quad 95\,R\,7$$
$$-36$$
$$\overline{27}$$
$$-20$$
$$\overline{7}$$

$$9 \overline{)588} \quad 64\,R\,12$$
$$-54$$
$$\overline{48}$$
$$-36$$
$$\overline{12}$$

Try these using the error:

$$6 \overline{)579} \qquad 8 \overline{)369}$$

$$(95\,R\,9) \qquad (45\,R\,9)$$

Error pattern: Remainder is greater than the divisor. The division was stopped too soon and is incomplete.

Remediation techniques: Stress estimation and use play money to focus on the size of the remainder.

Consider these problems:

$$31 \overline{)4696} \quad 15\,R\,3$$
$$-31$$
$$\overline{158}$$
$$-155$$
$$\overline{3}$$

$$21 \overline{)38874} \quad 184\,R\,10$$
$$-21$$
$$\overline{177}$$
$$-168$$
$$\overline{94}$$
$$-84$$
$$\overline{10}$$

Try these using the error:

$$16\overline{)4650}$$
$$-32$$
$$\overline{145}$$
$$-144$$
$$\overline{1}$$
$$(29r1)$$

$$8\overline{)5892}$$
$$-56$$
$$\overline{33}$$
$$-32$$
$$\overline{12}$$
$$-8$$
$$\overline{4}$$
$$(741r4)$$

Error pattern: Lost track of which numbers were left to "bring down."

Remediation techniques: This error happens because digits are not kept in proper alignment. Graph paper or notebook paper turned sideways offers a help for alignment.

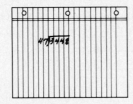

Consider these problems:

$$862$$
$$47\overline{)3448}$$
$$-32$$
$$\overline{24}$$
$$-24$$
$$\overline{8}$$
$$-8$$
$$\overline{0}$$

$$3236r1$$
$$24\overline{)6473}$$
$$-6$$
$$\overline{4}$$
$$-4$$
$$\overline{7}$$
$$-6$$
$$\overline{13}$$
$$-12$$
$$\overline{1}$$

Try these using the error:

$$37\overline{)9846} \qquad 43\overline{)2867}$$

$$(3282) \qquad (716r3)$$

Error pattern: Uses only the first digit of the divisor to divide.

Remediation techniques: Estimate the quotient. Use graph paper to record the answer with the size of the quotient indicated. Also, checking the answer may give a clue to the error.

Simultaneous Thought Processing

Students who process information from the whole to the part often benefit from the learning mats where they can see all the parameters from the start of the problem. These students are often better at performing subtraction and division algorithms than they are at working with the multiplication and addition algorithms. Perhaps it is because the nature of the task is one that requires the whole to be divided or reduced into its parts. The copy method of multiplication and division are also helpful because the rectangular area is easily seen. The copy method of division is shown here.

The problem $21\overline{)672}$

	32				
	10	10	10	1	1
10	100	100	100	10	10
10	100	100	100	10	10
1	10	10	10	1	1

(21 at left)

Explanation of How It Works with Base 10 Blocks:

1. The divisor, 21, is placed down the left side.
2. The dividend, 672, is represented by base 10 blocks going out to the right of the divisor to form a rectangle, showing how 672 is divided into 21 sets.
3. The quotient, 32, is represented on the top of the dark line (shown here by the dotted base 10 blocks of 3 sets of ten and 2 ones. The quotient must fit exactly over the top line of the dividend. It shows how many things are in each group of the 21 groups.
4. The largest number of the base 10 blocks (in this case 6 sets of hundreds) is represented first as the dividend is modeled from left to right at the right of the dark line, then comes tens and ones. The representation is just like the written numeral representation:

$$672 = 600 \ (6 \times 100) \qquad \text{six sets of } 100$$
$$70 \ (7 \times 10) \qquad \text{seven sets of } 10$$
$$2 \ (2 \times 1) \qquad \text{two sets of } 1$$

5. In some examples, the dividend may need to be regrouped to smaller powers of ten so that its configuration will fit exactly to the right of the divisor with no parts "hanging down" past the end of the divisor (as shown by the arrow in the last example).

Successive Thought Processing

Students who process information from details to the whole often benefit from the traditional algorithms that take one column at a time, find the answer, and then move on to the next column. These students are often better at performing addition and multiplication algorithms rather than subtraction or division. Perhaps it is because both addition and multiplication are joining activities ending in the whole at the culmination of the algorithm. The important thing is to use manipulatives where exploration is possible. Work of students has been included in the exercises so that you can practice analyzing and planning instruction for typical solutions to students' difficulties.

Summary

This chapter emphasized the rules for the step-by-step procedures of multi-digit addition, subtraction, multiplication, and division problems. Proportional and non-proportional models were illustrated throughout the chapter to give teachers an idea of the vast number of models that can be used to help children conceptualize what children perceive to be a long process.

The teaching strategies included the use of calculators and estimation activities for each algorithm and integers to show how students will use the skills to prepare for adult life. Simultaneous and successive thought processing were discussed in diagrams as relevant for each operation.

The next chapter shows many numerical relationships. Understanding these theories will give the students a much better knowledge of base 10 numbers and provide time-saving techniques for estimation and mental arithmetic.

Exercises

Directions: Read all questions *before* answering any one exercise. Frequently the last question in one category leads to the first question in the next category.

A. **Memorization and Comprehension Exercises**
Low-Level Thought Activities

1. Explain the two types of division and give an example of each type.

2. Explain how the computer spreadsheet program can help students learn about algorithms. Tell what the activity did for your own learning.

3. Explain the role of estimation and rounding in the study of algorithms.

4. Explain why it is recommended that the division algorithm not be taught as a subject of drill over long periods of time.

5. Explain the difference between representational, non-representational, proportional, and non-proportional models. Give an example of each.

B. **Application and Analysis Exercises**
Middle-Level Thought Activities

1. Think of an addition, multiplication, subtraction, and division algorithm modeled in this chapter. Show the same algorithm with a different manipulative material than the one used in the chapter.

2. Apply what you have learned about the copy method of multiplication and division to work a problem in division with a remainder. Try a variation of the problem:

$$21\overline{)672}$$

Make it 6 7 8.

See what happens.

3. To reinforce math across the curriculum:
   a) List five mathematical questions/problems that come from popular storybooks for children. Show how they are related to the topics of this chapter.
   b) List five ways to use the topics of this chapter when teaching lessons in reading, science, social studies, health, music, art, physical education, or language arts (writing, English grammar, poetry, etc.).

4. Search professional journals for current articles on research findings and teaching ideas with algorithms or integers.

Follow the form below to report on the pertinent findings:

Journal Reviewed: _____

Publication Date: _____ Grade Level: _____

Subject Area: _____

Author(s): _____
_____

Major Finding:

Study or Teaching Procedure Outlined:

Reviewed by: _____

Some journals of note are listed below:
Journal for Research in Mathematics Education
Arithmetic Teacher
Mathematics Teacher
School Science and Mathematics

5. Computer applications for classroom use are found in the journals listed here. Their programs are by no means the only ones.

Search the periodical section of the library for more.
Teaching and Computers
Classroom Computer Learning
Electronic Learning
Computers in the Schools
The Journal of Computers in Mathematics and Science Teaching
The Computing Teacher

6. Analyze the following algorithm. It was created by a bright fourth grader. Why does it work? Try other numbers. Does it work all the time? What would tell you so?

$$
\begin{array}{r}
34 \; R5 \\
23\overline{)787} \\
60 \\
18 \\
9 \\
97 \\
80 \\
17 \\
12 \\
5
\end{array}
$$

C. Synthesis and Evaluation Exercises
High-Level Thought Activities

1. After analyzing the student's algorithm in 6 above, design your own algorithm. Prove that it works with several examples.

2. Using the chip trading mat below, work the following operations in the different bases as indicated. Let A through F stand for the numerals beyond nine before the first regrouping in bases eleven through fifteen. Notice that the first one is answered for you in numerals. Show the corresponding regrouping on the chip trading mat. The second one is to be worked on your own. Remember to use the written word for the number base when working with young people so they understand they have no "12" as such in base twelve.

3. Create a number line in base seven using as a model the multi-base blocks seen in the introduction. Create an algorithm problem to do with the multi-base blocks

R	G	B	Y

+

(Show your work with chips above)

Base Twelve

$$
\begin{array}{r}
76A \\
+ \; 4B8 \\
\hline
1066
\end{array}
$$

Answer in numerals

R	G	B	Y

−

(Show your work with chips above)

Base Fifteen

$$
\begin{array}{r}
4C02 \\
- \; 370E
\end{array}
$$

Answer in numerals

in base seven. Show your work in pictorial form. Write the symbolic form to the side.

4. Find a concept on the Scope and Sequence Chart that you would like to teach.

    a. Search professional journals for current articles on research findings and teaching ideas with algorithms and integers. Follow the report form found in B4. The journals noted there are still pertinent to the study of algorithms. Look for computer applications.

    b. Use the current articles from the professional journals to help plan the direction of the lesson. Document your sources.

    c. Evaluate at least one computer software program as a part of the lesson. Use the software evaluation in Appendix A. Show how it will be integrated with the rest of the lesson.

    d. Develop behavioral objectives. Show how you will use these to evaluate the lesson.

    e. Use the research-based lesson plan format of i) review, ii) development, iii) controlled practice, and iv) seatwork.

5. Below is the work of a student. Create a remediation plan to help the student overcome the misconceptions when working with algorithms. Your plan should include:

    a. Analysis of the error pattern.

    b. An evaluation of the thought processing required in the examples and the thought processing displayed by the student.

    c. The sequential development of concrete, pictorial, and symbolic models to reteach the concept.

. . . With Subtraction

. . . With Multiplication

6. Prepare a session utilizing a computer spreadsheet in a cooperative learning activity to determine how long it will take a faster moving vehicle to pass a slower moving vehicle. Make up a problem using distance = speed × time. A spreadsheet example, for a truck traveling 45 miles per hour and a car traveling 60 miles per hour with the truck leaving two hours before the car is as follows:

Time and Distance Traveled

Speed	
Truck	45
Car	60

Distance Traveled

Vehicle	Start	1 hr	2 hr	3 hr	4 hr	5 hr	6 hr	7 hr	8 hr	9 hr	10 hr
Truck	90	135	180	225	270	315	360	405	450	495	540
Car	0	60	120	180	240	300	360	420	480	540	600

Have the student prepare the spreadsheet and make up problems for solution utilizing the spreadsheet. Encourage the students to generalize their findings.

7. Consider the importance of computational proficiency and the recommendations in the NCTM curriculum and evaluations standards (Commission on Standards for School Mathematics, 1987) that a de-emphasis should be placed on tedious paper-and-pencil computation. You are a classroom teacher using textbooks that include multi-digit multiplication and long division. You decide to skip that material and allow time for new topics such as number and spatial sense, relations, and functions. Some parents have challenged you. Write a position paper on how you would justify your decision.

8. Evaluate some ways to help students remember the essential steps required to answer multi-digit problems in mathematics and understand the reasoning behind the answers they obtain. Which ones mentioned in the chapter seem most valid to you? Justify your answer.

Bibliography

Abel, Jean, Glenn D. Allinger, and Lyle Andersen. "Popsicle Sticks, Computers, and Calculators: Important Considerations." *Arithmetic Teacher* 34 (May 1987): 8–12.

Ball, Stanley. "From Story to Algorithm." *School Science and Mathematics* 86 (May/June 1986): 386–394.

Battista, Michael T. "A Complete Model for Operations on Integers." *Arithmetic Teacher* 30 (May 1983): 26–31.

Beattie, Ian D. "Modeling Operations and Algorithms." *Arithmetic Teacher* 33 (February 1986): 23–28.

Bidwell, James K. "Using Grid Arrays to Teach Long Division." *School Science and Mathematics* 87 (March 1987): 233–238.

Bledsoe, Gloria J., and V. Carl Bledsoe. "Add-Lo." *Mathematics Teacher* 79 (September 1986): 434–437.

Campbell, Patricia F. "What Do Children See in Mathematics Textbook Pictures?" *Arithmetic Teacher* 28 (January 1981): 12–16.

Capps, Lelon R., and Mary M. Hatfield. "Mathematical Concepts and Skills: Diagnosis, Prescription, and Correction of Deficiencies." In *Instructional Planning for Exceptional Children*, edited by Edward L. Meyen, pp. 279–293. Denver: Love Publishing, 1979.

Chang, Lisa. "Multiple Methods of Teaching the Addition and Subtraction of Integers." *Arithmetic Teacher* 33 (December 1985): 14–19.

Cheek, Helen N. and Melfried Olson. "A Den of Thieves Investigates Division." *Arithmetic Teacher* 33 (May 1986): 34–35.

Commission on Standards for School Mathematics of the National Council of Teachers of Mathematics. *Curriculum and Evaluation Standards for School Mathematics*. Working Draft. Reston, Va.: The Council, 1987.

Davis, Robert B. *Explorations in Mathematics: A Text for Teachers*. Palo Alto, Calif.: Addison-Wesley, 1967.

Dossey, John A., Ina V. S. Mullis, Mary M. Lindquist, and Donald L. Chambers. *The Mathematics Report Card, Are We Measuring Up? Trends and Achievement Based on the 1986 National Assessment.* Princeton, N.J.: Educational Testing Service, June 1988.

Grossman, Anne S. "A Subtraction Algorithm for the Upper Grades." *Arithmetic Teacher* 32 (January 1985): 44–45.

Hamic, Eleanor J. "Students' Creative Computations: My Way or Your Way?" *Arithmetic Teacher* 34 (September 1986): 39–41.

Hendrickson, A. Dean. "Verbal Multiplication and Division Problems: Some Difficulties and Some Solutions." *Arithmetic Teacher* 33 (April 1986): 26–33.

Huber, John. "Stabilizing Archimedes' Algorithm for Pi." *School Science and Mathematics* 82 (May/June 1982): 380–382.

Kamii, Constance and Linda Joseph. "Teaching Place Value and Double-Column Addition." *Arithmetic Teacher* 35 (February 1988): 48–52.

Kouba, Vicky L., Catherine A. Brown, Thomas P. Carpenter, Mary M. Lindquist, Edward A. Silver, and Jane O. Swafford. "Results of the Fourth NAEP Assessment of Mathematics: Number, Operations, and Word Problems." *Arithmetic Teacher* 35 (April 1988): 14–19.

Labinowicz, Ed. *The Piaget Primer: Thinking, Learning, Teaching.* Menlo Park, Calif: Addison-Wesley Publishing, 1980.

Madell, Rob. "Children's Natural Processes." *Arithmetic Teacher* 32 (March 1985): 20–22.

Milne, William J. *Standard Arithmetic.* New York: American Book Co., 1892.

Peck, William G. *United Course: Complete Arithmetic.* New York: A. S. Barnes and Co., 1877.

Royal Council of Public Instruction. *New Treatment of Arithmetic for the Decimal and Metric System.* 39th ed. Trans. by Nancy Tanner Edwards. Poussielgue-Rusand, Paris: Royal Council of Public Instruction, 1848.

Stone, John C., Virgil S. Mallory, and Foster E. Grossnickle. *A Higher Arithmetic.* New York: Benjamin H. Sanborn and Company, 1930.

Swart, William L. "Some Findings on Conceptual Development of Computational Skills." *Arithmetic Teacher* 32 (January 1985): 36–38.

Thompson, Charles S., and A. Dean Hendrickson. "Verbal Addition and Subtraction Problems: Some Difficulties and Some Solutions." *Arithmetic Teacher* 33 (March 1986): 21–25.

Van de Walle, John, and Charles S. Thompson. "Partitioning Sets for Number Concepts, Place Value, and Long Division." *Arithmetic Teacher* 32 (January 1985): 6–11.

Vest, Floyd. "Using Physical Models to Explain a Division Algorithm." *School Science and Mathematics* 85 (March 1985): 181–188.

Weiland, Linnea. "Matching Instruction to Children's Thinking About Division." *Arithmetic Teacher* 33 (December 1985): 34–35.

Wentworth, George, and David E. Smith. *Complete Arithmetic.* New York: Ginn and Company, 1909.

———. *Essentials of Arithmetic: Advanced Book.* New York: Ginn and Company, 1915.

9

Number Theory

Key Question: *What is it about the concepts in number theory that continue to fascinate mathematicians and young students alike?*

Introduction

Mathematicians throughout the ages have been fascinated with number patterns, and many interesting relationships have been developed during the evolution of mathematics knowledge. Children are usually exposed to these ideas beginning in the third grade and continuing through middle school. The units may be independent of the sequential skill development normally associated with the study of numeration, basic facts, and algorithms. The Scope and Sequence Chart shows a typical plan for the introduction of number theory concepts. Frequently textbooks use a discovery approach to simulate activities similar to those experienced by early mathematicians. Some children enjoy this approach and are challenged to find new patterns, but not all students will find patterns fascinating.

According to the NCTM curriculum standards (Commission on Standards for School Mathematics, 1987), looking for patterns is the essence of inductive reasoning. As students explore problem situations with patterns, conjectures are made which must be validated. This process encourages students to develop supporting logical arguments. Cooperative learning may help disinterested students maintain concentration during the study of number theory concepts.

Number theory explores the relationships (patterns) between and among counting numbers and their properties. Specific definitions and properties of each concept will be introduced when appropriate in the following section on teaching strategies. Beginning concepts include odd and even numbers, prime and composite numbers, greatest common factor, and least common multiple.

Scope and Sequence

Second Grade
- Odd and even numbers.

Third Grade
- Odd and even numbers.

Fourth Grade
- Common multiples.
- Understand multiples.
- Odd and even numbers.

Fifth Grade
- Find patterns in number sequences.
- Greatest common factor.
- Prime and composite numbers.
- Find least common multiple.
- Prime and composite numbers.
- Use factor trees.

Sixth Grade
- Prime and composite numbers.
- Prime factorization.

- Divisibility rules.
- Least common multiple.
- Greatest common factor.
- Least common denominator.

Seventh Grade
- Least common multiple and greatest common factor.
- Prime factorization.
- Divisibility rules.
- Use exponential form.
- Prime and composite numbers.
- Proper factors and twin primes.

Eighth Grade
- Review of all concepts.
- Relatively prime.
- Figurate numbers.
- Perfect, abundant, deficient numbers.

Teaching Strategies

Beginning concepts

Odd and Even Numbers

Children learn to tell the difference between odd and even numbers by pairing manipulative materials. *A number is defined as even if it can be divided by two with no remainder. A number is defined as odd if it has a remainder of one when divided by two.* Figure 9.1 shows some of the materials children can use in the primary grades to discover odd and even numbers through pairing (which is dividing a counting number by two).

Children should be given the opportunity to draw the pictorial representations of many num-

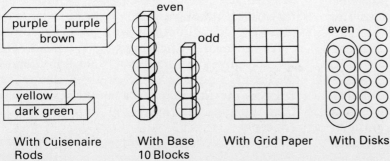

With Cuisenaire Rods   With Base 10 Blocks   With Grid Paper   With Disks

Figure 9.1

bers while deciding if the numbers are even or odd. Children will form an accurate definition of even and odd numbers if allowed to compare and contrast the paired and non-paired sets as seen in Figure 9.1. The numeral for each even number, *e*, can be designated on the number line. The same procedure can follow for each odd number, *o*. Children will be able to predict what larger numerals will be even or odd once the pattern is established.

	e		e		e		e		e		e
1	2	3	4	5	6	7	8	9	10	11	12
o		o		o		o		o		o	

Note that zero is considered an even number since it would fit the pattern shown on the number line. Children can experiment with odd and even numbers to generalize a rule about the properties of such numbers in addition and multiplication.

Activity

Addition with Odd and Even Numbers

Directions:

1. Using a calculator, make a list of what happens when you add one even number to another even number. Do the same for odd numbers. Then combine both odd and even numbers and record the results.

2. Set up a chart like the one below. Create more examples, following the pattern as established.

3. Generalize a rule for each set.

Even + Even	**Odd + Odd**	**Even + Odd**
2 + 4 =	7 + 3 =	2 + 11 =
106 + 38 =	201 + 55 =	348 + 29 =
1346 + 2794 =	2403 + 9825 =	7374 + 5689 =

(Create some more examples of large and small numbers.)

Rule: **Rule:**   **Rule:**

_____   _____   _____
_____   _____   _____
_____   _____   _____

Activity

Multiplication with Odd and Even Numbers

Directions:

1. Follow the same procedure for multiplication of even and odd numbers as you did in the previous calculator exploration with addition.

2. Use multi-digit numbers as well as single-digit numbers.

3. Show your work and rules below:

Even × Even	**Odd × Odd**	**Even × Odd**
_____	_____	_____
_____	_____	_____

(Multiply several more examples and record results.)

Rule: **Rule:**   **Rule:**

_____   _____   _____
_____   _____   _____
_____   _____   _____

Students enjoy comparing charts of classmates to see if the same rule will apply when other number combinations are used. Teachers should encourage students to see that the digit on the right of any multi-digit number is the one that determines if a number is odd or even. Students can problem solve further possibilities.

Activity _____

PROBLEM SOLVING

Odd and Even Numbers

Will the same rules apply if more than two odd or even numbers are added or multiplied together? Think what will happen in the following cases:

Even + Even + Even = ?   Odd + Odd + Even + Even = ?

Even + Odd + Odd = ?   Odd × Odd × Odd = ?

(Create some more examples and problem solve the rules that would apply in each case.)

If you know the rule for "Even + Odd + Odd," do you need to test the rule for "Odd + Even + Odd"? Why or why not? What mathematical properties can help with problems like these?

What other combinations do you already know without the need to test them?

Students who enjoy problem-solving explorations like the preceding one can be challenged to find many more combinations on their own time and share them with the class.

Prime and Composite Numbers

A prime number is an integer that is evenly divisible only by itself and one; i.e., only itself and one are factors of the number. Zero and one are not considered prime numbers. *A composite number is a number that has more than itself and one as factors.* Therefore, seven is a prime number, whereas eight is a composite number.

7	8
7 × 1	8 × 1
1 × 7	1 × 8
	2 × 4
	4 × 2
	2 × 2 × 2
	etc.

Activity _____

MENTAL MATH

The Mystery Even Number Prime

There is only one even number that fits the definition for a prime number. Think of many even numbers. Visualize their factors in your mind. Which factor, beside one and itself, does every even number have?

Did you think of two? The definition of an even number means that two will always be a factor of every even number. What about the two itself? Visualize the factors of two. What are they? The factors of two can only be one and itself (two). Therefore, two meets the criteria for a prime number, and it is the only even number that can be prime.

How can we know if a number is prime? The ancient Greeks had a method to find primes of a given number and through the centuries, it has become known as the Sieve of Eratosthenes.

The Sieve of Eratosthenes: Over 2000 years ago the Greek mathematician, Eratosthenes, cre- ated a process to help sieve (filter out) the composite numbers, leaving only the prime numbers. There are many interesting patterns to be found by using the Sieve of Eratosthenes. Teachers can lead students to discover many of the patterns on their own by asking a few leading questions such as the ones found in the problem-solving exploration that follows.

Activity

PROBLEM SOLVING

Number Patterns with the Sieve of Eratosthenes

Directions:

1. You are asked to use six colors on the Sieve of Eratosthenes (see chart at end of step 4) so that you can see the number patterns more easily. The colors of orange, green, purple, blue, red, and black are suggested but any eye-catching colors will do.

2. On the chart in step 4, find the first prime number, which is 2, and circle it with an orange marker. Now slash (/) all the multiples of 2 using the orange marker. You have just eliminated all even numbers because they are composites. Note the color pattern you have just created.

3. Now find the next prime which is 3 and circle it with a green marker. Sieve (/) all the multiples of 3 using the green marker. Some numbers now have an orange and green slash. You can easily see the numbers which are multiples of both 2 and 3. Note the color pattern created by the green slashes.

4. Follow the same procedure using the color code:

Primes	Color
five	= purple
seven	= blue
eleven	= red
all others	= black

1	2	3	4	5	6	7	8	9	10
11	12	13	14	15	16	17	18	19	20
21	22	23	24	25	26	27	28	29	30
31	32	33	34	35	36	37	38	39	40
41	42	43	44	45	46	47	48	49	50
51	52	53	54	55	56	57	58	59	60
61	62	63	64	65	66	67	68	69	70
71	72	73	74	75	76	77	78	79	80
81	82	83	84	85	86	87	88	89	90
91	92	93	94	95	96	97	98	99	100

(continued)

Questions Based on the Sieve of Eratosthenes:

1. How many of the first 100 numbers are prime?

2. Of the first 25 numbers, what percent were multiples of three numbers?

3. Of the second 25 numbers, what percent were multiples of three numbers?

4. Would you expect this number to increase or decrease as we continue through the number system? Why?

5. The number of primes in the first 50 numbers is what fractional portion of the total primes in the first 100 numbers?

6. The number of primes in the second 50 numbers is what fractional portion of the total primes in the first 100 numbers?

7. Will the relationship found in questions 5 and 6 hold for the second 100 numbers? What could be done to find out?

8. When is the first time that a multiple of 7 has not been crossed out by a multiple of a smaller prime?

9. What do you notice about multiples of 11? Predict at which number the multiple of 11 will be crossed out for the first time with no smaller primes as its multiple.

10. Generalize a rule for what you discovered after answering questions 8 and 9.

A person can figure out which multi-digit numbers are prime by using the Sieve of Eratosthenes. The first two examples which follow are performed the way mathematicians and upper grade students figured the problem of primes for centuries. Then the same process is placed in a BASIC computer program to show what Eratosthenes and others could have done with a computer if one had been available.

Let's discover if the number 201 is a prime or composite number. When working with any large number, n, you want to find the greatest prime number that could be a factor or multiple of n, thus proving that n is a composite number, having more numbers than itself and one as factors. Think of the greatest two multiples that n could have. It would be that multiple squared. We find that $14 \times 14 = 196$ and $15 \times 15 = 225$. Since 201 falls between 196 and 225 we know that we need not look for a prime number above 15. The highest prime number below 14 is 13. Now we can see if 201 is divisible by the primes that are 13 or lower. (To be considered divisible by a number, 201 must yield a whole number with no fractional parts in the answer.)

Divisible by:	Answer:	Is 201 a composite?
13)201	= 15.46	NO
11)201	= 18.27	NO
7)201	= 28.71	NO
5)201	= 40.2	NO
3)201	= 67	YES

Therefore, 201 is a composite number because 3 and 67 are multiples as well as one and itself. A student only has to find one pair of multiples this way to declare the number a composite although the number may have more factors than those first found.

Activity

Large Numbers with the Sieve of Eratosthenes

Directions:

1. Follow the same procedure as described for the number 201. This time find if 1373 is a prime or composite number.

2. Roughly we know 1373 is between 30 × 30 and 40 × 40 which is 900 to 1600.

3. Try between the squared numbers of 33 to 39. Find between which set of numbers 1373 falls.

4. Now look at the Sieve of Eratosthenes and find the next lowest prime number. Start dividing 1373 into each prime number through the number 2 until a multiple is found. If no multiple is found, what will 1373 be considered . . . prime or composite? Why?

5. Record the calculator work here:
   Divisible by: Answer: Is it composite?

You can see that a calculator can be used with large primes like the previous one. Early mathematicians used the long division algorithms to work such problems. A computer could also be used to check on a number. Children can use the following BASIC program to check on large numbers without the need to use the Sieve of Eratosthenes in the detail we have just gone through. A person exposed to both methods can appreciate the power of computers.

A student who has experimented with the Sieve of Eratosthenes can see the patterns and will not waste time checking numbers like 30, 125, 144, or 895. Those numbers can be noticed right away if students have been encouraged to look for easy patterns when working with the Sieve of Eratosthenes. Some experimenting can be done with these computer activities, and a teacher can create more programs using these as a start.

Activity

Prime Numbers in BASIC

Directions:

1. This program is saved on the BASIC disk under the name, PRIME NUMBERS.

2. Type LOAD PRIME NUMBERS and follow the directions in step 3.

   (If you have no disk, type in the following program.)

```
10 REM THIS PROGRAM FINDS PRIMES
20 PRINT "TYPE IN A NUMBER FROM 30 to 840."
30 PRINT "LET'S SEE IF IT IS A PRIME."
40 INPUT A
50 IF A < 30 THEN 300
60 IF A > 840 THEN 340
70 PRINT A;"/2 IS ";A/2
80 PRINT A;"/3 IS ";A/3
```

(continued)

```
 90 PRINT A;"/5 IS ";A/5
100 PRINT A;"/7 IS ";A/7
110 PRINT A;"/11 IS ";A/11
120 PRINT A;"/13 IS ";A/13
130 PRINT A;"/17 IS ";A/17
140 PRINT A;"/19 IS ";A/19
150 PRINT A;"/23 IS ";A/23
160 PRINT A;"/29 IS ";A/29
170 PRINT
180 PRINT "LOOK AT THE LIST ABOVE."
190 PRINT "IS ANY ONE OF THE ANSWERS A"
200 PRINT "WHOLE NUMBER?"
210 PRINT "TYPE Y OR N."
220 INPUT B$
230 IF B$= "Y" THEN GOTO 380
240 PRINT "THEN ";A;" IS A PRIME NUMBER."
250 PRINT
260 PRINT "DO YOU WISH TO DO ANOTHER ONE?"
270 INPUT C$
280 IF C$= "Y" THEN GOTO 20
290 GOTO 420
300 PRINT
310 PRINT "YOU HAVE TYPED A NUMBER SMALLER"
320 PRINT "THAN 30. TRY AGAIN."
330 GOTO 20
340 PRINT
350 PRINT "YOU HAVE TYPED A NUMBER LARGER"
360 PRINT "THEN 840. TRY AGAIN."
370 GOTO 20
380 PRINT "THEN ";A;" IS A COMPOSITE NUMBER."
390 PRINT "IT IS NOT A PRIME NUMBER."
400 PRINT
410 GOTO 260
420 END
```

3. Type RUN and do several examples before ending the program.

4. *How the Program Works:*
   a. Look at the printout of the program (printed in step 2).
   b. Lines 70 to 160 show the test used to find primes. From what is already known about the Sieve of Eratosthenes, it is possible to tell what the BASIC program is doing.
   c. There are shorter commands in BASIC which will do the same thing as lines 70 to 160. However, students will not be able to see the process the way it is shown here.

Activity

COMPUTERS

PROBLEM
SOLVING

Large Numbers with the Sieve of Eratosthenes

If you wanted to check numbers from 900 to 1600 to see which numbers are prime, what lines in the BASIC program would you add and/or change? Why? Make the changes and run your new program. Do you need to make corrections or did your program perform as you had wished?

One interesting feature of our number system is known as the Unique Factorization Theorem or the Fundamental Theorem of Arithmetic. *The Fundamental Theorem of Arithmetic states that every composite number can be expressed as a product of its primes.* Applying that to what we learned about the number 201 earlier in this section, it can be seen that:

$$201$$
$$\diagup \ \ \diagdown$$
$$3 \quad 67$$

By checking the Sieve of Eratosthenes, we see that 67 is a prime as is the familiar prime 3. Therefore, the number 201 can be expressed as 3×67 and will only have this one prime factorization.

Another investigation of number theory is finding the prime factorization of a number. *When a composite number is divided into its primes (not including 0 or 1) so that only primes are present as its factors, the process is known as prime factorization.* The following shows two methods for finding the prime factorization of the number 24.

Factor Tree Method *Dividing Method*

```
        24                    3
       /  \                2)6
      2    12              2)12
          /  \             2)24
         2    6
             / \
            2   3
```

(reading the left number in each branch, the answer is $2 \times 2 \times 2 \times 3$)

(reading up the list of divisors and final quotient, the answer is $2 \times 2 \times 2 \times 3$)

Both methods start with the smallest prime and continue its use until it is no longer a factor or divisor of the number. Then the next appropriate prime is found and the process continues until all the factors are prime numbers (factor tree method) or until the final quotient is a prime number (dividing method). Both show $2 \times 2 \times 2 \times 3$ as the prime factors of the number 24. From the associative property, we know that any arrangement of the four prime numbers will yield the unique number 24 when multiplied as factors, but it is customary to write the prime factors from least to greatest.

Least Common Multiple and Greatest Common Factor: The Least Common Multiple (LCM) and the Greatest Common Factor (GCF) can help students factor numbers efficiently. The GCF can also be called the Greatest Common Divisor, but most elementary textbooks choose the GCF terminology.

The least common multiple of two counting numbers is the smallest (least) non-zero number that is a multiple of both numbers. Since every number has an infinite number of multiples, every pair of numbers has an infinite number of common multiples. The smallest of the common multiples is known as the LCM. For example, the numbers 4 and 5 have many multiples, the least (smallest) common one is 20.

4   8   12   16   **20**   24   28   30
5   10   15   **20**   25   30   35   40

The key word is *multiple*, which implies that the resulting number will be greater than either of the two numbers. The LCM is useful in finding the common denominator in adding and subtracting fractions, for example:

$$\tfrac{1}{3} \text{ and } \tfrac{1}{4}$$

The common denominator is 12, which is the least common multiple of 3 and 4.

The greatest common factor (GCF) of two counting numbers is the greatest counting number that is a factor of each of the numbers. The key word is *factor*, which implies that the resulting number will be less than either of the two numbers. The numbers 18 and 24 have several common factors (1, 2, 3, 6) and the greatest common factor is 6.

A visual way to remember the idea is to think of the placement of each in relation to a diagram. Using the numbers 8 and 12:

```
LCM (8, 12)        GCF (8, 12)
   8)LCM            GCF) 8
  12)LCM            GCF)12
```

Both the LCM and GCF may be extended to more than two numbers, as seen in Figure 9.2 where the LCM and GCF of three numbers are illustrated.

Finding the LCM and GCF of Numbers

Both the LCM and the GCF start by factoring each number to be compared into its primes (as shown in the following factor diagrams, known as factor trees). The LCM is found by looking for

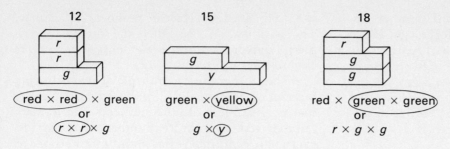

The most of any color is . . . (circled)
2 red × 1 yellow × 2 green
LCM = 2 × 2 × 5 × 3 × 3

Figure 9.2

LCM = the number of primes used most often in each composite number. Therefore, 2 × 2 is the most twos used (seen in the number 12): whereas 3 × 3 is the most threes used (seen in the number 18). Five is used only once but it still applies because it is the most times it is used in any number. The LCM is 2 × 2 × 3 × 3 × 5 or 180. The GCF is found by looking for the greatest divisor or factor that all numbers have in common. There is only one factor common to all the numbers and that is 3. Therefore, 3 is the GCF.

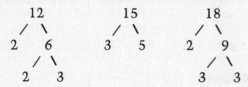

Figure 9.2 shows how the numbers 12, 15, and 18 can be factored using Cuisenaire rods. Each set of primes is made into a "prime tower" with the prime rods stacked one on top of the other. The LCM is found by asking which is the most color any prime tower uses. The answer is two reds, two greens, and one yellow, which equates the same as 2 × 2 × 3 × 3 × 5. The GCF is found by asking which is the largest color rod common in all three prime towers. The answer is the green rod or the numeral 3.

Venn diagrams (Fig. 9.3) are another useful tool for finding the LCM, and they may appeal to students who are comfortable with set diagrams. The LCM is 30. While 60 and 90 are also common multiples, they are not the least common multiple.

Number theory is an integral part of studying mathematics in the upper grades and middle school years. The following intermediate concepts are most often explored as sixth, seventh,

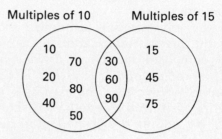

Figure 9.3

and eighth grade material. Students are introduced to the concepts much as the topics are introduced here; however, time for individual exploration should be provided. These intermediate concepts are: divisibility rules, polygonal or figurate numbers, Pythagorean triples, Fibonacci numbers, and Pascal's triangle.

Intermediate concepts

Divisibility Rules

One number is said to be divisible by another number if and only if the first number divides evenly into the second, leaving no remainder. Sometimes a person needs to estimate quickly if one number will divide evenly into a greater multi-digit number. Knowing some simple divisibility rules can help a person conquer the time factor on standardized achievement and aptitude tests. While other students are trying to perform laborious calculations taking up valuable time, the person who knows divisibility rules can answer more rapidly. Since calculators are not allowed in such tests, estimation and knowledge of number patterns become crucial. The following activity can be completed quickly after learning a few divisibility rules.

Activity

MENTAL MATH

Achievement Test Simulation

Directions: Which of the following problems results in a whole number?

a. $3\overline{)2537887}$

b. $9\overline{)7812477}$

c. $5\overline{)1259954}$

d. $8\overline{)7891302}$

e. $4\overline{)3520938}$

Divisibility by 2, 5, and 10: From working with the Sieve of Eratosthenes, students can "see" patterns for the divisibility of numbers by 2, 5, and 10. They will be able to generalize that any number ending with an even number in the ones place will be divisible by 2. Any number ending with a 0 or 5 in the ones place will be divisible by 5, and any number ending with a 0 in the ones place will be divisible by 10. Other divisibility rules are not that easily seen and require more questions on the part of the teacher if rules are to be learned by students.

Divisibility by 4 and 8: Since our number system is based on regrouping at ten, the powers of ten can help our understanding of divisibility rules. Look at the following divisibility pattern for 4.

Division by Powers of Ten	Does it Divide Evenly?
$10^1 \div 4$	NO
$10^2 \div 4$	YES . . . and all following powers of ten will be divided evenly also

Rationale for the Rule: Some numbers in the tens and ones place (10 to the first power) are divisible by 4, such as $4 \times 4 = 16$, $4 \times 5 = 20$.

Therefore, if we look at the last two digits to the right in a multi-digit number and find them divisible by 4, we know the entire number will be divisible by 4.

The rule for divisibility by 4 is *a number is divisible by 4 if and only if the last two digits of the number are divisible by 4.*

Following the same reasoning, divisibility by 8 will be examined:

Division by Powers of Ten	Does it Divide Evenly?
$10^1 \div 8$	NO
$10^2 \div 8$	NO
$10^3 \div 8$	YES . . . and all the following powers of ten will divide evenly also.

Rationale for the Rule: Some numbers in the ones, tens, and hundreds place are divisible by 8, such as $8 \times 1 = 8$, $8 \times 7 = 56$, $8 \times 13 = 104$, etc. Therefore, if we look at the last three digits to the right in a multi-digit number and find them divisible by 8, we know the entire number will be divisible by 8.

The rule for divisibility by 8 is *a number is divisible by 8 if and only if the last three digits of the number are divisible by 8.*

Activity

CALCULATORS

Divisibility by 4 and 8

Directions: Which numbers are divisible by 4 and 8?
(Students should be encouraged to fill in the right side of the chart as they work with the problems. The chart has been filled in for you.)

Problems	Divisible by 4 and 8?	Explanation
4)197836	YES	Because 36 is divisible by 4
8)765872	YES	Because 872 is divisible by 8
8)197836	NO	Because 836 is not divisible by 8

Check with a calculator to see if you were correct. Make up others and check with your calculator.

Students should also notice that any number divisible by 8 is also divisible by 4, but that not every number that is divisible by 4 will be divisible by 8. Have students discuss in cooperative learning groups why this fact is true. How does this fact relate to divisibility by 2?

Divisibility by 3 and 9:   Looking once again at the powers of ten with the division by 9:

$$10^1 \div 9 = 9 + 1$$
$$10^2 \div 9 = 99 + 1$$
$$10^3 \div 9 = 999 + 1$$

Notice: All these numbers have a pattern of addition along with a set of numbers that are divisible by 9.

Rationale for the Rule: Any number divisible by 9 is also divisible by 3, *and any number will always be divisible by 9 or 3 if and only if all the digits added together equal a number that is divisible by 9 or 3.*

Children may find the previous explanation more difficult to understand than the following activity card. Both explanations are based on the concept of regrouping numbers after 9 in each digit.

Activity

MENTAL MATH

Divisibility by 3 and 9

Guided Questions:

1. Look at the Sieve of Eratosthenes and find every number that is a multiple of 9. All the multiples of 9 have what color in common? (Answer: Green, which is used to mark all the multiples of 3.)

2. This color pattern tells a general rule about 9: What is it? (Answer: Every multiple of 9 is also a multiple of 3.)

3. Think what you have learned about the basic facts with 9 as a factor. The following pattern may be used to refresh students' memories:

   $$9 \times 1 = 9$$
   $$9 \times 2 = 18 \; (1 + 8 = 9)$$

$9 \times 3 = 27 \; (2 + 7 = 9)$
$9 \times 4 = 36 \; (3 + 6 = 9)$
$9 \times 5 = 45 \; (4 + 5 = 9)$
$9 \times 6 = 54 \; (5 + 4 = 9)$
$9 \times 7 = 63 \; (6 + 3 = 9)$
$9 \times 8 = 72 \; (7 + 2 = 9)$
$9 \times 9 = 81 \; (8 + 1 = 9)$

4. What is a general rule that can be developed from the above pattern? (Answer: If the digits of the answer are added together and equal 9, then the number is divisible by 9.)

5. What can be said about multiples of 3? (Answer: Since all multiples of 9 are also multiples of 3, the rule applies to multiples of 3.)

6. By extending the rule to large numbers, it would be: A multi-digit number is divisible by 9 or 3 respectively if and only if the digits of an answer are added together to make a number that is divisible by 9 or 3 respectively.

7. Which ones are divisible by 3 and 9?

Can do

$3 \overline{)8122572} = 8 + 1 + 2 + 2 + 5 + 7 + 2 = 27$
and 27 is divisible by 3
so 8122572 is divisible by 3

$9 \overline{)8122572} = 8 + 1 + 2 + 2 + 5 + 7 + 2 = 27$
and 27 is divisible by 9
so 8122572 is divisible by 9

$3 \overline{)7382937} = 7 + 3 + 8 + 2 + 9 + 3 + 7 = 39$
and 39 is divisible by 3
so 7382937 is divisible by 3

Cannot Do

$9 \overline{)7382937} = 7 + 3 + 8 + 2 + 9 + 3 + 7 = 39$
but 39 is NOT divisible by 9
so 7382937 is NOT divisible by 9

Any number divisible by 9 is also divisible by 3, but not every number divisible by 3 is divisible by 9. Since $3 \times 1 = 3$ and $3 \times 2 = 6$ are smaller than 9, it follows that some numbers will be multiples of 3 that will not be multiples of 9.

Now go back to the Achievement Test Activity Card and see how quickly you can answer the questions applying the rules you have just learned. Students beginning middle school can benefit from such practice with divisibility rules.

Polygonal or Figurate Numbers

Polygonal Numbers (also known as figurate numbers) are numbers that take the shape of polygons or figures. We will explore several kinds in this section. There are polygonal or figurate numbers called triangular numbers, square numbers, pentagonal numbers, hexagonal numbers, octagonal numbers, and so on. The most common are known as square numbers and take the form of a square figure.

The following Logo program will show figurate numbers known as perfect squares. See if you can deduce the formula for finding perfect squares as you work with the Logo program.

Activity

COMPUTERS

PROBLEM SOLVING

Perfect Squares in Logo

Directions:

1. This program is saved under PERFECTSQUARES.

2. Type `LOAD "PERFECTSQUARES` and follow the directions in step 3.

   (If you have no disk, type the following:)

   ```
   TO SQUARE :N
   REPEAT 4 [FD :N * 10 RT 90]
   HT
   MAKE "P :N * :N
   TYPE "SIDE TYPE "= TYPE CHAR 32 PR :N
   TYPE "AREA TYPE "= TYPE CHAR 32 PR :P
   END
   ```

3. *What the program does:* The procedure, SQUARE, will show the number pattern associated with perfect squares.

4. Start by typing `SQUARE 1` to see the first square. It will compute the area and side measurements of the square. Record the measurements in the chart below. Then type `SQUARE 2`, etc. and complete the chart.

Square #	Side	Area
SQUARE 1		
SQUARE 2		
SQUARE 3		
SQUARE 4		
SQUARE 5		
SQUARE 6		
SQUARE 7		
SQUARE 8		
SQUARE 9		
SQUARE 10		

5. The pattern is easy to see and students can predict what SQUARE 11 and SQUARE 12 would look like and what the side and area measurements would be. Then they can run the program to check their predictions.

6. *How the program works in Logo:*
   a. The first command draws a square with the side being any size given to the variable :N. Logo uses the "* 10" command to multiply the millimeter turtle step by 10 to make the centimeter measurement of the sides.

b. The HT hides the turtle so the squares can be seen more easily without the obstruction of the turtle head.

c. The variable :P is computed by squaring the side measurement (:N * :N), thus becoming the area.

d. The next two lines tell the computer to print the words, SIDE = and AREA =, along with the computed measurements.
The command, TYPE'', is used when you want to print all the words and variables on one line.
The command, TYPE CHAR 32, tells the computer to leave a space between words if they are printed on the same line with each other.

7. If students have worked with Logo programming in early elementary grades, they will be able to figure out the pattern for perfect squares by looking at the Logo procedure and only need to run the program for verification by the middle school grades.

Squares can be drawn whose sides and areas are not whole numbers. For instance, a square can be drawn whose sides are 2.1 × 2.1, which yields a square area of 4.41, and a square with an area of 3 will yield sides whose measure is an irrational number, 1.7320508 (unending decimal). These are not considered perfect squares because the sides in both cases are not whole numbers. Therefore, *a perfect square is a square whose sides result in a whole number with no fractional parts. An imperfect square is a square whose sides result in a whole number plus some fractional part.* How many imperfect squares are there with whole number areas ranging from 1 to 100? The next computer activity will show the answer. Predict the number mentally before you start the program. You should not need pencil and paper to figure out the answer if you have done the preceding Logo progam with perfect squares.

Activity

MENTAL MATH

COMPUTERS

Imperfect vs. Perfect Squares with Logo

Directions:

1. This program is already loaded in the computer if you did the previous Logo activity. If not, you will need to type LOAD "PERFECTSQUARES

(If you have no disk, type in the following program.)

```
TO AREA :A
CLEARSCREEN
REPEAT 4 [FD 10 * :A/SQRT :A RT 90]
HT
WAIT 185
CLEARTEXT
PR [THE SQUARE ROOT IS]
PR SQRT :A
WAIT 240
IF :A>100 [STOP]
CLEARTEXT
PR [HERE IS THE NEXT ONE.]
PR [THE AREA IS]
PR :A+1
AREA :A+1
END
```

(continued)

2. This program will show the difference between square numbers and numbers that are defined as PERFECT SQUARES.

3. Make a checklist of the areas whose square roots (the sides) are only whole numbers (with no decimals in the answer) and those whose square roots have decimals in the answer. A chart has been started to help you (see step 6).

4. The picture of each square number will appear on the screen for only a few seconds and then the square root (the measurement for each side) will appear. Mark the chart quickly before the square root disappears. The first two are done for you.

5. The program draws squares up to 100.

6. Start by typing AREA 1

CHECK ONE:

Area	Square root is a decimal	Square root is a whole number
1		X
2	X	
3		
4		
5		
6		

(Record the list all the way to 100.)

Questions:

1. How many squares did you find with a whole number for their square roots? These are called the *perfect squares*.

2. Did you predict correctly how many imperfect squares there would be?

3. What do you think would happen for squares with whole number areas from 101 to 200?

For those who like a challenge:

4. What part of the Logo program would you have to change to find squares up to an area of 200?

5. Change the program and test your prediction. Use Appendix B to follow the editing commands you will need to change the program.

There are many possibilities for figurate (polygonal) numbers. Pythagoras, a Greek mathematician and philosopher of 2500 years ago, started a society of persons interested in many issues of that day, including clever patterns dealing with numbers. Known as the Pythagorean Society, this secret society explored many possibilities with figurate numbers. Each figurate number is found by counting the number of units used to make the original figure (polygon) and the number of equivalent units needed to produce the polygon as it grows larger. Figure 9.4 shows the configurations for the triangular, square, and pentagonal numbers as the Pythagoreans discovered them.

Note the numbers shown below each polygon. Do you see a pattern that could predict how many units a hexagonal number would have as it grows larger? The following Logo program will help test your predictions. The polygons in Figure 9.4 were drawn by the same Logo program.

Activity

Figurate Numbers in Logo

Directions:

1. This program is saved under the name FIGURATE.

2. Type LOAD "FIGURATE and follow the directions in step 3.

   (If you have no disk, type the following procedures.)

```
TO C                       TO FIG :S :N
HT                         MAKE "M "1
FD 15 CIRCLER 2            F
END                        END

TO F
REPEAT :S [REPEAT :M [C] RT 360 / :S]
MAKE "M :M + 1
IF :M > :N [STOP]
F
END
```

3. To see the polygons in Figure 9.4, you would type:

   FIG 3 4   to see the triangle to four levels

   FIG 4 4   to see the square to four levels

   FIG 5 4   to see the pentagon to four levels

   Note: Type CS to clear screen after each figure group is drawn. The program may appear slow to you because the figure is drawing over itself as it proceeds to make larger figures.

4. What are your predictions for the hexagonal figure? Fill in the chart. What variables would you type in FIG to see a hexagon to four levels?

5. If you chose the command, FIG 6 4, you were right. Count the unit points and see if your prediction was correct.

6. Predict the number of points in:

   a heptagonal number
   an octagonal number
   a nonagonal number

   Test your predictions to six levels of each number sequence.

7. Test other figurate numbers to various levels.

8. *How the Logo program works:*
   The procedure C makes the points of each figure, and the procedure F makes the shape unit by unit (the M variable) along with the angle size in right turns.

   Note: The angle size (:S variable) of a figure is important. Notice that it takes 360 degrees to complete an entire figure. When making triangles (requiring 3 angles), each turn was RT 120 degrees, 360 / :S. When making pentagons (requiring 5 angles), each turn was RT 72 degrees, 360 / :S.

   The variable :N tells the computer how many levels to build each figurate number. These facts should help you in planning programs in the following chapters.

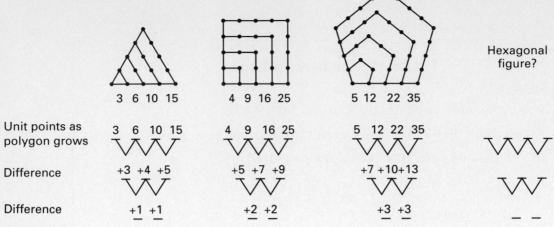

Unit points as polygon grows	3 6 10 15	4 9 16 25	5 12 22 35	
Difference	+3 +4 +5	+5 +7 +9	+7 +10 +13	
Difference	+1 +1	+2 +2	+3 +3	– –

Hexagonal figure?

Figure 9.4

Pythagorean Triples

Pythagoras is thought to have created the Pythagorean theorem which states that

$$A^2 + B^2 = C^2$$

where A and B are legs of a right triangle and C is the hypotenuse. The hypotenuse is the side opposite the right angle.

A visual representation of the theorem is shown in Figure 9.5. Pythagoras found sets of three whole numbers, known as triples, that would fit this pattern. One triple given frequently as an example is the triple of 3, 4, and 5 as the sides of the right triangle (seen in Fig. 9.5). If asked to find other triples, middle school students are most likely to see multiples of the 3, 4, 5 triple as possibilities.

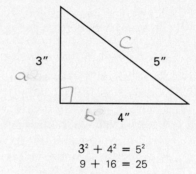

$$3^2 + 4^2 = 5^2$$
$$9 + 16 = 25$$

Figure 9.5

Activity

CALCULATORS

Triples

Directions:

1. Think of a pattern that might create triples from what you know about the triple 3, 4, 5.

2. Follow the theorem.
$$A^2 + B^2 = C^2$$

3. Use the calculator to see if the triple fits the theorem.

Triples	Theorem
3 4 5	9 + 16 = 25
	25 = 25
6 8 10	? + ? = ?
	? = ?
? ? ?	? + ? = ?
	? = ?

4. Create four more triples, following the same pattern.

5. Generalize a rule for finding Pythagorean triples using the table you created in step 3.

There are other ways to create Pythagorean triples. Analyze the computer programs in BASIC or Logo to discover another way of finding triples. Both programs follow the same steps to generate the next set of triples.

Activity

COMPUTERS

PROBLEM
SOLVING

Pythagorean Triples

Directions:

1. This program is saved under the name TRIPLES in both BASIC and Logo.

   Remember: You must work in one computer language at a time, and reboot (restart) when you switch from one language to another.

2. To work in the BASIC program, type   LOAD TRIPLES

3. To work in the Logo program, type   LOAD "TRIPLES

   (If you have no disk, type the following programs.)

BASIC

```
10 PRINT "THIS PROGRAM PRINTS"
20 PRINT "PYTHAGOREAN TRIPLES"
30 PRINT "IT STARTS WITH 3,4,5"
40 LET N = 1
50 LET A = 3
60 LET B = 4
70 LET C = 5
80 LET N = N + 1
90 LET A = A + 2
100 LET B = B + (4 * N)
110 LET C = B + 1
120 PRINT "THE NEXT TRIPLE IS"
130 PRINT A;", ";B", ";C
140 IF N > 9 THEN 160
150 GOTO 70
160 END
```

Logo

```
TO TRIPLES
TEXTSCREEN
PR [THIS PROGRAM PRINTS]
PR [PYTHAGOREAN TRIPLES.]
PR [IT STARTS WITH 3,4,5.]
MAKE "A "3
MAKE "B "4
MAKE "C "5
MAKE "N "1
TRIPLEFORMULA
END

TO TRIPLEFORMULA
MAKE "A :A + 2
MAKE "N :N + 1
MAKE "B :B + (4 * :N)
MAKE "C :B + 1
PR [THE NEXT TRIPLE IS]
PR :A
PR :B
PR :C
WAIT 120
IF :N > 9 [STOP]
TRIPLEFORMULA
END
```

Guided Questions:

1. Type the word RUN to start the BASIC program.

2. Type the word TRIPLES to start the Logo program.

3. Analyze the formulas seen in the program printouts as typed above, and generalize a rule that would produce the triples seen in the computer programs.

(continued)

4. *How It Works:*

    a. To generate variable A in each new triple, 2 is added to the preceding number for A.

    b. The variable B in each triple is found by adding the preceding number for B to the quantity, 4, multiplied by the number of times a new set of triples has been found.

    c. The third variable, C, is found by adding 1 to each new number generated for the variable B.

      Therefore, the next triple after 3 4 5 would be 5 12 13.

5. Check the programs again if you are having trouble seeing the pattern.

For those who like a challenge:

6. If you wanted the programs to print 20 sets of triples, what lines would you change? What would the change be? See if the programs did as you wished. See Appendix B for the instructions to edit programs.

7. Check the twentieth triple and see if it fits the Pythagorean theorem.

Fibonacci Number Sequence

The Fibonacci number sequence is most often seen in the field of science but also occurs in other fields including music. *A Fibonacci number sequence is defined as a sequence of numbers where any number is generated by adding the two previous numbers together.* Figure 9.6 shows a diagram of the number pattern as it might occur in growing things. The numbers to the right show the number pattern as it progresses through seven cycles.

Using the model of a tree branch (Fig. 9.6), Fibonacci numbers must meet the following criteria:

1. An old branch can generate a new branch only once.

2. The old branch continues to generate itself at each new level.

3. A new branch can generate itself (becoming an old branch) and a new one (at the next level of procreation).

When the number pattern becomes apparent, one can predict the number of new and old branches for endless levels yet to come. It may be easier to see the number pattern if presented horizontally with numbers only:

1 1 2 3 5 8 13 25 38   etc.

Notice that 13 is derived from adding the two numbers directly before it. Do you get 8 the same way? Do you get the other numbers the same way? Now you have established the number pattern of Fibonacci numbers.

Look at the three criteria needed in the preceding branch model. Substitute the word, variable, for the word, branch, to generalize the Fibonacci sequence to all areas. This general rule for finding Fibonacci numbers will help you perform the higher-ordered thought activity of creating your own computer program at the end of this chapter.

Activity

MATH ACROSS THE CURRICULUM

Music and Science with Fibonacci Numbers

Directions:

1. Since the Fibonacci sequence of numbers occurs in many fields, including music and science, have students choose a science or music topic and illustrate how the Fibonacci sequence is seen.

2. Find models in the science book and from nature.
A hint to start: Rabbits help us see the sequence every few weeks.

3. If students are creative, they may be able to compose their own examples of musical arrangements with the Fibonacci sequence.

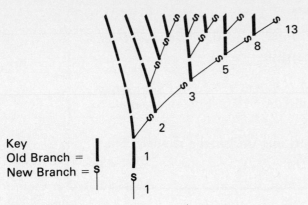

Key
Old Branch = |
New Branch = S

Figure 9.6

Pascal's Triangle

Pascal's triangle is a special arrangement of numbers discovered by Blaise Pascal, a seventeenth century mathematician. Pascal's triangle is in the configuration of a triangle with many number patterns visible as each row is added to the triangle. This arrangement of numbers relates to the number of possible outcomes in

probability. It relates later to algebra and solving equations. A formal definition is left for the reader to discover during the problem-solving activity that follows.

1		start
1 1		level 1
1 2 1		level 2
1 3 3 1		level 3
1 4 6 4 1		level 4
1 5 10 10 5 1		level 5
1 6 15 20 15 6 1		level 6
		etc.

It is easy to see the symmetry of Pascal's triangle. It makes an excellent bulletin board, especially if a new row is added daily, starting after the initial pattern is established at level 6. Students find themselves anticipating what the next row will be. They should also be encouraged to look at emerging number patterns as the triangle develops. The following exploration presents the kind of questioning that will help students begin to develop problem-solving strategies using the triangle.

Activity

PROBLEM SOLVING

Pascal's Triangle

Directions:

1. This is the same as Pascal's triangle seen above. Some numbers have been removed to help you see some of the number patterns more clearly.

```
            1
          1   1
        1   2   1
      1   3   3   1
    1   4   6   4   1
  1   5   10  10  5   1
 1   6  15  20  15   6   1
1  7  21  35  35  21  7   1
```

How many spaces will continue on the next two levels?
Do you see a pattern developing?

2. The spaces give us an idea of the numbers required on each side to make the symmetry occur. By looking at the first diagram completed to level 6, we see that two 10s come in the first two blank spaces after the numeral, 5. Which numbers in the previous row help to produce the 10s in level 5? In level 6, how were the numbers 15 and 20 chosen for that level? What numbers on the previous level helped? What will the numbers be on level 7, 8, and 9?

3. Add the numbers horizontally on each level. What pattern do you notice?

4. Turn the pattern into a definition of Pascal's triangle by explaining how the formula is derived.

Activity

CALCULATORS

Pascal's Triangle

1. Predict what the sums will be on level 7, 8, and 9. You can use a calculator to do this quickly. Test your prediction by finishing the triangle to level 9.

2. Find at least two more interesting patterns in Pascal's triangle. Share what you discovered with others in class.

The patterns in number theory can be very helpful to students as they work with the number system. The unique features can help students calculate more quickly, estimate reasonable answers, perform mental computation *(mental math)*, and problem solve where expanded answers can lead to new discoveries. The latter supports Bruner's idea of discovery learning as a mark of a true education.

Diagnosis and Remediation

Successive Thought Processing

Throughout the examples in the teaching strategies section, material was presented in chart or table form where one detail could build upon another. Charts and tables, rather than diagrams, seem more helpful to students who find successive processing more to their liking. The step-by-step progression of the Sieve of Eratosthenes helps because primes are divided out one at a time and factors are combined in the end to see a pattern. The number pattern involved in the Fibonacci sequence may be seen more clearly if the numbers (minus the diagram) are presented as seen below:

1 1 2 3 5 8 13 21 34 55

The diagram seems to "get in the way" of seeing the number sequence. If you found yourself confused rather than helped by the diagram, you may have been experiencing the same frustrations that successive thinkers feel at such times.

The dividing method for prime factorization is often more helpful than the stylized method of factor trees. BASIC programs that develop any number pattern one step at a time may be more easily understood than Logo programs that produce a picture as each explanation develops.

Simultaneous Thought Processing

Throughout the teaching strategies section, material was also presented in a diagram or stylized picture form. It is thought that such an approach may be more beneficial to the simultaneous thinker who can see patterns more easily if expressed in relationship to a whole picture or idea from the beginning. Factor trees, figurate (polygonal) numbers, Venn diagrams, and the branch drawing of the Fibonacci sequence are some of the teaching materials that can help students develop specific patterns in number theory. Logo computer programs that show the whole pattern in a drawing may prove more helpful to simultaneous learners than the BASIC programming language.

Correcting Common Errors

- **Jumping to Premature Conclusions**
  Frequently students will decide on an answer before the real pattern can be discerned. In their desire to find a pattern, they have been too hasty in their conclusions. A good example is the false pattern emerging from this attempt to find the solution for Pascal's triangle.

```
        1              start
      1   1            level 1
    1   2   1          level 2
  1   3   3   1        level 3
1   4   5   4   1      level 4
1  5   6   6   5  1    level 5
```

Students who are asked to find the number pattern of Pascal's triangle after only seeing the first three levels will quite naturally jump to the wrong conclusion about the pattern. The best remediation technique is a preventive one—a teacher must anticipate the points at which a misinterpretation could easily occur and make sure that enough examples of the pattern have been given to ensure a correct analysis of the pattern. As was mentioned earlier, it is good to present at least six levels of Pascal's triangle before asking children to find the pattern involved.

- **Careless Errors**

Some students discover the correct pattern but a teacher may not know it because the students make careless errors that do not show what they know. In the following Fibonacci example, the student understands the pattern but would be marked down by a teacher who was looking only at the answers in the blanks. Examples like this continue to emphasize the need for the teacher and student to dialogue about the pattern. The teacher must analyze the step-by-step work of each student when answers appear to be incorrect.

1 1 2 3 5 **8** **14** **25** **39**

- **Not Seeing the Pattern**

There are some students that cannot see a pattern and/or will not search for one if it is not seen easily. Teachers must ask more probing questions in such cases. It is tempting to tell the answer rather than construct more and more decisive questions. If problem solving and self-discovery of patterns are the important components of the lesson, then probing questions and clues are the best helps a teacher can give. Notice that clues appear throughout this chapter, especially when you were asked to problem solve a pattern in number theory that could be quite difficult for the non-mathematics major. Such clues did not provide the answer but should have helped you discover a way to the solution.

Summary

The chapter discussed odd and even numbers, prime and composite numbers, plus several number theories. Computer programs were provided to explore several of the theories. The Fundamental Theorem of Arithmetic, the Least Common Multiple, and the Greatest Common Factor were developed using concrete models. Divisibility rules for 2, 3, 4, 5, 8, 9, and 10 were provided. The Fibonacci number sequence and Pascal's triangle were introduced with simulations of what it might have been like to be the first mathematician to discover each pattern. It called for the use of good problem-solving skills.

The divisibility rules used in this chapter will help students find common denominators in the next chapter on rational numbers (common fractions). These topics are carefully developed with concrete models rather than memorized algorithms.

Exercises

Directions: Read all questions *before* answering any one exercise. Frequently the last question in one category leads to the first question in the next category.

A. Memorization and Comprehension Exercises
Low-Level Thought Activities

1. State each divisibility rule in your own words. Start each rule: "A multi-digit number is divisible by __ if and only if _____."

2. Test this number for divisibility by 2, 5, 10, 4, 8, 3 and 9.

$$? \overline{)946872}$$

3. Change the number in exercise 2 so it is divisible by the two numbers not represented in the previous exercise.

4. Tell the difference between prime and composite numbers and show the difference using Cuisenaire rods to demonstrate

the multiples of a prime and a composite number.

5. Using manipulatives, how could you show the rules for addition and multiplication of odd and even numbers found in the first calculator exploration at the beginning of the Teaching Strategies section? Model each rule using at least two different manipulative devices.

6. Define LCM and GCF in your own words. Give an example of both.

B. Application and Analysis Exercises
Middle-Level Thought Activities

1. Find the LCM and the GCF using Cuisenaire rods for the numbers 8, 6, and 12. Check your answers using numerals. Did you achieve the same answer using the concrete and symbolic methods?

2. Choose two more numbers and find the LCM and the GCF using pictorial models (Venn diagrams, number lines, or any other device) seen in the Teaching Strategies section.

3. Using the BASIC computer program that finds prime numbers, make another BASIC program to find primes beyond 1600.

4. Give examples of multi-digit numbers that are and are not divisible by 2, 5, 10, 4, 8, 3, and 9. These can become the basis of teacher-planned worksheets for students in your own classroom.

5. Find at least three more number patterns in Pascal's triangle. Explain each one.

6. Apply what you have already learned about the Sieve of Eratosthenes to decide if 19711 is a prime number.

7. Find divisibility rules for 6, 7, and 11.

8. Computer programs and computer reviews for the use of number theory in the classroom are found in the journals listed below. These articles are by no means the only ones. Search the periodical section of the library for more.
Teaching and Computers
Classroom Computer Learning
Electronic Learning
Computers in the Schools
The Journal of Computers in Mathematics and Science Teaching
The Computing Teacher

9. List five ways to integrate number theory into other areas of the curriculum.

C. Synthesis and Evaluation Exercises
High-Level Thought Activities

1. Evaluate the bases between two and ten. In which bases is 10 an odd number? Sketch the multi-base blocks in the different bases to help you. You will need to remember the definition of an odd and even number to pick the correct arrangements of multi-base blocks.

2. Evaluate prime numbers in the base two number system. List all the prime numbers between 0 and 10000. (Clue: Any even number except 10 can be excluded because it can be divided by 10 as well as itself and one.)

3. Will the divisibility rules found in this chapter apply to different bases other than the base 10 system? Why or why not? If not, what would need to be changed so the principle of divisibility could remain intact? (Clue: Create a divisibility rule for base 4 numbers, applying the same principles as the base 10 divisibility rule for 9.)

4. Create two cooperative learning activities for teaching one concept in number theory. Follow these steps:
   a. Look at the Scope and Sequence Chart. Find a grade level and a concept within that grade level. Trace the previous concept development of the topic taught in earlier grades so you will know where to begin your instruction.
   b. Include at least one way that the topic could be used for another academic area, supporting math across the curriculum.
   c. Use current articles from professional journals to help plan the direction of the lesson. Document your sources.
      Some journals of note where excellent articles may be found:
   Journal for Research in Mathematics Education
   Arithmetic Teacher
   Mathematics Teacher
   School Science and Mathematics

5. Create a BASIC or Logo computer program for Fibonacci numbers. The following set of steps will help structure your thinking process as you start your plan.

Activity

COMPUTERS

BASIC and Logo with Fibonacci Numbers

Directions:

1. Create a computer program that will generate Fibonacci numbers for at least 14 levels.

2. Decide which computer language you will use.

3. Look at the three criteria needed in the branch model. Substitute the word variable for the word branch to generalize Fibonacci numbers to all areas.

4. What stipulations will you place on each variable as you create the program? Name each variable with a letter name and describe its limits mathematically.

5. Place each variable in the sequential order needed to allow the program to fit the number pattern indicative of Fibonacci numbers.

6. Initiate a statement that will allow the program to stop after the Fibonacci numbers have reached the fourteenth level.

---

6. Create BASIC or Logo computer programs that would yield:
   a. The next even number in base 10.
   b. The next odd number in base 10.
   c. The next prime number in base 10.

7. Prepare a lesson for students to explore using a spreadsheet to produce:
   a. Fibonacci numbers.
   b. Pascal's triangle.

8. What is it about the concepts in number theory that continue to fascinate mathematicians and young students alike?

Bibliography

Bezuszka, Stanley J. "A Test for Divisibility by Primes." *Arithmetic Teacher* 33 (October 1985): 36–42.

Commission on Standards for School Mathematics of the National Council of Teachers of Mathematics. *Curriculum and Evaluation Standards for School Mathematics.* Working Draft. Reston, Va.: The Council, 1987.

Dearing, Shirley, and Boyd Holtan. "Factors and Primes with a T Square." *Arithmetic Teacher* 34 (April 1987): 34.

Dossey, John A., Ina V. S. Mullis, Mary M. Lindquist, and Donald L. Chambers. *The Mathematics Report Card, Are We Measuring Up? Trends and Achievement Based on the 1986 National Assessment.* Princeton, N.J.: Educational Testing Service, June 1988.

Edwards, Flo McEnery. "Geometric Figures Make the LCM Obvious." *Arithmetic Teacher* 34 (March 1987): 17–18.

Kouba, Vicky L., Catherine A. Brown, Thomas P. Carpenter, Mary M. Lindquist, Edward A. Silver, and Jane O. Swafford. "Results of the Fourth NAEP Assessment of Mathematics: Number, Operations, and Word Problems." *Arithmetic Teacher* 35 (April 1988): 14–19.

Litwiller, Bonnie H., and David R. Duncan. "The Extended Subtraction Table: A Search for Number Patterns." *Arithmetic Teacher* 33 (May 1986): 28–31.

———. "Pentagonal Patterns in an Addition Table." *Arithmetic Teacher* 32 (April 1985): 36–28.

Peera, Zehra. "Numbér Patterns and Bases." *Arithmetic Teacher* 30 (October 1982): 52–53.

Thompson, Alba G. "On Patterns, Conjectures, and Proof: Developing Students Mathematical Thinking." *Arithmetic Teacher* 33 (September 1985): 20–23.

Tierney, Cornelia C. "Patterns in the Multiplication Table." *Arithmetic Teacher* 32 (March 1985): 36.

Whitin, David J. "More Magic with Palindromes." *Arithmetic Teacher* 33 (November 1985): 25–26.

———. "More Patterns with Square Numbers." *Arithmetic Teacher* 33 (January 1986): 40–42.

10

Rational Numbers– Common Fractions

Key Question: *How can teachers use materials in meaningful ways to convey the real meaning behind abstract concepts like rational numbers?*

Introduction

This chapter will explore the development of rational numbers as they relate to common fractions. Decimals are also rational numbers but they will be discussed in Chapter 11 along with percents, ratios, and proportions—all special cases of rational numbers.

A rational number is a number that can be expressed as a ratio (a/b), or as a quotient of two integers when the divisor is not zero. The word *rational* comes from the word *ratio*. Rational numbers can be named by a fraction in the form *a/b* for which *a* and *b* are integers, but *b* is not zero. In this form, *a/b*, *a* is called the numerator and *b* is called the denominator. All rational numbers except zero have an inverse for multiplication. Therefore, all rational numbers are closed with respect to division except for division by zero, which is undefined as explained in Chapter 7.

We commonly refer to rational numbers as fractions. The system of rational numbers includes all the fractional numbers and their opposite negative numbers. This means it consists of $\frac{1}{2}, \frac{1}{5}, \frac{3}{4}, \frac{7}{12} \ldots$ and also $-\frac{1}{2}, \frac{1}{5}, -\frac{3}{4}, -\frac{7}{12} \ldots$ In most textbooks, the concept of negative rational numbers is presented in seventh or eighth grade. Negative integers and negative rational numbers are abstract concepts that usually are taught when students are beyond the age of 12. Hopefully, they will have begun the period of cognitive development which Piaget calls Formal Operational Thinking.

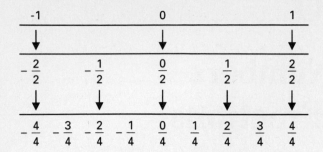

Rational numbers can be expressed as common fractions, decimal fractions, or decimals. *A common fraction is a fraction having any integer except zero in the denominator* such as $\frac{3}{4}$, $\frac{2}{3}$, $\frac{10}{12}$, $\frac{7}{10}$. *A decimal fraction is a fraction having only powers of 10 as denominators*, such as $\frac{7}{10}$, $\frac{19}{100}$, $\frac{120}{1000}$. Another way to name rational numbers is through our decimal system. This topic is covered in Chapter 11. In this text, the word fraction will be used in place of the term common fraction.

In the rational number, a/b, b is considered the complete unit representing a whole to which a portion, a, is compared. Each a is considered to be an equal portion of the complete unit. This complete unit may be referred to as the basic unit of wholeness or oneness.

Since b represents the basic unit of wholeness, b cannot equal 0 because it would mean that there was no complete unit with which to compare other portions. Thus, $a/0$ is considered impossible or undefined based on the same argument as presented in Chapter 7 for division by zero.

A number is a rational number if and only if given two rational numbers, a/b and c/d, the operations of equality, addition, and multiplication can be represented such that $ad = bc$. Substituting numbers for letters, $\frac{1}{2}$ and $\frac{3}{6}$ are considered rational numbers if and only if $1 \times 6 = 2 \times 3$. This allows for the existence of equivalencies, where different numbers can represent the basic unit of wholeness. Other examples could be:

$$\frac{1}{3} = \frac{4}{12} \qquad \frac{2}{5} = \frac{6}{15}$$

A further explanation of this idea is found under the concept of equivalence classes.

There are three distinct meanings that can be represented with rational numbers: part of a complete set, division, and ratio. Each meaning will be explored.

Part of a Complete Set and the Complete Set

In the fraction, a/b, the denominator states the number of equal portions in which the whole set or complete unit is divided. The a states the portion of complete set, b, that is to be considered at any given time. For example, the following manipulatives show portions of a complete unit or set.

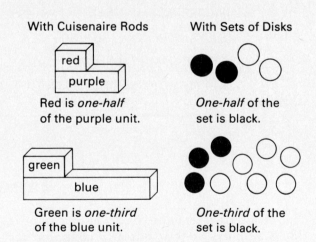

With Cuisenaire Rods

Red is *one-half* of the purple unit.

Green is *one-third* of the blue unit.

With Sets of Disks

One-half of the set is black.

One-third of the set is black.

Activity

Cuisenaire Rods as Fractional Parts of a Complete Set

Find as many Cuisenaire rods that represent $\frac{1}{3}$ as you can. How can you prove that each is exactly one-third of the basic unit of wholeness?

Now find Cuisenaire rods that represent $\frac{2}{3}$.

How can you prove that each is exactly two-thirds of the basic unit of wholeness? (HINT: If you proved one-third, proving two-thirds should follow easily.)

Division

Fractions denote division of one set by the other. This concept is presented after much work with fractions has taken place in the elementary school. Many elementary texts introduce this concept after decimal representations are presented. The division concept of fractions will be presented in Chapter 11 of this text. Here are the equivalent forms for $\frac{10}{6}$:

$$\frac{10}{6} = 6\overline{)10} = 10 \div 6$$

Ratio

A fraction such as $\frac{3}{4}$ can also represent a ratio which means that 3 elements of one set are present for every 4 elements of another set. A discussion of ratio is presented in Chapter 11.

Fractions are symbols to represent rational numbers. Different fractions may represent the same rational numbers. For example, $\frac{1}{3}$ and $\frac{3}{9}$ are different fractions because they have different numerators and denominators; however, both fractions represent the same rational number or the same amount. They are called equivalent fractions.

Equivalence classes mean different names for the same amount. For example:

Equivalence on a Number Line

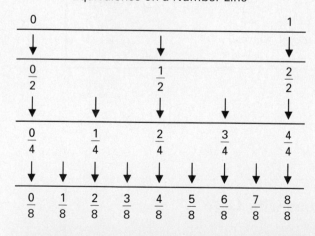

Hence, $\frac{1}{2}$, $\frac{2}{4}$, and $\frac{4}{8}$ are called an equivalence class. Notice $\frac{0}{2}$, $\frac{0}{4}$, and $\frac{0}{8}$ all mean no parts of the whole, whether it is divided into halves, fourths, or eighths.

All of the following Cuisenaire rods show different names for the same amount of the basic unit being measured.

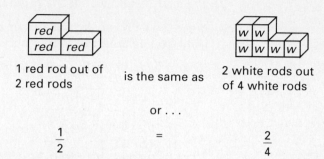

1 red rod out of 2 red rods is the same as 2 white rods out of 4 white rods

or . . .

$$\frac{1}{2} \quad = \quad \frac{2}{4}$$

If many names describe the same amount of a whole then the equality should remain if the numerator of one fraction is multiplied by the denominator of the other fraction, in other words, cross multiplication. Hence,

$$\frac{1}{2} = \frac{2}{4}$$
$$1 \times 4 = 2 \times 2$$
$$4 = 4$$

The following programs in BASIC and Logo are based on the preceding formula and can be used by elementary and middle school students to check if two fractions are equivalent or nonequivalent. Students can do programs like these after they have been introduced to the concept by the teacher using concrete manipulatives as their first exposure.

A significant line is missing in each computer program. Analyze each program and add the missing line. Change each program on the computer and see if you have problem solved correctly. The exercises at the end of the chapter will give you one way to arrive at the correct answers.

Notice: The Cuisenaire Company represents fractions with the numerator to the left of the denominator as illustrated in Figure 10.1. It coincides well with the way older math textbooks used to represent fractions in the form of 2/4 written side by side. Now most elementary textbooks represent fractions as: $\frac{2}{4}$. Therefore, this text will represent Cuisenaire rod fractions with the numerator placed on top and the denominator placed below it.

Figure 10.1

Activity

COMPUTERS

Equivalent Fractions with BASIC

Directions:

1. Load the program from the disk by typing LOAD EQUIV FRACTIONS

2. Type LIST and watch the program appear on the monitor.

3. Now type 110 at the bottom of the list and add the pertinent information after 110 on the same line.

   (If you have no disk, type the following:)

```
10  PRINT "FIND IF TWO FRACTIONS"
20  PRINT "ARE EQUIVALENT."
30  PRINT "TYPE FIRST NUMERATOR."
40  INPUT A
50  PRINT "TYPE FIRST DENOMINATOR."
60  INPUT B
70  PRINT "TYPE SECOND NUMERATOR."
80  INPUT C
90  PRINT "TYPE SECOND DENOMINATOR."
100  INPUT D
110
120  PRINT "THE FRACTIONS ARE"
130  PRINT "NOT EQUIVALENT."
140  GOTO 170
150  PRINT "THE FRACTIONS ARE"
160  PRINT "EQUIVALENT."
170  END
```

4. Now type RUN before trying each set of fractions to see if the program finds equivalent fractions.

5. Make adjustments to the program until it works.

Activity

COMPUTERS

Equivalent Fractions with Logo

Directions:

1. Load the program by typing LOAD "EQUIVFRACTIONS

2. Add the pertinent information following the word IF in line 11. Use the edit commands found in Appendix B to help you.

3. Now type FRACTIONS and see if the program finds equivalent fractions.

4. Type FRACTIONS before trying each new set of fractions.

5. Keep editing the program until it works.

   (If you have no disk, type the following:)

```
TO FRACTIONS
TEXTSCREEN
PR [TYPE FIRST NUMERATOR.]
MAKE "A FIRST READLIST
PR [TYPE FIRST DENOMINATOR.]
MAKE "B FIRST READLIST
PR [TYPE SECOND NUMERATOR.]
MAKE "C FIRST READLIST
PR [TYPE SECOND DENOMINATOR.]
MAKE "D FIRST READLIST
IF
IF :A * :D > :B * :C [PR [THE FRACTIONS ARE NOT EQUIVALENT.]]
IF :A * :D < :B * :C [PR [THE FRACTIONS ARE NOT EQUIVALENT.]]
END
```

Properties

In the study of rational numbers, three important properties need to be taught to students. The first is the property of denseness. *The property of denseness* is defined as: *for any two rational numbers, a and b, there is at least one other rational number that is the average of the two numbers.* This can be seen easily if a strip of paper is folded in half and then that $\frac{1}{2}$ strip is folded in half again and so on (Fig. 10.2).

How far do you think you can continue to fold the paper until there is no rational number between two others? Most adults can envision

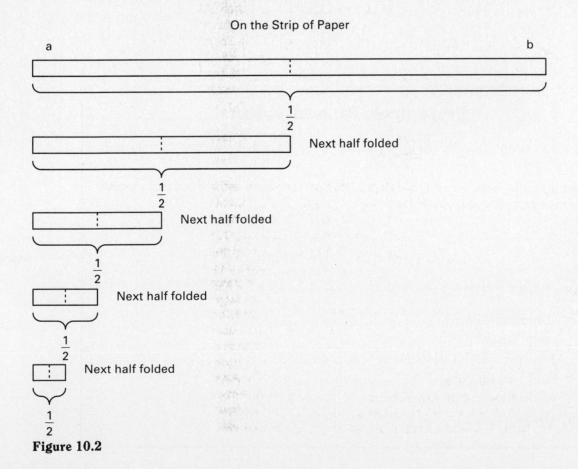

Figure 10.2

the answer in their mind but children cannot. They will have to test the property of denseness by folding a long strip of paper in half and continuing in half again and again before they can see that the possibilities are infinite. Computer paper with its long border strips (normally discarded) can be a handy manipulative material to show the density of rational numbers. This activity is a good test to see which students are still in Piaget's concrete operational period and those students who are ready for formal operational thought. Students who are strictly in the concrete operational period will not see that the possibilities are infinite and will count only the folds that they are able to make with the paper. Formal operational thinkers can generalize that the folds continue even when their fingers can no longer perform the activity.

The second important property of rational numbers, *the reciprocal property*, is important in developing algorithms for multiplication and division. Also known as the *multiplicative inverse, the reciprocal of any rational number, a, is $\frac{1}{a}$ and conversely, the reciprocal of $\frac{1}{a}$ is a.* When a rational number and its reciprocal are multiplied together, the result is 1 (or it brings the rationals back to the basic unit). Therefore,

$$\frac{1}{a} \times a = 1 \quad \text{and} \quad a \times \frac{1}{a} = 1$$

Proof of this is presented in the teaching strategies under the multiplication operation since an abstract statement such as the one above needs physical proof for students to comprehend it.

The third important property relates to *the equivalency rule for fractions. It says that if a and b are integers, then multiplying the numerator and the denominator of a fraction by the same number produces an equivalent fraction.* The same is true if the numerator and denominator are divided by the same number. Therefore,

$$\frac{a}{b} = \frac{a \times c}{b \times c} \quad \text{OR} \quad \frac{a}{b} = \frac{a \div c}{b \div c}$$

Substituting numbers for letters:

$$\frac{2}{3} = \frac{2 \times 4}{3 \times 4} = \frac{8}{12} \quad \text{OR} \quad \frac{8}{12} = \frac{8 \div 4}{12 \div 4} = \frac{2}{3}$$

This property becomes important when studying addition and subtraction of fractions when an equivalent form of a fraction must be found.

Activity

PROBLEM SOLVING

Properties of Rational Numbers

Do the properties of whole numbers apply to the properties of rational numbers?

Set up equations using 1/a, 1/b, and 1/c to represent unique rational numbers. Verify equality by letting

$$\frac{1}{a} = \frac{1}{2}, \quad \frac{1}{b} = \frac{1}{3}, \quad \frac{1}{c} = \frac{1}{4}$$

Test for the following properties:

Commutative

Associative

Distributive (multiplication over addition)

Additive Identity

Multiplicative Identity

Closure

Scope and Sequence

Kindergarten
- Readiness for equal parts.

First Grade
- Meaning of equal parts.
- One half, one third, and one fourth of a region.
- Parts of a set.

Second Grade
- Recognizing $\frac{1}{2}, \frac{1}{3}, \frac{1}{4}$ of a region or set.
- Identifying and writing fractional parts.

Third Grade
- Part of regions or sets for common fractions.
- Writing common fractions.
- Finding equivalent fractions with pictures.
- Knowing equivalent to one.
- Mixed numbers—identify and write.
- Adding and subtracting like denominators.

Fourth Grade
- Numerator/denominator as terms.
- Comparing unit or common fractions using < or > symbols.
- Ordering fractions.
- Adding like and unlike denominators.
- Subtracting like and unlike denominators.
- Using least common multiple.
- Changing mixed numbers to improper and improper to mixed.

- Finding fractional part of a number.
- Adding and subtracting mixed numbers.

Fifth Grade
- Rounding fractions.
- Cancelling and cross products.
- Adding three fractions.
- Equivalent fractions.
- Multiplying fractions and whole numbers.
- Comparing and ordering fractions.
- Multiplying mixed numbers.
- Using the reciprocal.
- Subtracting fractions and mixed numbers.
- Dividing fractions.

Sixth Grade
- All four operations with fractions.
- All four operations with mixed numbers.
- Using fractions in ratios, decimals, percents, probabilities.
- Review all other concepts of fractions.

Seventh Grade
- Review all previous concepts.
- Complex fractions.
- Using fractions with other components of mathematics.

Eighth Grade
- Review all previous concepts.
- Pre-algebra concepts with rational numbers.
- Using fractions in other areas.

Teaching Strategies

Fractions are difficult to understand for several reasons. Frequently models are not explored and used when concepts are developed. Many teachers demonstrate with pictorial models rather than concrete aids. The study of fractional numbers should begin with a variety of models and shapes. Manipulative models should be used throughout the middle school years. Unless this practice is followed, fractions will continue to be difficult to learn and internalize. Too often

the rule is presented before the underlying foundation has been laid. Many terms are unclear and ambiguous such as "reduce to lowest terms." Operations on whole numbers and on fractional numbers are not the same. For example, multiplying whole numbers yields a product larger than either factor, while multiplying two proper fractions results in a product smaller than either factor. Proper, concise language used when teaching operations with fractions plays an important role in the correct interpretation of the numbers.

The NCTM curriculum and evaluation standards (Commission on Standards for School Mathematics, 1987) specify a de-emphasis on paper-and-pencil computation with fractions. More time should be spent on creating a number sense for fractions through estimation with familiar fractions and mental computation. In the Standards for grades 5–8 mathematics curriculum, it is recommended that all operations with fractions and mixed numbers be limited to fractions with denominators involving numbers 2, 3, 4, 5, 6, 7, 10, 16, and 100. Instruction involving symbols should relate to models and simple real-world situations.

Whole-to-Part Activities

Early activities with fractions should begin with the whole-part meaning of fractions. Having children divide concrete objects such as a paper, egg cartons, or scored candy bars provides direct experiences to avoid such misconceptions as wanting the "biggest half."

Activity

COMPUTERS

Explorations with Logo

This Logo program helps students evaluate when a diagram is a true fraction and when it is divided unequally. As the program develops, students will create their own fractional pieces with different shapes representing one whole.

Directions:

1. This program is saved on the disk under the name TRUEFRACTIONS.

2. Type LOAD "TRUEFRACTIONS and follow the directions starting with step 3, following the printout of the program parts.

   (If you have no disk, type the following:)

```
TO BOX1
REPEAT 4 [FD 40 RT 90]
END
```

```
TO BOX2
REPEAT 2 [FD 40 RT 90 FD 60 RT 90]
END
```

```
TO BOX3
REPEAT 2 [FD 40 RT 90 FD 20 RT 90]
END
```

```
REPEAT 3 [BOX1 LT 90 BK 40 RT 90]
LT 90
END
```

```
REPEAT 3 [BOX1 RT 90]
LT 90
END
```

```
TO REC3C
BOX1 RT 60 BOX1 RT 60 BOX1
END
```

```
TO REC3D
BOX1 RT 45 BOX1 RT45 BOX1
END
```

```
TO REC3E
LT 90 FD 80 BOX1 RT 90 BOX2
LT 90 BK 60 RT 90 BOX3
END
```

```
TO FILLBOX1
REPEAT 39 [LT 90 FD 1 RT 90
FD 40 BK 40] HT
END
```

```
TO FILLBOX2
REPEAT 59 [RT 90 FD 1 LT 90
FD 40 BK 40] HT
END
```

```
TO FILLBOX3
REPEAT 19 [RT 90 FD 1 LT 90
FD 40 BK 40] HT
END
```

3. Call up the following 11 procedures to see what each one will do. Sketch the design next to the corresponding procedure in the printout (shown in step 2) so you can remember more easily what each one does. Remember to clear the screen (CS) between each procedure. Type:

```
        BOX1    BOX2    BOX3
   REC3A  REC3B  REC3C  REC3D  REC3E
      FILLBOX1  FILLBOX2  FILLBOX3
```

4. Using BOX1 type in commands that will show four equal portions where ¼ can be seen easily.

   Helpful Hints:
   a. The procedures, REC3A and REC3B and FILLBOX1, may help you set up the commands you need.
   b. To prevent "wrap around," move the turtle left 30 spaces before you start creating your new figure.
   For those who like a challenge:
   If you have your own copy of the disk, save the work you have just created to use with children. Write two procedures named REC4A and REC4B with the commands you have just used. See Appendix B to edit procedures.

5. Look at the following figure and answer these questions:

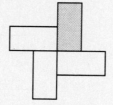

   a. Is this in equal portions?
   b. What would the shaded area be called?
   c. Type in commands using the appropriate BOX procedure that could produce the Logo design above.

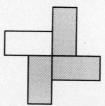

6. Look at the following figure:
   a. Change the procedure to make the Logo design.
   b. If all the portions are equal, then the portions represent fractional numbers and show the meaning of ¾.

7. Create several commands that will show the true meaning of fractions in the designs and those that will not show fractional designs. Sketch the ones you created. If you have a printer that will print the designs, make a "hard copy" (print out your designs) and label them: Fractions and Not Fractions.

An important, and often disregarded, consideration is to discuss what is represented by a whole. *The whole is whatever is designated as the unit.* Take a sheet of 8½x11 paper. Tear it in half. Each of the two congruent regions represent one-half. Show a notepad made of half-sheets of paper. Tear off one sheet and ask if this sheet is a whole piece of paper. It is in relationship to the pad of paper, but it is one-half when compared to the former whole sheet. Do the same for a pad consisting of paper the size of one-fourth the original sheet. Again ask the preceding questions and refer to the importance of knowing what is designated as the unit. This is important in terms of understanding operations with fractions discussed later in this chapter.

Fractions can be seen as parts of sets of objects. In this model, the unit represents the entire set.

Show a set of soft drinks.
How many cans are in the set? 6
Remove 2 cans. Suppose we drank 2 cans.
What part of the set did we drink? $\frac{2}{6}$
What does the 6 represent?
What does the 2 represent?
Show an egg carton with 12 eggs.
How many eggs are in the set? 12
Suppose I use 5 eggs to make a cake. Remove them.
What part of the set did I use? $\frac{5}{12}$
What does the 12 represent?
What does the 5 represent?
Repeat with Connecting cubes, colored tiles, wooden cubes, and other concrete materials.

Have children construct sets of their own and interpret the fraction represented.

Partitioning Activities

Initial work with fractions should be based on the natural activities children experience with

sharing. The sharing may be a region such as a candy bar or a set such as a bag of marbles. *Partitioning relates to dividing parts into equal shares.* In the concept development of fractions, emphasis should be on unit fractions (where the numerator is one) and on common fractions.

The region model (a circle, square, or rectangle partitioned into parts) is usually the child's introduction to work with fractions and is found in most children's elementary school textbooks. It is important not to limit the shape to the "pie shape" or circle. Various regions should be used by exploring fractions with shapes on a geoboard, with pattern blocks, Cuisenaire rods, and paper strips. Have children subdivide regions into equal-sized sections. In this way, they will understand the meaning of the subdivisions (Fig. 10.3).

An effective technique to develop the concept of comparing unit fractions is to take rectangular strips of construction paper, 4x18 inches and fold each into various pieces. The activity below explains the procedure required to make the concrete material seen in Figure 10.4.

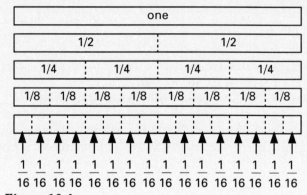

Figure 10.4

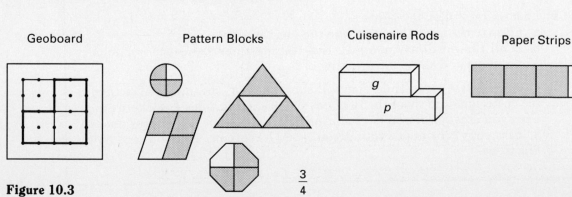

Geoboard   Pattern Blocks   Cuisenaire Rods   Paper Strips

Figure 10.3

Activity

MANIPULATIVES

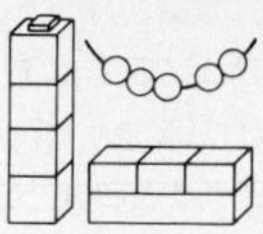

Making a Rectangular Fraction Kit

Materials: Five 4x18 inch strips of construction paper, each strip a different color.
Scissors
Pen or pencil

Procedure:

1. Take one of the colored strips (indicate a specific color). Label it "one whole." Take another strip. Fold it in half lengthwise. How many equal pieces result? 2

2. Cut on the fold. What is each piece in terms of the whole (1 out of 2 or $\frac{1}{2}$). Label each piece "$\frac{1}{2}$."

3. Take a third strip. Fold in half and fold in half again. Open it out and count how many equal pieces. 4
What is each piece in terms of the whole? ($\frac{1}{4}$) Cut on the fold. Label each piece "$\frac{1}{4}$."

4. A fourth strip is folded into 8 equal pieces and labeled as "$\frac{1}{8}$" and the fifth strip is folded into 16 equal pieces and labeled as "$\frac{1}{16}$."

Use the rectangular fraction kit to compare unit fractions and to understand that the larger the denominator, the smaller the fractional size since the whole has been partitioned into more parts. A common misconception is that the larger number means more even when it appears as the denominator in a fractional number. The teacher can select three unit fractions and ask children to order from least to greatest using the fraction kit as a concrete model.

The fraction strip chart in Figure 10.5 also serves as a powerful visual tool for comparing

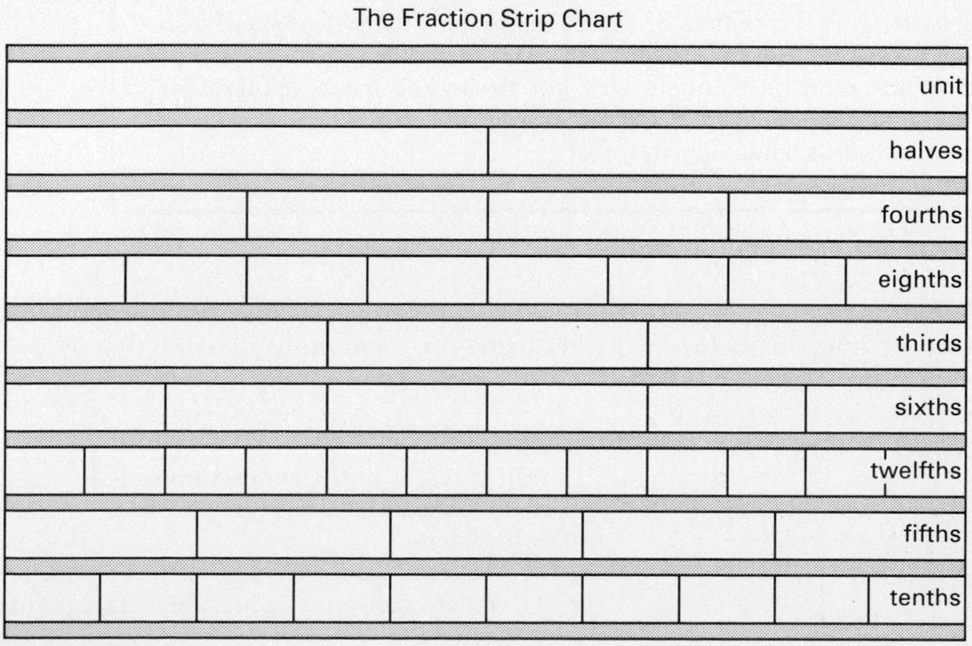

The Fraction Strip Chart

Figure 10.5

fractions. It is made essentially the same way as the rectangular fraction strips except all the strips are pasted on a 9x12 piece of construction paper. Every strip can be compared with the whole unit, 1, without the need to move many strips around.

Equivalence Activities

Equivalence is another key concept associated with fractions. Equivalence relates to the various ways that show equal shares of the same portion. Some educators believe that equivalence plays such a key role in the understanding of fractions that all work with the operations

should be delayed until after the fourth grade. The NCTM curriculum standards (Commission on Standards for School Mathematics, 1987) suggest that paper-and-pencil computation with fractions should be delayed until middle grades. This practice would allow ample time to develop the concept of equivalence and would set the stage for later work with unlike denominators and mixed numbers. As part of equivalence, the identity element of multiplication, or property of one for multiplication, should be emphasized. The aforementioned fraction kit or fraction strip chart are ways to show how the one whole strip can be renamed. The problem-solving exploration below shows how the fraction strip chart, Figure 10.5, can be used.

Activity

PROBLEM SOLVING

Equivalent Fractions: Fraction Strip Chart

Materials: Fraction Strip Chart

Procedure: *For one whole*—Move down each row of the chart. How many pieces have the whole piece been partitioned into? Write the fractional name for each strip such as $\frac{2}{2}$, $\frac{3}{3}$, $\frac{4}{4}$. Discuss the pattern for these names for 1. What rule can we generate about how a name for 1 looks (the numerator and the denominator are the same number)? Why is this the case? What other names for 1 can we develop? Make a chart of them.

For one-half—Place a straightedge so that the edge is perpendicular to the mark showing $\frac{1}{2}$. Look down the rows of strips for pieces that also are equivalent to $\frac{1}{2}$ such as $\frac{2}{4}$, $\frac{3}{6}$, $\frac{5}{10}$. Make lists of all the equivalent names for $\frac{1}{2}$ and also a list of all the pieces that do not equal $\frac{1}{2}$. Discuss the pattern of the numbers on each list. What kind of numbers will not be on the list of equivalent names? Repeat this procedure with all unit fractions on the chart as well as other common fractions when appropriate.

Many experiences are needed to internalize the concept of equivalence of fractions before the symbolic level holds meaning. The identity element for multiplication should be related to whole numbers as well as to fractional numbers. For example:

What happens when you take any fraction and multiply it by 1?

$$\frac{1}{2} \times 1 = ? \qquad \frac{3}{4} \times 1 = ?$$

What happens if you substitute some equivalent name for 1 and multiply a fraction by it?

$$\frac{1}{2} \times \frac{2}{2} = ? \qquad \frac{3}{4} \times \frac{4}{4} = ?$$

Will it result in the same number?

Is the outcome an equivalent name for that fraction? Why?

Try several fractions and different names for 1. What can you generalize from this experience?

Multiple bars (Fig. 10.6) offer a helpful way to show equivalent fractions. Many children are hindered with the concept development of fractions when they are not familiar with multiples. The multiple bars are cut into horizontal strips and used as explained in the following activity.

1	2	3	4	5	6	7	8	9	10
2	4	6	8	10	12	14	16	18	20
3	6	9	12	15	18	21	24	27	30
4	8	12	16	20	24	28	32	36	40
5	10	15	20	25	30	35	40	45	50
6	12	18	24	30	36	42	48	54	60
7	14	21	28	35	42	49	56	63	70
8	16	24	32	40	48	56	64	72	80
9	18	27	36	45	54	63	72	81	90
10	20	30	40	50	60	70	80	90	100

Figure 10.6

Activity

PROBLEM SOLVING

Equivalent Fractions: Multiple Bars

Materials: Multiple bars

Procedure: Place the 1 bar above the 8 bar to show a set of equivalent fractions for $\frac{1}{8}$. Read various equivalent names aloud. What name for 1 was used to multiply by $\frac{1}{8}$ to form $\frac{4}{32}$? ($\frac{4}{4}$)

Write this number sentence. $\frac{1}{8} \times \frac{4}{4} = \frac{4}{32}$

Repeat for other equivalent fractions. Place the 3 bar above the 4 bar. What fraction is formed? ($\frac{3}{4}$) Read other equivalent names for $\frac{3}{4}$. Tell what name for 1 was used to form each one.

1	2	3	4	5	6	7	8	9	10
8	16	24	32	40	48	56	64	72	80

Multiple bars can be used to see how dividing a fraction by 1 or an equivalent name for 1 produces another equivalent name for that fraction. If the resulting fraction cannot be divided again, the fraction is said to be "expressed in lowest terms" or "expressed in simplest form." These terms are preferred by the authors because such terms are more meaningful to students than the phrase "reduced to lowest terms" and do not hold the misconception that reducing results in a reduction in the size of the region.

Practice activities should not always focus on finding an equivalent fraction given the numerator or denominator. Some activities should include problems such as:

$$\frac{2}{3} \times \frac{?}{?} = \frac{12}{18}$$

What name for 1 has been multiplied by $\frac{2}{3}$?

$$\frac{12}{20} \div \frac{?}{?} = \frac{3}{5}$$

What name for 1 has $\frac{12}{20}$ been divided by?

The authors believe that multiple embodiments should always be used to help build concept development. However, in the interest of space, most of the remaining teaching strategies will be covered using fraction kits made from circular regions. Try using Cuisenaire rods, multiple bars, and attribute pieces (patterns in Appendix C) to teach the same strategies as those modeled here with circular fractions. You may prefer certain materials over others. Generally, students learn best when teachers feel comfortable using the materials they have chosen.

Texas Instruments has created a fraction calculator, the Math Explorer, that can simplify fractions automatically or by entering the greatest common multiple. Students can perform the four basic operations using the built-in fraction algorithm. This calculator presents many possibilities for students to test their knowledge of fractions with the power of immediate feedback. Estimation should also be included with calculator activities.

The directions for making a circular fraction kit are described here for those who feel that they can understand more fully by actually manipulating the materials. All the items are inexpensive and can be used with students in many grade levels. The fraction kit consists of 4-inch circular regions as the whole units plus wholes partitioned into the following pieces: $\frac{1}{2}$, $\frac{1}{4}$, $\frac{1}{3}$, $\frac{1}{6}$, $\frac{1}{8}$, $\frac{1}{12}$. To have adequate pieces with which to work in a flexible manner, there should be four wholes and the subdivisions of halves, thirds, fourths, sixths, and twelfths to create four wholes (e.g., 8 halves = 4 wholes, 48 twelfths = 4 wholes). Making the various fractional parts from different colored paper helps with easy recognition of the pieces. The fractional names can also be written on each piece to aid in quick recognition.

Using the Circular Fraction Kit

Concept: *Equivalent Fractions*

Procedure:
- Take a whole.
- Place the halves on it. How many does it take to cover it?
- Record this information in a chart.
- Repeat this procedure using the other pieces in the kit.
- How many thirds does it take to make the whole?
- How many fourths does it take to make the whole?

Name of fractional part	Number of parts to make 1 whole
Halves	2
Fourths	4
Thirds	3
Sixths	6
Eighths	8
Twelfths	12

How many ways can one-half be covered with other fractional parts? Record the findings in a chart.

Name of fractional part	Number of parts to make $\frac{1}{2}$
Fourths	2
Thirds	Cannot be done
Sixths	3
Eighths	4
Twelfths	6

What other parts can be found that are equivalent? Record findings in a chart. Discuss the relationship between the numbers and the unit pieces. For example, $\frac{1}{2} = \frac{4}{8}$. The one-half piece has been divided into four congruent regions so you have multiplied it by four. Areas are still congruent so the fractions are equivalent.

Concept: *Comparison of Unit Fractions*

Procedure:
- Which is more $\frac{1}{3}$ or $\frac{1}{4}$?
- Place two wholes on the desk.
- Put one-half on top of one whole.
- Put one-third on top of the other whole.
- What part is more? Compare the two pieces directly on top of each other if necessary.
- Use a similar procedure to compare other unit fractions.

Refer to the denominator as a label naming the number of parts in the whole. The numerator tells how many parts there are.

Many experiences are necessary in order for children to feel totally comfortable with fractional equivalencies. Activities with a variety of materials are helpful as well as playing games in a comfortable, non-threatening environment. The following two games are from Marilyn Burn's *The Math Solution* (1984). They are to be played using the rectangular strip fraction kit described in Figure 10.4.

Activity

Game 1—Cover Up

Materials: The fraction kit of rectangular strips
Number cube marked with these sides: $\frac{1}{2}, \frac{1}{4}, \frac{1}{8}, \frac{1}{8}, \frac{1}{16}, \frac{1}{16}$

Procedure: This game is to help children understand equivalent fractions and to compare fractional size of unit fractions $\frac{1}{2}, \frac{1}{4}, \frac{1}{8}, \frac{1}{16}$. The object of the game is to be the first player to completely cover the whole strip with the other fractional pieces from the kit. No overlapping is allowed.
Roll the number cube and compare the fractional size of the fraction for each player. The player rolling the least fractional number goes first. The first player rolls the number cube. This fraction tells what size piece that the student puts on the whole strip from the kit. Each student builds the fractions on his or her own whole strip. The number rolled must be the fraction piece placed on the whole strip. If an overlap is the result, the player loses a turn and must wait until that exact piece is the fraction named on the number cube. For example, if the player needs only a small piece, such as $\frac{1}{8}$ or $\frac{1}{16}$, to cover up and win, rolling a $\frac{1}{2}$ or $\frac{1}{4}$ will not work. The student must roll exactly what is needed. The first to cover up the entire strip wins. Students will eventually begin to estimate the fraction needed for the remaining piece as well as to realize that the larger the number in the denominator, the smaller the size of the fractional piece.

Activity

Game 2—Uncover

Materials: The fraction kit of rectangular strips
Number cube marked with sides: $\frac{1}{2}, \frac{1}{4}, \frac{1}{8}, \frac{1}{8}, \frac{1}{16}, \frac{1}{16}$

Procedure: This game is to help children understand fractional equivalencies. The object of the game is to be the first player to uncover the whole strip completely. The player must roll exactly what is needed to uncover the strip in order to win. Each player has the whole strip covered with the two $\frac{1}{2}$ pieces. These are the pieces that must be removed to uncover the whole strip to win.

Children roll the number cube and the player rolling the least fractional number plays first. The first player rolls the number cube. The fraction on the cube tells how much the player can remove. An exchange or trade may be necessary before the fraction can be removed. This is the object behind the game as it focuses on equivalencies. For example, if a student rolls a $\frac{1}{8}$ on the first roll, one of the $\frac{1}{2}$ pieces needs to be exchanged in order to remove a $\frac{1}{8}$. A number of exchanges are possible. Students begin to estimate the variety of possibilities as the game becomes more familiar.

Some sample plays are described below. As with Cover Up, the player must roll exactly what is needed to uncover the strip before being declared the winner. It is important to have the players verbalize the exchanges aloud and other players must agree before the fractional piece is removed. It is also advised that students make the exchanges, then remove the piece rather than do the subtraction mentally. This practice will allow greater transfer to the symbolic level and will reinforce the connecting level which will come next.

Sample plays for player one: Start with the whole covered with the two halves.

First roll of the game—$\frac{1}{4}$. Rename the $\frac{1}{2}$ as $\frac{2}{4}$. Remove $\frac{1}{4}$.

Second roll of the game—$\frac{1}{8}$. Rename the $\frac{1}{4}$ as $\frac{2}{8}$. Remove $\frac{1}{8}$.

Third roll of the game—$\frac{1}{8}$. Remove $\frac{1}{8}$.

Fourth roll of the game—$\frac{1}{4}$. Rename the $\frac{1}{2}$ as $\frac{2}{4}$. Remove $\frac{1}{4}$.

Fifth roll of the game—$\frac{1}{16}$. Rename the $\frac{1}{4}$ as $\frac{4}{16}$. Remove $\frac{1}{16}$ leaving $\frac{3}{16}$ to uncover and win.

Sixth roll of the game—$\frac{1}{2}$. No play possible because $\frac{1}{2}$ is greater than $\frac{3}{16}$.

Seventh roll of the game—$\frac{1}{8}$. Remove $\frac{2}{16}$.

Eighth roll of the game—$\frac{1}{8}$. No play possible because $\frac{1}{8}$ is greater than $\frac{1}{16}$.

Ninth roll of the game—$\frac{1}{16}$. Remove $\frac{1}{16}$ and WIN (if you are the first one to remove all pieces from the whole).

After students have become familiar with this game and are ready to be introduced to the symbols, it is important for them to record the steps on paper or small individual chalkboards as they do the action.

For example:

$\frac{1}{8}$ is rolled. On the strip, there is the $\frac{1}{4}$ piece. The problem is then $\frac{1}{4} - \frac{1}{8}$. The exchange is to trade $\frac{1}{4}$ for $\frac{2}{8}$.

This is recorded as

$$\begin{array}{r} \frac{1}{4} = \frac{2}{8} \\ -\frac{1}{8} = \frac{1}{8} \\ \hline \frac{1}{8} \end{array}$$

If the recordings are done as the exchanges are made, transfer from the concrete level to the symbolic level is much greater. These two games provide excellent readiness for addition and subtraction of unlike denominators.

Operations with Fractions

Exploratory fraction work should introduce the use of physical materials to solve simple real-world problems involving fractions. Data from the fourth mathematics assessment of the national assessment of educational progress (Kouba et al., 1988) indicate that many students have learned computations with fractions as procedures without developing the underlying conceptual knowledge about fractions. About one-third of the seventh-grade students demonstrated "extremely limited knowledge of fractions and could not perform simple computational procedures with fractions" (Brown et al., 1988, p. 246). More teaching should involve the use of models. About 50 to 60 percent of the seventh-grade students in the NAEP study (Kouba et al., 1988) could perform computations with simple mixed numbers involving subtraction or multiplication. When subtraction involved regrouping, only one-third of the students correctly worked the problem. The circular fraction kit can be used to show the operations with common fractions. Each activity, presented here with addition and subtraction, will show the sequence with which the operations should be taught. Notice that the activities grow gradually more complex.

Addition and Subtraction with the Circular Fraction Kit

Each step in the procedures below is outlined in detail. The actions should be replicated over and over again with students verbalizing what is happening in each step. Manipulatives model these concepts, but all action must be accompanied with verbal descriptions, followed by the written steps, to link the algorithm with the words. There must be a link and match between what is done, said, and written. The teacher reading this text is encouraged to model each of the examples. Just reading the explanations may not be enough to teach the concepts adequately to children.

Concept: *Adding Fractions with Like Denominators*

Procedure:

The problem: $\frac{1}{4} + \frac{1}{4}$

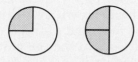

- Place a whole on the desk. Put a fourth on it.
- Add another fourth to it. How many in all?
- Record that one-fourth + one-fourth = two fourths.
- At the symbolic level: $\frac{1}{4} + \frac{1}{4} = \frac{2}{4}$

Repeat for other unit fractions and for fractions with like denominators. Remember that only the numerators are being added.

Concept: *Subtracting Fractions with Like Denominators*

Procedure:

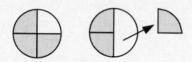

- Place a whole on the desk. Place three-fourths on it.
- To subtract one-fourth from three-fourths, simply take away the one-fourth piece.
- Record at symbolic level: $\frac{3}{4} - \frac{1}{4} = \frac{2}{4}$.

Repeat for other fractions with like denominators and record. The operation of subtraction is being done only with numerators. With subtraction, the take-away method is most commonly used.

Concept: *Adding Mixed Numerals*

Procedure:

The problem: $1\frac{1}{3} + 1\frac{1}{3}$

- Place two groups of one whole and one-third on the desk.
- Always begin with the fractional units first

(as with whole numbers, the column on the right is where addition or subtraction begins).

- Combine the fractional units and get $\frac{2}{3}$.
- Combine the wholes and there are 2. Answer: $2\frac{2}{3}$
- Record answer in symbolic form and discuss.

Concept: *Adding Fractions with Unlike Denominators*

Procedure:

The problem: $\frac{1}{2} + \frac{1}{4}$

- Denominators are not the same so the first step is to see if one number is a multiple of the other.

- If it is, use the least common denominator. In other words, is one of the denominators a multiple of the other?
- Exchange $\frac{1}{2}$ for $\frac{2}{4}$.
- Record this step at the symbolic level.
- Proceed in the same manner as adding like denominators.
- Record the answer: $\frac{3}{4}$

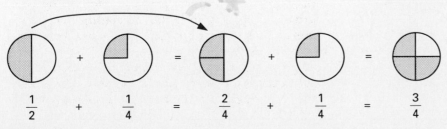

$$\frac{1}{2} + \frac{1}{4} = \frac{2}{4} + \frac{1}{4} = \frac{3}{4}$$

(continued)

WARNING! Do not allow students to short-cut this step! If they join the one-fourth to the one-half and give an answer of three-fourths which is easily observed, the transfer and linking to the symbolic, written level will be lost. When more difficult problems are encountered, the actions and sequenced steps will not have been developed and learning may be slower.

Concept: *Subtracting Fractions with Unlike Denominators*

Procedure:

The problem: $\frac{2}{3} - \frac{1}{4}$

- Denominators are not the same. Place the whole piece with two-thirds on top.
- In order to subtract one-fourth, rename the two-thirds and the one-fourth with the least common multiple of 3 and 4.

- With the fraction kit, make an exchange of thirds and fourths with twelfths.
- Exchange $\frac{2}{3}$ for $\frac{8}{12}$ and $\frac{1}{4}$ for $\frac{3}{12}$.
- Record this step at the symbolic level.
- Now the denominators are the same so subtract the numerators.
- With the fraction kit, take away three twelfths from the eight.
- Record the answer: $\frac{5}{12}$

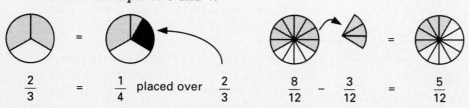

$$\frac{2}{3} = \frac{1}{4} \text{ placed over } \frac{2}{3} \qquad \frac{8}{12} - \frac{3}{12} = \frac{5}{12}$$

Concept: *Adding Mixed Numerals with Regrouping*

Procedure:

The problem: $2\frac{2}{3} + 1\frac{1}{2}$

- Form the two numbers with the fraction kit pieces.
- To combine the fractional units.
- Is one number a multiple of the other? (No)

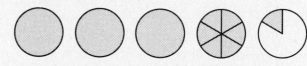

- Change both pieces for the least common denominator (thirds and halves can become sixths).
- Exchange $\frac{2}{3}$ for $\frac{4}{6}$ and $\frac{1}{2}$ for $\frac{3}{6}$.
- Record with symbols.

- Combine the sixths (7). Six-sixths can be regrouped as a whole with one-sixth left.
- Combine the wholes (4) and record final answer: $4\frac{1}{6}$

Concept: *Subtracting Mixed Numerals with Regrouping—Like Denominators*

Procedure:

The problem: $2\frac{1}{3} - 1\frac{2}{3}$

- Place the two whole pieces and the one-third on the desk.

- In order to take away two-thirds from one-third, exchange one whole for three-thirds.
- Now there are four-thirds in all.
- Record this step at the symbolic level.
- Take away two-thirds leaving two-thirds. Take away one whole.
- Record the answer: $\frac{2}{3}$

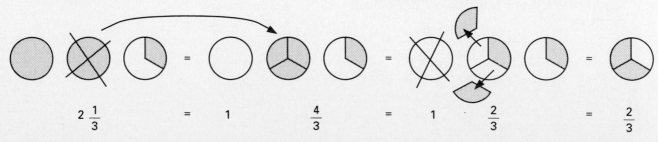

$$2\frac{1}{3} = 1 \quad \frac{4}{3} = 1 \quad \frac{2}{3} = \frac{2}{3}$$

Activity

PROBLEM
SOLVING

Addition and Subtraction with the Circular Fraction Kit

Directions:

1. Problem solve the following combinations:

$$2\tfrac{3}{4} + 1\tfrac{1}{4}$$
$$2\tfrac{5}{6} - 1\tfrac{2}{6}$$
$$2 - \tfrac{7}{8}$$
$$3\tfrac{1}{8} - 2\tfrac{3}{4}$$

2. Write down the actions in words as it has been done in the examples above.
3. Draw a pictorial representation as you write the words.
4. Make up more examples on your own.

Present situations where students must estimate if the fraction is closer to 0, to $\tfrac{1}{2}$, or to 1. Have students name some fractions near 0, $\tfrac{1}{2}$, and 1. Develop a sense of relative magnitude of a fraction such as $\tfrac{3}{4}$ is large compared to $\tfrac{1}{3}$, about the same as $\tfrac{5}{8}$, and closer to 1 than to 0. It is important to establish the relationship of the numerator and denominator (that the numerator is one-half of the denominator) to help children become comfortable with this activity. They should also see that any fractional number close to 1 will have a numerator and a denominator that approach the same number. Develop activities where students must figure mentally fractional parts such as 9 of the 20 people at the picnic were men. Was this about 0, about $\tfrac{1}{2}$, or about 1? If 12 of the 15 children in Mrs. Henry's class are girls, can we say over half the class is girls? Can you mentally picture a candy bar with $\tfrac{4}{10}$ of it cut off?

These activities are also helpful in estimating sums and differences of fractions and in giving children some checkpoint about the correctness of their answers in computational problems. You may need to write the problems on large cards to hold before the class in a limited time period or to write them on the overhead projector and control the time period to be shown. Children (and adults) have many insecurities about estimating with fractions because they have not had sufficient opportunities to practice this skill.

Activity

ESTIMATION

Estimation with Fractions

Problem: Look quickly at the fractional number or numbers and decide if they are about 1, about $\tfrac{1}{2}$, or about 0. Determine the sum or difference.

$$\tfrac{7}{12} + \tfrac{3}{6} = \qquad \tfrac{4}{10} + \tfrac{11}{10} =$$
$$\tfrac{2}{9} + \tfrac{7}{8} = \qquad \tfrac{9}{20} + \tfrac{4}{5} =$$
$$\tfrac{8}{9} - \tfrac{4}{5} = \qquad \tfrac{10}{9} - \tfrac{1}{12} =$$

Problems can also include mixed numbers. Ask students to round to the nearest whole number and add or subtract.

$$2\tfrac{4}{5} + 2\tfrac{9}{10} = \qquad 3\tfrac{1}{9} + 8\tfrac{8}{10} =$$
$$7 - 2\tfrac{6}{7} = \qquad 4\tfrac{8}{12} - \tfrac{7}{9} =$$

When adding and subtracting mixed numbers, students can benefit from knowing how to change a mixed number to an improper fraction. The terms, proper and improper fractions, are ones that may be difficult for students to remember. *A proper fraction is one whose numerator is of a lower degree (less) than its denominator and an improper fraction is one whose numerator is of a greater degree (more) or is equal to its denominator.* Hence,

Proper	Improper
$\frac{7}{12}$	$\frac{12}{7}$ $\frac{8}{8}$

Elementary textbooks also show students how to change improper fractions to an equivalent proper fraction. This necessitates changing the improper fraction to a *mixed number which is a whole number and a proper fraction.*

Biehler and Snowman (1986) point out that ridiculous associations can help people remember terms that would otherwise be forgotten very quickly. This association has helped elementary students in the past to remember the terms proper and improper fractions:

- It would be *improper* (not polite) for a person of less weight to carry a person of greater weight, but

- It is *proper* (polite) for a person of greater weight to carry a person of less weight.

Therefore,

- An improper fraction is one where a number of lesser degree has "to carry" a number of greater degree on "its back."

- A proper fraction is one where the number of greater degree has "to carry" a number of lesser degree on "its back."

Some disturbing data from NAEP results (Kouba et al., 1988) show that about 80 percent of the seventh-grade students could change a mixed fraction to an improper fraction, but fewer than one-half recognized $5\frac{1}{4}$ was the same as $5 + \frac{1}{4}$. Kouba et al. suggest that it may be a lack of understanding about the relationships between the fractional parts and the whole. Just as students should have the ability to rename whole numbers in many ways ($54 = 5$ tens $+ 4$ ones, 54 ones, 4 tens $+ 14$ ones, $5 \times 10 + 4$) so should they have the ability to rename fractions in a set of equivalence values ($5\frac{1}{4} = 5 + \frac{1}{4}$, $\frac{20}{4} + \frac{1}{4}$, $\frac{16}{4} + \frac{5}{4}$, $\frac{21}{4}$).

Activity

MANIPULATIVES

Changing Mixed Numbers to Fractional Numbers

Materials: Circular fraction kit

Procedure:

Problem: $2 + \frac{3}{4}$

Get two wholes and on a third whole place three-fourths. How can this number be changed to all fourths? Exchange each whole for fourths. How many fourths will this take? 8 How many fourths in all? $8 + 3 = 11$ fourths.

$$2\frac{3}{4} = 4 \text{ fourths} + 4 \text{ fourths} + 3 \text{ fourths} = 11 \text{ fourths}$$

Changing Fractional Numbers to Mixed Numbers

Materials: Circular fraction kit

Procedure:

Problem: $\frac{13}{4} = ?$

Get thirteen fourths. How many fourths make one whole? (4) Form the fourths into as many wholes as possible. How many can be made? (3 with one of the fourths left.) The mixed number is $3\frac{1}{4}$.

$$\frac{13}{4} = 4 \text{ fourths} + 4 \text{ fourths} + 4 \text{ fourths} + 1 \text{ fourth} =$$
$$1 \quad + \quad 1 \quad + \quad 1 \quad + \quad \tfrac{1}{4} \quad = 3\tfrac{1}{4}$$

Multiplication and Division of Fractions with the Circular Fraction Kit

Multiplication and division of fractions are areas that pose many difficulties for children primarily because the foundation for understanding the underlying concepts is not fully developed. The case too often is that the teacher uses the textbook examples and verbally explains them using a chalkboard or the overhead. Students are moved too quickly to the symbolic level and are given the easy-to-remember but difficult-to-comprehend rules. Errors result when students move symbols around meaninglessly as they rotely learn rules for procedures such as:

- when multiplying fractions, multiply the numerators and the denominators.
- when dividing fractions, use the reciprocal (often referred to as "invert the second number and multiply the two numbers").

Many teachers do not use manipulative materials in intermediate grades or the middle school. There are a limited number of materials available to model various problems. Therefore, the effectiveness of manipulatives is limited. Many examples are needed for adequate practice to solidify one's understanding of these operations with fractions.

An important aspect of multiplication is understanding the term "of" for multiplication and how it relates to whole number multiplication. It is important to build upon the prior knowledge students have about whole numbers whenever possible when teaching fractions. Although there are several differences that occur with fractions and whole numbers, it is important to address them and to create bridges of understanding when desirable. The following teaching sequence is suggested.

Write this equation: $3 \times 4 = 12$

Ask students to give the mathematical language and draw the set pictures that model this equation. It says: 3 groups of 4. It means there are 4 in each group and there are 3 groups. This becomes a review of the meaning of multiplication seen in Chapter 7. If teachers use the word "of" from the beginning of multiplication, the transfer of learning will help in problems like those that follow.

Activity

MANIPULATIVES

MATH
ACROSS THE
CURRICULUM

Multiplying Fractions—Whole Number by a Fractional Number

Materials: Circular fraction kits

Procedure:

Problem: $2 \times \frac{1}{2}$

Situational Problem: Lesley wants to double a recipe of cookies. The recipe calls for $\frac{1}{2}$ cup sugar. How much sugar will Lesley need to make 2 batches?

Take two groups of the one-half piece and place on a whole. These equal the whole.

In the problem of multiplying with a fraction, the language that accompanies the equation is that there are 2 groups of one-half.

Record the answer.

Write this problem: $\frac{1}{2} \times 2$

Situational Problem: Andy's recipe calls for 2 cups of flour. He wants to only use half of the recipe. How much flour will be needed?

Ask if the commutative principle works for fractions. What is the mathematical language that accompanies this problem? It says: One-half a group of 2. What does it mean? It means there are two wholes and you take one-half of that amount. How do you model this problem? Was the answer the same? Why?

Example #1

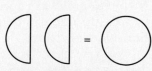

2 of the halves

$2 \times \dfrac{1}{2} =$ one whole

Example #2

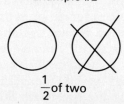

$\dfrac{1}{2}$ of two

$\dfrac{1}{2} \times 2 =$ one whole

The next step is to multiply one fractional part by another, no longer relying on a whole number to make the concept easier to understand. Most textbooks represent this concept with an array model. It is difficult for many students to understand. An area interpretation can also show multiplication with fractions. The procedure for making an array is: Take a unit region and divide it according to the fractional parts being multiplied. For example:

$$\tfrac{3}{4} \times \tfrac{2}{3}$$

Divide the unit horizontally into thirds and shade 2 of them. Next divide the unit region vertically into fourths and shade 3 of them. Count the number of equal parts in the unit region to represent the denominator and the number of parts that are shaded for the numerator.

Situational problem: Whole wheat bread takes $\frac{2}{3}$ hour to bake. It takes gingerbread $\frac{3}{4}$ as long. How long does it take gingerbread to bake?

The answer is $\frac{1}{2}$ hour. The model can be seen in the figure.

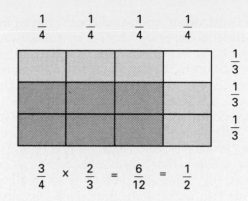

$$\frac{3}{4} \times \frac{2}{3} = \frac{6}{12} = \frac{1}{2}$$

12 parts and 6 parts are shaded twice

Activity

MANIPULATIVES

Multiplying Fractions—Fractional Number by Fractional Number

Procedure:

MATH ACROSS THE CURRICULUM

Problem: $\frac{1}{2} \times \frac{1}{4}$

Situational Problem: Lynn's lawn mower needs $\frac{1}{4}$ pint of oil to be mixed with gasoline. If only half of the mixture is made, how much oil is needed?

What is the mathematical language that accompanies this problem? It says: One-half group of $\frac{1}{4}$. To represent it with the fraction kit, do we start with halves or fourths? Remember that the second factor tells the number in the group and the first factor tells how many groups. To model this problem, you must take the whole piece and place the one-fourth piece on it.

Here it becomes critical to have the whole piece as the representative unit. *The one-fourth is one-fourth only in relationship to the whole circular region.* To take one-half of the one-fourth, the one-fourth piece needs to be partitioned into two equal parts. What fractional piece will do this? $\frac{1}{8}$.

Place the two one-eighths pieces on the $\frac{1}{4}$. How much is half of this group? $\frac{1}{8}$.

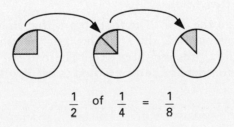

$$\frac{1}{2} \quad \text{of} \quad \frac{1}{4} = \frac{1}{8}$$

Think about the answer. The one-eighth piece is one-half of the one-fourth, but it is one-eighth in relationship to the whole. Use the commutative property and have children see the other way to consider this problem and that the answer is still the same.

It is important to provide many examples so that students become familiar with how multiplying two fractional numbers will result in a smaller amount than either of the two proper factors. Give practice estimating with fractional numbers, adapting the estimation activity seen at the end of the addition and subtraction strategies. Test the students' understanding by evaluating answers to see if the answer is reasonable. This will help when numbers become too cumbersome for manipulatives to work well, and students still need to check if their answers are in the correct range of possibilities. Multiplying two mixed numbers is an example of a cumbersome set to manipulate. The fraction kits do not work well with such problems. If sufficient prior work has been done with multiplying in ways mentioned above, it is hoped that this process can be understood and estimations can be made to check for accuracy.

Division with fractions is one of the most difficult concepts to understand. Again, it is important to relate the operation of division with whole numbers and discuss what mathematical language and models are appropriate. Consider the problem: $\frac{12}{3} = 4$. What does it mean? There

are 12 objects to be divided into groups of 3. The question is "How many groups of 3 are there in 12"? The answer tells you the number of groups of 3 you can get from 12. This is the measurement approach to division. You know the amount in each group and must determine the number of groups that can be made. This is the approach that will be used in division with fractional numbers. Although not every division situation, i.e., division by a whole number, can be interpreted as the measurement concept of division, it fits many problems as noted in this section.

When students first work with division of fractions, we advocate using a common denominator approach. This keeps the operation as division rather than using the reciprocal, which converts the process into multiplication. Students are familiar with finding common denominators to add and subtract fractions, and the algorithm ties into that prior knowledge. The steps for division are the same as for addition and subtraction: find common denominators, then perform the operation (division) on the numerators.

Activity

MANIPULATIVES

MATH
ACROSS THE
CURRICULUM

Division of Fractional Numbers—
Whole Number by a Fraction

Procedure:

Problem: $2 \div \frac{2}{3}$

Situational Problem: Mrs. Morozzo needs 2 cups of evaporated milk for her recipe. The milk comes in cans that contain $\frac{2}{3}$ cup. How many cans of evaporated milk must she buy?

What is the mathematical language for this problem? How many groups of two-thirds are in two wholes?

Place 2 wholes on the desk. The first step is that in order to decide how many two-thirds there are in 2, the 2 wholes must be renamed or exchanged for thirds. How many thirds are there in 2? There are 6. Record this step to show that the common denominator method is being used.

$$2 \div \frac{2}{3} = \frac{6}{3} \div \frac{2}{3}$$

How many groups of two-thirds can be formed from six thirds? Record the answer.

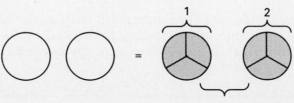

Therefore, when the denominators are the same (thirds), the operation is performed on the numerator. That means 6 divided by 2 is 3.

$$2 \div \frac{2}{3} = \frac{6}{3} \div \frac{2}{3} = \frac{6 \div 2}{3 \div 3} = \frac{3}{1} = 3.$$

To interpret the final answer, it says: There are 3 groups of two-thirds in 2. Mrs. Morozzo must buy 3 cans of evaporated milk to get 2 cups of milk.

Problem: $2 \div \frac{4}{12}$

The problem asks: How many groups of four-twelfths are there in 2? First step is to rename two whole pieces as twelfths. Take the 2 wholes and rename as 24 twelfths. Record this step.

$$2 \div \frac{4}{12} = \frac{24}{12} \div \frac{4}{12}$$

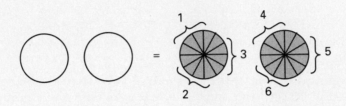

$$2 \div \frac{4}{12} = \frac{24}{12} \div \frac{4}{12} = \frac{24 \div 4}{12 \div 12} = \frac{6}{1} = 6$$

How many groups of four-twelfths can be made from the 24 twelfths? Look at the numerators and perform the division of $\frac{24}{4} = 6$. The answer says that there are 6 groups of 4 twelfths in 24 twelfths.

Be sure students have many experiences with this kind of problem where the division results in an equal number of groups. Have them verbalize the steps and then record the symbolic numbers and model the procedures with the fraction kit. The important thing to remember is that using the common denominator method keeps the operation of division clearly in mind rather than the traditional "invert and multiply." For most children and many teachers, the latter approach can better be interpreted as "Ours is not to reason why, simply invert and multiply." The rule is easy to learn, but students have no idea about the reasonableness of answers when problem solving.

Here are some more examples with mixed numbers as the final answer. See how the procedure works with them.

Activity

MANIPULATIVES

Using the Common Denominator Method

Procedure:

Problem: $2 \div \frac{3}{4}$

What is the mathematical language for this problem? How many groups of three-fourths are there in 2? The first step is to rename the two wholes as fourths. How many are there? 8. Now the problem is rewritten as

$$2 \div \frac{3}{4} = \frac{8}{4} \div \frac{3}{4}$$

(continued)

The question is how many groups of three-fourths are there in eight-fourths? Take the eight fourths and place into groups of 3. There are two complete groups of 3 that can be made and a part of a group or a remainder. The big question is how to express and interpret this remainder. There are 2 pieces of fourths left. How many fourths made each group? 3. How many are left? 2. There are two-thirds of a complete group left. The final answer is $2\frac{2}{3}$.

$$2 \div \frac{3}{4} = \frac{8}{4} \div \frac{3}{4} = \frac{8 \div 3}{4 \div 4} = \frac{8 \div 3}{1} = \frac{8}{3} = 2\frac{2}{3}$$

Problem: $2 \div \frac{5}{6}$

How many groups of five-sixths are there in two wholes? Rename the two wholes as sixths. Record this step: $\frac{12}{6} \div \frac{5}{6}$. It asks how many groups of five-sixths are there in twelve sixths? The denominators are the same so the operation of division occurs on the numerators. 12 divided by 5 = $2\frac{2}{5}$.

To show this step with the kit, you have 12 sixths and putting them into groups of 5 you can make 2 complete groups with 2 pieces leftover. These are a part of the group—2 out of the 5 that make a complete group. Look at the picture and the symbolic level together.

With Symbols Included

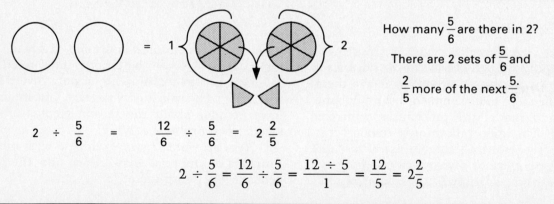

How many $\frac{5}{6}$ are there in 2?

There are 2 sets of $\frac{5}{6}$ and $\frac{2}{5}$ more of the next $\frac{5}{6}$.

$$2 \div \frac{5}{6} = \frac{12}{6} \div \frac{5}{6} = 2\frac{2}{5}$$

$$2 \div \frac{5}{6} = \frac{12}{6} \div \frac{5}{6} = \frac{12 \div 5}{1} = \frac{12}{5} = 2\frac{2}{5}$$

The common denominator method can be used also with fractional numbers divided by fractional numbers. The following activities show how it can be done with the circular fraction kit. The most difficult problems to show and correctly interpret are those where the divisor is larger than the dividend. The fraction kit can be used to model this process and some ex-amples are given in this text. Silvia (1983) uses graph paper to illustrate the algorithm for the division of fractions. To solidify your understanding, you may want to read and compare both approaches.

One important consideration—keep the numbers simple and within the realm of being conceptually understood. The NCTM curriculum standards (Commission on Standards for School Mathematics, 1987) recommend that only common familiar fractions should be used: none should be more difficult than $1\frac{1}{4} \div \frac{1}{2}$. We must be reasonable about what problem-solving situations would require complicated fractional numbers. We cannot justify including difficult numbers that require a lot of valuable instructional time but have limited, or no, real-life application.

Activity

Division of Fractional Numbers— Fraction by a Fraction

Procedure:

Problem: $\frac{1}{2} \div \frac{1}{4}$

What is the language for the problem? How many groups of one-fourth are there in one-half? Place the one-half piece on a whole. The first step is to rename the half as fourths in order to determine how many fourths can be made from the half.

$$\tfrac{1}{2} \div \tfrac{1}{4} = \tfrac{2}{4} \div \tfrac{1}{4}$$

How many groups of one-fourth can be made from two-fourths? There are 2 groups of one-fourth in two-fourths.

Problem: $\frac{3}{4} \div \frac{1}{2}$

How many groups of one-half can be made from three-fourths? First rename the one-half as fourths. Record the step.

$$\tfrac{3}{4} \div \tfrac{1}{2} = \tfrac{3}{4} \div \tfrac{2}{4}$$

How many groups of two-fourths can be made from three-fourths? This is the same as 3 divided by 2. The answer is $1\frac{1}{2}$ or that one group of two-fourths can be made and one-half of a group is left.

$$= \quad 1\frac{1}{2}$$

$$\frac{1}{2} \text{ of the next half}$$

$$\frac{3}{4} \div \frac{1}{2} = \frac{3}{4} \div \frac{2}{4} = \frac{3 \div 2}{1} = \frac{3}{2} = 1\frac{1}{2}$$

Problem: $\frac{2}{3} \div \frac{3}{4}$

How many groups of three-fourths are in two-thirds?
Place the two-thirds on a whole.

(continued)

Next change to like denominators—

Can thirds be exchanged for fourths? (No)

So what is the lowest common denominator? (12) Exchange $\frac{2}{3}$ for $\frac{8}{12}$.

Write problem with like denominators.

$$\tfrac{2}{3} \div \tfrac{3}{4} = \tfrac{8}{12} \div \tfrac{9}{12}$$

How many groups of nine-twelfths are in eight twelfths? Not even a whole-group can be made. Therefore, the fractional part is $\frac{8}{9}$ of a group.

$$\frac{2}{3} \div \frac{3}{4} = \frac{8}{12} \div \frac{9}{12} = \frac{8 \div 9}{1} = \frac{8}{9}$$

When middle school students are ready to work with the reciprocal, teachers must be ready to explain why the reciprocal works. This is difficult for students to conceptualize. In the previous examples, the complete answers were always given as multiplication problems. For example:

There are 2 groups of one-fourth.

$$2 \times \tfrac{1}{4}$$

There are 1 and $\frac{1}{2}$ of the two-fourths.

$$1\tfrac{1}{2} \times \tfrac{2}{4}$$

The multiplicative inverse is possible because answers to the division of fractions are inherently the inverse operation of multiplication. It was also shown that the reciprocal relationship is present, such as

2 of $\frac{1}{2}$ $\frac{1}{2}$ of 2

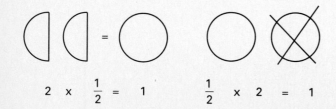

$$2 \quad \times \quad \frac{1}{2} \quad = \quad 1 \qquad\qquad \frac{1}{2} \quad \times \quad 2 \quad = \quad 1$$

Notice that each example comes back to the same basic unit of wholeness or 1. When the numerator and the denominator can be inverted and multiplied by a corresponding number in such a way as to maintain the same basic unit of wholeness, it is said that one fractional representation is in a reciprocal relationship with the other. Putting the two ideas together,

1. Complete answers to division of fractions are really multiplication problems, and
2. Using a reciprocal of a fraction does not change the basic unit of wholeness, one has a justification for the rule "invert and multiply." In addition, the example has shown another property of the multiplication of fractions,

$$2 \times \tfrac{1}{2} \text{ and } \tfrac{1}{2} \times 2$$

is referred to as the commutative property.

An important technique to help children get a sense about their answers is to encourage estimation. Experiences should be provided where students estimate quotients. The procedure is to round each mixed number to the nearest whole number or to round fractions to more convenient numbers to allow for mental computation. Try the following.

Activity

ESTIMATION   MENTAL MATH

Rounding Mixed Numbers in Division

$5\frac{3}{4} \div 2\frac{2}{3}$ Round to $6 \div 3$ so the answer is about 2.

$3\frac{1}{4} \div \frac{5}{6}$ Round to $3 \div 1$ so the answer is about 3.

$\frac{7}{8} \div \frac{1}{4}$ Round to $1 \div \frac{1}{4}$ so the answer is about 4.

Using Other Manipulatives

A brief discussion of the teaching strategies using other manipulative materials is presented in Tables 10.1–10.6 to show the power of using multiple embodiments. Always provide students with free exploration time using the various materials to help them become familiar with the relationships between the components as well as to satisfy their innate need to explore new materials. This will ensure greater participation during the instructional time.

All four operations are presented together, using the same two fractions for each of the four operations. This has been done so the reader can see how the same rational numbers are affected when the operations change.

Table 10.1 Cuisenaire Rods—Traditional Algorithm

This method supports the way most textbooks teach the algorithm for addition and subtraction of fractions. It is the build up of the basic unit of wholeness for one. This method finds the least common denominator without the need to find the lowest terms.

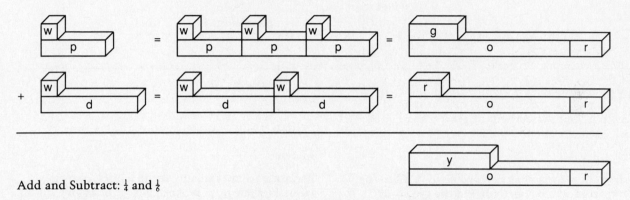

Add and Subtract: $\frac{1}{4}$ and $\frac{1}{6}$

Step 1: Represent each fraction in its original representation.

Step 2: Build up the denominator until both are the same length. Notice that each denominator retains its own uniqueness "purple is still purple," "dark green is still dark green"; now both fractions have the same unit of wholeness.

Step 3: The numerators are increased by the same respective magnitude as the denominators.

Step 4: Both numerators and denominators are re-represented in the equivalent rods, making an exchange for the least amount of rods to show any one number.

Step 5: For addition, add numerators and represent the sum with the appropriate rod. Show the denominator as the part that stands for one. $\frac{5}{12}$

Step 6: For subtraction, compare the green rod with the red rod and show the difference, one white rod. $\frac{1}{12}$ The answer in rods is shown below:

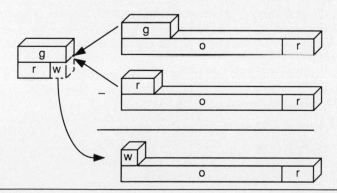

Table 10.2   Pattern Blocks

Pattern blocks may be made from various colors for various shapes. The shapes have an interrelationship with each other as the drawings show below. Find all the different ways to build the yellow hexagon with different combinations of other blocks—using all the same color.

If yellow hexagon = 1, what are the fractional values for other combinations?

       1 green = ?   1 blue = ?   1 red = ?

If red trapezoid = 1, find all the fractional values of the other blocks

 1 green = ?   1 blue = ?   1 yellow = ?   1 red = ?

This activity can be repeated letting other blocks equal 1.

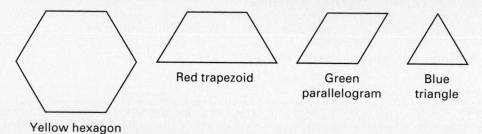

Red trapezoid   Green parallelogram   Blue triangle

Yellow hexagon

To add or subtract with the blocks, you must decide which block to assign the value of 1. It is good to *estimate* if your answer will be greater than or less than 1, greater than or less than one-half.

Add: $\frac{1}{2} + \frac{1}{3}$

Let the yellow hexagon have a value of 1. Then the red trapezoid will be one-half and the green parallelogram will be one-third.

Here's how it looks using the blocks. When both are added together, there is five-sixths of the yellow hexagon covered. Answer in symbols:

$$\frac{1}{2} + \frac{1}{3} = \frac{5}{6}$$

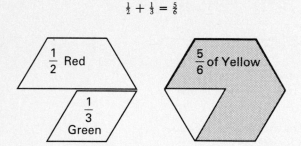

Subtract: $\frac{1}{2} - \frac{1}{3}$

The same values remain for the blocks. Now the green parallelogram ($\frac{1}{3}$ of the yellow hexagon) will be taken away from the red trapezoid ($\frac{1}{2}$ of the yellow hexagon). The area remaining is the blue triangle or one-sixth of the yellow hexagon.

Answer in symbols: $\frac{1}{2} - \frac{1}{3} = \frac{1}{6}$

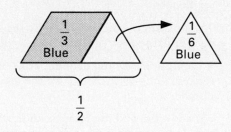

Multiply: $\frac{1}{2} \times \frac{1}{3}$

If yellow is the basic unit for one whole, then $\frac{1}{3}$ is the green parallelogram and $\frac{1}{2}$ of $\frac{1}{3}$ is a half of the green parallelogram which is the blue triangle which, in turn, is $\frac{1}{6}$ of the whole unit (yellow hexagon).

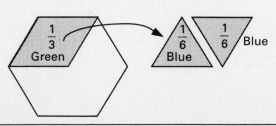

Divide: $\frac{1}{2} \div \frac{1}{3}$

How many green parallelograms ($\frac{1}{3}$) are there in the red trapezoid ($\frac{1}{2}$)? Answer: There are 1 and $\frac{1}{2}$ more green parallelograms in the red trapezoid.

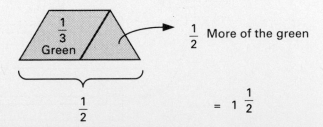

Table 10.3   Multiple Bars

Multiple bars were first introduced in Teaching Strategies. Review the use of multiple bars to show equivalencies. Then do the following activities:

Add:

$$\tfrac{2}{5} + \tfrac{1}{3}$$

2	4	6	8	10	12	14	16	18	20
5	10	15	20	25	30	35	40	45	50

1	2	3	4	5	6	7	8	9	10
3	6	9	12	15	18	21	24	27	30

Get the 2 bar and the 5 bar.

Place the 2 bar above the 5 bar to form the fraction $\tfrac{2}{5}$ at the left of the bars.

All other fractions formed across the bars are equivalent fractions to this one.

Form the second fraction with the 1 bar and the 3 bar.

Since the denominators are unlike, the first step is to locate the least common denominator for both of them.

2	4	6	8	10	12	14	16	18	20
5	10	15	20	25	30	35	40	45	50

1	2	3	4	5	6	7	8	9	10
3	6	9	12	15	18	21	24	27	30

Move one pair of the multiple bars until you have the same denominator lined up.

Once the denominators are alike, the operations can be performed with the numerators.

In this problem, the least common denominator is 15 and the two equivalent fractions are $\tfrac{2}{5} = \tfrac{6}{15}$ and $\tfrac{1}{3} = \tfrac{5}{15}$.

The two numerators are 6 and 5 so, if these numbers are added, the result is 11 with the denominator or label of 15.

$$\tfrac{6}{15} + \tfrac{5}{15} = \tfrac{11}{15}$$

Subtract:

$$\tfrac{2}{5} - \tfrac{1}{3}$$

The same procedure is used to subtract with the multiple bars, only once the numerators have been found, the operation of subtraction is performed on them.

Remember: Move one pair of the multiple bars until you have the same denominator (the least common denominator) for both fractions. Line up the bars. Now you can subtract the numerators.

$$\tfrac{2}{5} = \tfrac{6}{15}$$
$$\underline{-\tfrac{1}{3} = \tfrac{5}{15}}$$
$$\tfrac{1}{15}$$

Table 10.4 Fraction Strip Chart

Each student is given a chart like the one pictured here. The strips have already been pasted on the chart. Each student also receives loose strips of paper measuring the same as the "whole unit" shown on the chart.

The picture below each example shows the actions that students would perform to find the answer using the four operations. The whole is clearly visible at all times; therefore, a student does not lose sight of the basic unit to which all fractions must be compared.

The Fraction Strip Chart

	unit
	halves
	fourths
	eighths
	thirds
	sixths
	twelfths
	fifths
	tenths

Add: $\frac{2}{3} + \frac{1}{4}$

Find $\frac{2}{3}$. It is automatically $\frac{2}{3}$ of the basic unit. Place a strip of paper over the $\frac{2}{3}$ and draw a line to mark it on the strip. Now move to the row of fourths and find $\frac{1}{4}$. Place the edge of the $\frac{1}{4}$ on the edge where $\frac{2}{3}$ ended so you are adding the $\frac{1}{4}$ to the $\frac{2}{3}$.

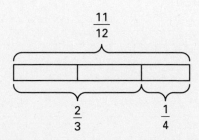

With the strip of paper at the top left edge, move the strip down the rows until it matches an equivalent fraction, which is $\frac{11}{12}$.

The $\frac{2}{3}$ can be easily seen as equivalent to $\frac{8}{12}$ and the $\frac{1}{4}$ can be easily seen as $\frac{3}{12}$ with this method.

Subtract: $\frac{2}{3} - \frac{1}{4}$

Find $\frac{2}{3}$. Place a strip of paper over the $\frac{2}{3}$ and draw a line to mark it on the strip. Now move to the row of fourths and find $\frac{1}{4}$. On the $\frac{2}{3}$ strip, mark off the $\frac{1}{4}$ measured originally on the line below.

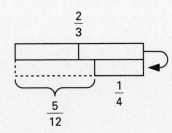

With the strip of paper at the top left edge, move the strip down the rows until it matches an equivalent fraction, which is $\frac{5}{12}$.

The $\frac{2}{3}$ can be easily seen as equivalent to $\frac{8}{12}$ with $\frac{3}{12}$ taken away from it, leaving $\frac{5}{12}$.

Multiply: $\frac{2}{3} \times \frac{1}{4}$

Read the problem as $\frac{2}{3}$ of $\frac{1}{4}$. Then the $\frac{1}{4}$ is found first and it is divided into thirds and two of the thirds are found.

With the strip of paper at the top left edge, run the strip down until it matches an equivalent fraction, which is $\frac{1}{6}$. This way it is seen in its simplified form first.

Note one drawback: Since this method bypasses the $\frac{2}{12}$ which children would find first in the symbolic form, it makes the bridge from concrete to symbolic difficult to see.

Divide: $\frac{2}{3} \div \frac{1}{4}$

Read the problem as how many $\frac{1}{4}$s in $\frac{2}{3}$. Find the $\frac{2}{3}$. Place a finger there and look at the $\frac{1}{4}$ strip. Compare the $\frac{2}{3}$ strip to the $\frac{1}{4}$ strip. It can be seen that there are 2 and $\frac{2}{3}$ of the $\frac{1}{4}$ strip in $\frac{2}{3}$.

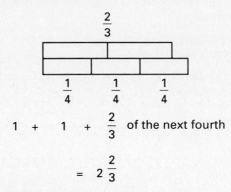

$$1 \;+\; 1 \;+\; \frac{2}{3} \text{ of the next fourth}$$

$$= \; 2\,\frac{2}{3}$$

If the comparison is too difficult to do by just "eyeing" the strips, a $\frac{1}{4}$ strip may be marked from one of the loose strips and then moved down right on top of the $\frac{2}{3}$ strip for comparison.

Note: This procedure works well when the denominator of the second fraction is smaller than the denominator of the first fraction.

Table 10.5 Cuisenaire Equivalency Chart

The chart is essentially the same as the multiple bars except it is *not* cut apart. Transparent strips are used to locate the appropriate fractions, yellow strips for numerators, blue strips for denominators. The entire chart looks like this:

The strips are placed over the appropriate fractions to "highlight" them. The following illustration shows the chart on the left with the highlighted part enlarged.

	1	2	3	4	5	6	7	8	9	10
1	1	2	3	4	5	6	7	8	9	10
2	2	4	6	8	10	12	14	16	18	20
3	3	6	9	12	15	18	21	24	27	30
4	4	8	12	16	20	24	28	32	36	40
5	5	10	15	20	25	30	35	40	45	50
6	6	12	18	24	30	36	42	48	54	60
7	7	14	21	28	35	42	49	56	63	70
8	8	16	24	32	40	48	56	64	72	80
9	9	18	27	36	45	54	63	72	81	90
10	10	20	30	40	50	60	70	80	90	100

	1	2	3	4	5	6	7	8	9	10
1	1	2	3	4	5	6	7	8	9	10
Yellow — 2	2	4	6	8	10	12	14	16	18	20
Blue — 3	3	6	9	12	15	18	21	24	27	30
4	4	8	12	16	20	24	28	32	36	40
5	5	10	15	20	25	30	35	40	45	50
6	6	12	18	24	30	36	42	48	54	60
7	7	14	21	28	35	42	49	56	63	70
8	8	16	24	32	40	48	56	64	72	80
9	9	18	27	36	45	54	63	72	81	90
10	10	20	30	40	50	60	70	80	90	100

(continued)

All Four Operations: $\frac{2}{3}$ and $\frac{1}{4}$

Step 1: The original fractions are represented by the numbers found in the first column to the left and the first row across the top.

Step 2: The important numbers will be where the strips intersect.

Step 3: All operations are possible using the intersection squares of 2, 3, 8, 12:

Note: The colors change where the strips intersect.
2 = dark yellow
3 = green
8 = green
12 = dark blue

For addition:
1. Add the green squares (8 and 3) as the numerators (8 + 3 = 11).
2. The dark blue square, the 12 square, will be the denominator; therefore, answer is $\frac{11}{12}$.

For subtraction:
1. Subtract 8 and 3 as the numerator 8 − 3 = 5.
2. The denominator is the same as above; therefore, answer is $\frac{5}{12}$.

For multiplication:
1. Record the dark yellow square, 2, as the numerator.
2. The denominator is the same as above; therefore, answer is $\frac{2}{12}$.

For division:
1. Top green number is the numerator, 8, and the bottom green number, 3, is the denominator; therefore, answer is $\frac{8}{3}$.

If the first fraction is recorded down the side and the second fraction along the top, then the division problem requiring the 8 to be on top may be easier for students to see.

Students may do a short form of the chart by sketching out their own chart. This method does not automatically give answers in lesser terms as the Cuisenaire traditional algorithm does. See how the next problem works:

$$\frac{5}{8} \text{ and } \frac{4}{7}$$

	4	7	
			add $\frac{67}{56}$
5	20	35	subtract $\frac{3}{56}$
8	32	56	multiply $\frac{20}{56}$
			divide $\frac{35}{32}$

Note: By writing the first fraction down the side, its numerator will always appear on the top row and the numerator of the second fraction will appear in a row below it. This helps students who think of the largest number "on top" in subtraction problems, and prevents a negative amount as a result.

Diagnosis and Remediation

Successive Thought Processing

Some students benefit from the gradual build-up of fractions from the parts to the corresponding whole. Materials like the fraction kits, the multiple bars and the Cuisenaire rod algorithm for adding and subtracting fractions (explained in the preceding section on teaching strategies) are all very helpful. These materials start with one fraction, find its equivalencies, and then compare them to the whole unit before performing the mathematical operation. If students do not respond well to one of these materials or seem to forget where they are in the sequence of building to the solution, materials supporting simultaneous thought processing should be presented.

Simultaneous Thought Processing

Some students benefit from having a "whole" unit constantly within sight as they work on fraction relationships. The whole and its parts are more clearly seen in materials like the fraction strip chart or the Cuisenaire equivalency chart. Both materials allow for problems to be presented by partitioning from the whole to its parts. Students are not required to handle as many small pieces or make as many manipulations as they may be required to do in successive processing models.

Correcting Common Errors

Adding or Subtracting Denominators as Whole Numbers: The denominator of a fraction serves as a label naming the number of parts into which the whole unit has been divided. Sometimes it is helpful to write the denominator in words to help focus on the numerator when teaching the symbolic level of adding and subtracting fractions. Since a common error is to add or subtract *both* the numerators and the denominators, this technique emphasizes that different parts or labels are involved and until the parts or labels are the same, the numerators cannot be added or subtracted.

Common Errors:

$$\frac{2}{3} + \frac{1}{4} = \frac{3}{7}$$

But:

$$\frac{2}{thirds} + \frac{1}{fourths} = \frac{?}{?}$$

Get labels or denominators the same, then perform the operation on the numerators.

$$\frac{8}{twelfths} + \frac{3}{twelfths} = \frac{11}{twelfths}$$

Changing Mixed Numbers to Improper Fractions: Students forget the sequence of steps and which part is added or multiplied. For the mixed number, $5\frac{1}{2}$, the following answers may occur when it is changed to an improper fraction:

$$\frac{5}{2} \quad \frac{1}{10} \quad \frac{6}{2} \quad \frac{6}{10} \quad \frac{8}{2}$$

Analyze what happened to obtain each answer. There are other possible variations also.

One of the most efficient means of remediation is to model the movement with manipulatives using concrete or pictorial materials. Remember findings from the NAEP study indicate that about half of the seventh-grade students were unable to show that $5\frac{1}{4} = 5 + \frac{1}{4}$. More exploratory experiences are needed to build a conceptualization and visualization of the problem. The circle fraction kit works well because the student can show five circles and one-half more of the next circle. Then the student can count the number of halves to find the eleven halves. After the physical model is seen, the teacher can turn the mistake into a problem-solving experience.

Activity

PROBLEM SOLVING

Diagnostic Help

The teacher asks:
What did you do to $5\frac{1}{2}$ to get $\frac{11}{2}$?
Look at some more examples. Model the representation. Generalize a procedure to get these answers every time.

$$2\frac{3}{4} = \frac{11}{4}$$
$$3\frac{4}{10} = \frac{34}{10}$$

Inverting and Multiplying the Incorrect Factors: This problem occurs when the student "inverts and multiplies" the first factor instead of the second when attempting to divide one fraction by another. Some students may invert *both* fractions, thinking the rule applies to both factors.

Original Example   *Student's Work*

$\frac{1}{2} \div \frac{1}{4}$ $\frac{2}{1} \times \frac{1}{4} = \frac{1}{2}$

Switching to an entirely new algorithm is advisable. Just restating the rule may leave the student still confused. The teaching strategy for division using the circular fraction kit would require a different set up and execution of the problem.

$$\frac{1}{2} \div \frac{1}{4} = \frac{2}{4} \div \frac{1}{4} = \frac{2 \div 1}{1} = \frac{2}{1} = 2$$

Learning a totally new algorithm helps the student to focus less on bad habits often associated with the earlier taught algorithm.

Multiplying and Dividing Mixed Numbers: Students know they can add or subtract the whole numbers after they have found the equivalent fractional parts. They assume logically that they can do the same when multiplying or dividing mixed numbers, yielding answers like the following:

Original Example   *Student's Answer*

$2\frac{1}{2} \times 4\frac{1}{4}$ $8\frac{1}{8}$

This type of problem is difficult to remediate. Some elementary textbooks insist on the change of mixed numbers to improper fractions as the first step in any addition or subtraction problem. Such texts hope that insistence on one procedure will eliminate problems with inappropriate transfer of learning later when multiplication and division of mixed numbers is taught. If students have learned different algorithms for addition and subtraction which are causing difficulties, it may be beneficial to go back to addition and subtraction, showing the improper fraction step first. After students have seen that it works with addition and subtraction, stress that the same procedure works for multiplication and division. For example:

With Addition

$$4\frac{1}{4} = \frac{17}{4} = \frac{17}{4}$$
$$+ \ 2\frac{1}{2} = \frac{5}{2} = \frac{10}{4}$$
$$\overline{\qquad\qquad\qquad}$$
$$\frac{27}{4} = 6\frac{3}{4}$$

With Multiplication

$$4\frac{1}{4} = \frac{17}{4} = \frac{17}{4}$$
$$\times \ 2\frac{1}{2} = \frac{5}{2} = \frac{10}{4}$$
$$\overline{\qquad\qquad\qquad}$$
$$\frac{170}{16} = 10\frac{10}{16} = 10\frac{5}{8}$$

Use the circular fraction kit to model the problem. Students should build their understanding from the mathematical representation of $2\frac{1}{2} \times 4\frac{1}{4}$. When students are involved in problem-solving experiences, they can later generalize ideas gained through models. Figure 10.7 illustrates that the problem states $2\frac{1}{2}$ groups of $4\frac{1}{4}$ which produces $10\frac{5}{8}$.

Another way to help students with multiplication of mixed numbers is to have them estimate to see if their answers are reasonable. For example:

To estimate the largest value possible:

$2\frac{1}{2} \approx 3, \quad 4\frac{1}{4} \approx 4, \quad$ so $3 \times 4 = 12 \neq 8\frac{1}{8}$

To estimate the smallest value possible:

$2\frac{1}{2} \times 4 = (2 \times 4) + (\frac{1}{2} \times 4) = 8 + 2 = 10 \neq 8\frac{1}{8}$

The answer must be between 10 and 12. The student's answer of $8\frac{1}{8}$ does not fit reasonably in the problem and this estimation technique alerts the student to check with a different algorithm like the one shown previously or to check with manipulative materials.

Regrouping Fractions As Whole Numbers: Students frequently confuse the regrouping process of whole numbers with the regrouping process required in the subtraction of fractions.

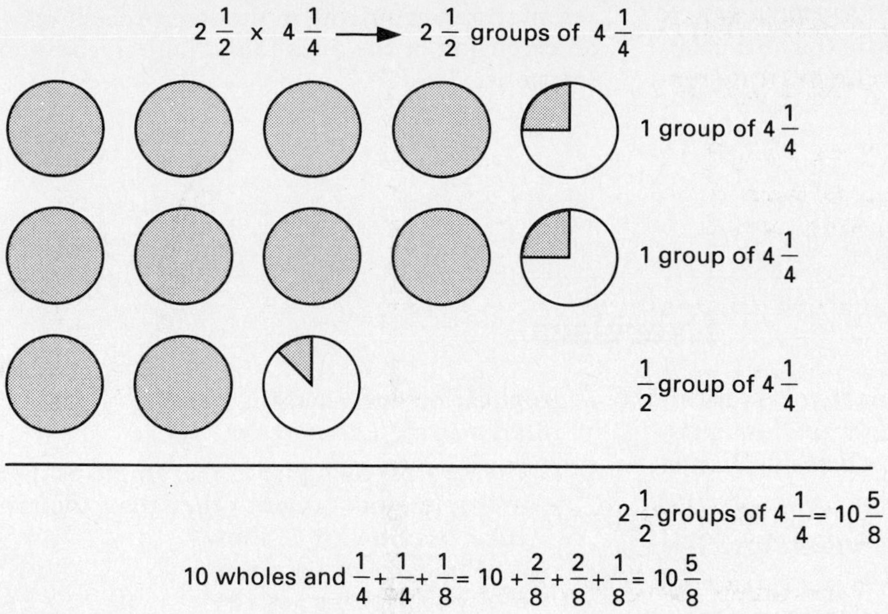

$$2\frac{1}{2} \times 4\frac{1}{4} \longrightarrow 2\frac{1}{2} \text{ groups of } 4\frac{1}{4}$$

1 group of $4\frac{1}{4}$

1 group of $4\frac{1}{4}$

$\frac{1}{2}$ group of $4\frac{1}{4}$

$$2\frac{1}{2}\text{ groups of } 4\frac{1}{4} = 10\frac{5}{8}$$

$$10 \text{ wholes and } \frac{1}{4} + \frac{1}{4} + \frac{1}{8} = 10 + \frac{2}{8} + \frac{2}{8} + \frac{1}{8} = 10\frac{5}{8}$$

Figure 10.7

Original Example		Student's Work
$5\frac{1}{4}$	$=$	$5^{4+1}\frac{2}{8} = 4\frac{12}{8}$
$-3\frac{7}{8}$		

The student regroups the 5 as 4 and 1 but "carries" the 1 over to the renamed two-eighths as if it were a whole number and not the numerator of a fraction. The student has forgotten that the whole number 1 must be renamed as:

$\frac{8}{8}$ and then $\frac{2}{8}$ to $\frac{10}{8}$
added to make

The strategies section stressed the need to ask, "What name is being used for one?" *con-stantly* as a student performs the activities. Students who forget this important step should write the question down and save it in a folder where they can refer to it as they work. The circular fraction chart and the physical manipulatives can help especially if the teacher did not spend much time with the concrete materials in the initial presentation of subtraction problems that require regrouping.

The end of chapter exercises will show the actual work of a student having trouble with one or more of the ideas presented here. You will be asked to apply what you have learned in this chapter to create a plan to reteach the needed skills with common fractions.

Summary

To provide a lasting sense of fraction and their relationships, learning must be experienced in an active way with concise language. In this chapter, rational numbers were developed using many concrete activities, modeling exact terminology to be used with children. The teacher was encouraged to read each activity to feel comfortable with the way to present each operation with common fractions. The fraction strip chart, circular fractions, and multiple bars were among the several manipulatives used to explore frac-tions and their operations. Estimation activities were introduced throughout the chapter as a means to teach the "reasonableness" of answers.

Children encounter more problems when working with fractions than when working with whole numbers. Specific suggestions for correcting common errors were given as a means of remedial help for the teacher to use with children.

The next chapter will discuss rational numbers (specifically as decimals), ratio, proportion, percent, and rate. The chapter includes money in a decimal representation. A calculator is strongly recommended to explore the next chapter as the concepts deal with more intricate rep-resentations, requiring multi-step procedures as an extension of the ones found here in common fractions.

Exercises

Directions: Read all questions *before* answering any one exercise. Frequently the last question in one category leads to the first question in the next category.

A. Memorization and Comprehension Exercises
Low-Level Thought Activities

1. Find all the Cuisenaire rods that represent $\frac{1}{4}$. Use the fractional rods up through two orange rods as the denominator.

2. List in order the recommended sequence for teaching fractions in the elementary and middle school.
   a. Operations with fractions
   b. Equivalence with fractions
   c. Whole-to-part representations of fractions
   d. Partitioning representations of fractions

3. State a better terminology than "reduce to lowest terms" to help children learn mental manipulations of fractions. Explain why "reduce to lowest terms" can be a problem for children.

4. The phrase "key to the basic unit as the whole or 1" has been used repeatedly throughout the chapter. Explain why it is crucial to mention continually to children.

5. Practice mental computation with estimation in the following examples:

Round to: Answer:

$2\frac{3}{5} \div \frac{3}{4}$ _____ = about _____

$1\frac{3}{8} \div \frac{1}{3}$ _____ = about _____

$5\frac{7}{9} \div \frac{5}{8}$ _____ = about _____

$\frac{3}{4} \div 7\frac{1}{2}$ _____ = about _____

B. Application and Analysis Exercises
Middle-Level Thought Activities

1. Create your own pictorial example using a multiple embodiment other than the circular fraction kit to show:
   a. Whole-to-part activity
   b. A partitioning activity
   c. An equivalence activity

2. Compare the common denominator method and the reciprocal method for division of fractions. Create a new number problem (one not seen in this chapter) to show both methods. Show a pictorial model to go along with your number problem. Practice with a model that was more difficult for you to understand as you read the chapter.

3. Analyze the following two lines. What are they asking the computer to do? This formula helps complete a mathematical definition. Which one? (CLUE: These are the lines omitted from the BASIC and Logo programs found in the introduction to this chapter.)

In BASIC
```
110 IF A * D = B * C THEN 150
```
In Logo
```
IF :A * :D = :B * :C [PR [THE
FRACTIONS ARE EQUIVALENT.]]
```

4. Look at one of the mathematics textbook series for the elementary and middle schools. Analyze how many different approaches to fractions are presented within one grade level. How many pages are devoted to concrete and/or pictorial models? What "extra" materials might you as a teacher need to use as supplementary work for your students?

5. Search professional journals for current articles on research findings and teaching ideas with fractions. You may wish to look at the first activity under Synthesis and Evaluation Exercises below and coordinate activities.

Follow the form below to report on the pertinent findings:

```
Journal Reviewed: _____

Publication Date: _____  Grade Level: _____

Subject Area: _____

Author(s): _____

          _____

Major Finding:



Study or Teaching Procedure Outlined:



Reviewed by: _____
```

Some journals of note are listed below:
Journal for Research in Mathematics Education
Arithmetic Teacher
Mathematics Teacher
School Science and Mathematics

6. Computer programs and computer reviews for classroom use with common fractions are found in the journals listed below. Their articles are by no means the only ones. Search the periodical section of the library for more. You may wish to look at the first activity under Synthesis and Evaluation Exercises below to coordinate activities.
Teaching and Computers
Classroom Computer Learning
Electronic Learning
Computers in the Schools
The Journal of Computers in Mathematics and Science Teaching
The Computing Teacher

C. Synthesis and Evaluation Exercises
High-Level Thought Activities

1. Create your own lesson plan for teaching one concept using common fractions. Follow the following steps:
   a. Look at the Scope and Sequence Chart. Find a grade level and a concept within that grade level. Trace the previous knowledge of the concept taught in earlier grades so you will know where to begin your instruction.
   b. Use current articles from professional journals to help plan the direction of the lesson. Document your sources.
   c. Include at least one computer software program as a part of the lesson. Show how it will be integrated with the rest of the lesson.
   d. Include behavioral objectives, concrete and pictorial materials, procedures, evaluation of student mastery.
   e. Use the research-based lesson plan format of i) review, ii) development, iii) controlled practice, and iv) seatwork.

2. Using the software evaluation form in Appendix A, evaluate a simulation, problem-solving, tutorial, or drill and practice program used to teach common fractions.

3. Here is the work of an actual student. Create a remediation plan to help the student overcome the misconceptions of addition and subtraction of fractions. Your plan should include:
   a. An analysis of the error pattern.
   b. An evaluation of the thought processing required in the examples and the thought processing displayed by the student.
   c. The sequential development of concrete, pictorial, and symbolic models to reteach the concept. (See Figure 10.8.)

4. Prepare a lesson using a calculator that has fraction capability. In the lesson, include objectives, how to solve fraction problems using the calculator, activities, and evaluations.

5. Evaluate how teachers can use materials in meaningful ways to convey the real meaning behind abstract concepts like rational numbers. Which materials seem the most valuable to you? Justify your position.

The symbolic problem was all that was given to the student.

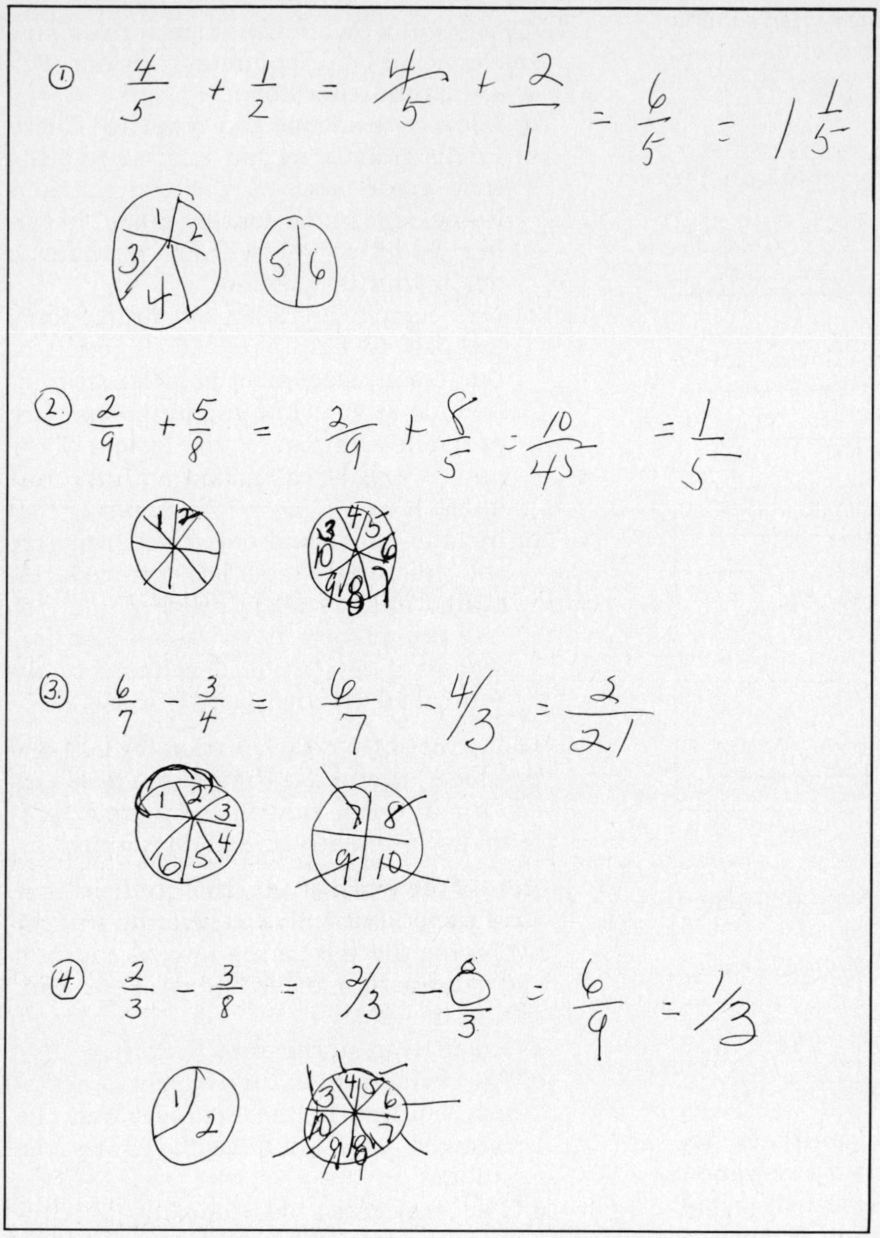

① $\frac{4}{5} + \frac{1}{2} = \frac{4}{5} + \frac{2}{1} = \frac{6}{5} = 1\frac{1}{5}$

② $\frac{2}{9} + \frac{5}{8} = \frac{2}{9} + \frac{8}{5} = \frac{10}{45} = \frac{1}{5}$

③ $\frac{6}{7} - \frac{3}{4} = \frac{6}{7} - \frac{4}{3} = \frac{2}{21}$

④ $\frac{2}{3} - \frac{3}{8} = \frac{2}{3} - \frac{8}{3} = \frac{6}{9} = \frac{1}{3}$

Figure 10.8

Bibliography

Biehler, Robert F., and Jack Snowman. *Psychology Applied to Teaching.* 5th ed. Boston: Houghton Mifflin Co., 1986.

Blaenuer, David A. "Fractions, Division, and the Calculator." *School Science and Mathematics* 84 (February 1984): 113–118.

Brown, Catherine A., Thomas P. Carpenter, Vicky L. Kouba, Mary Lindquist, Edward A. Silver, and Jane O. Swafford. "Secondary School Results for the Fourth NAEP Mathematics Assessment: Discrete Mathematics, Data Organization and Interpretation, Number and Operations." *Mathematics Teacher* 81 (April 1988): 241–248.

Burns, Marilyn. *The Math Solution: Teaching for Mastery Through Problem Solving.* Sausalito, Calif.: Marilyn Burns Education Associates, 1984.

Carnine, Douglas, Siegfried Engelmann, Alan Hofmeister, and Bernadette Kelly. "Videodisc Instruction in Fractions." *Focus on Learning Problems in Mathematics* 9 (Winter 1987): 31–52.

Commission on Standards for School Mathematics of the National Council of Teachers of Mathematics. *Curriculum and Evaluation Standards for School Mathematics.* Working Draft. Reston, Va.: The Council, 1987.

Dossey, John A., Ina V. S. Mullis, Mary M. Lindquist, and Donald L. Chambers. *The Mathematics Report Card, Are We Measuring Up? Trends and Achievement Based on the 1986 National Assessment.* Princeton, N. J.: Educational Testing Service, June 1988.

Driscoll, Mark. "What Research Says." *Arithmetic Teacher* 31 (February 1984): 34–36.

Effline, J. Fred. "A Uniform Approach to Fractions." *Arithmetic Teacher* 33 (March 1985): 42–43.

Grossnickle, Foster E., and Leland M. Perry. "Division with Common Fraction and Decimal Divisors." *School Science and Mathematics* 85 (November 1985): 556–566.

Hollis, L. Y. "Mickey." "Teaching Rational Numbers—Primary Grades." *Arithmetic Teacher* 31 (February 1984): 36–39.

Jacobson, Marilyn Hall. "Teaching Rational Numbers—Intermediate Grades." *Arithmetic Teacher* 31 (February 1984): 40–42.

Kalman, Dan. "Up Fractions! Up n/m!" *Arithmetic Teacher* 32 (April 1985): 42–43.

Kouba, Vicky L., Catherine A. Brown, Thomas P. Carpenter, Mary M. Lindquist, Edward A. Silver, and Jane O. Swafford. "Results of the Fourth NAEP Assessment of Mathematics: Number, Operations, and Word Problems." *Arithmetic Teacher* 35 (April 1988): 14–19.

Lester, Frank K., Jr. "Teacher Education: Preparing Teachers to Teach Rational Numbers." *Arithmetic Teacher* 31 (February 1984): 54–56.

McBride, John W., and Charles E. Lamb. "Using Concrete Materials to Teach Basic Fraction Concepts." *School Science and Mathematics* 86 (October 1986): 480–488.

Ockenga, Earl. "Chalk Up Some Calculator Activities for Rational Numbers." *Arithmetic Teacher* 31 (February 1984): 51–53.

Payne, Joseph N. "Curricular Issues: Teaching Rational Numbers." *Arithmetic Teacher* 31 (February 1984): 14–17.

Prevost, Fernand J. "Teaching Rational Numbers—Junior High School." *Arithmetic Teacher* 31 (February 1984): 43–46.

Scott, Wayne R. "Fractions Taught by Folding Paper Strips." *Arithmetic Teacher* 28 (January 1981): 18–21.

Silvia, Evelyn M. "A Look at Division with Fractions." *Arithmetic Teacher* 30 (January 1983): 39–41.

Steiner, Evelyn E. "Division of Fractions: Developing Conceptual Sense with Dollars and Cents." *Arithmetic Teacher* 34 (May 1987): 36–42.

Trafton, Paul R., and Judith S. Zawojewski. "Teaching Rational Number Division: A Special Problem." *Arithmetic Teacher* 31 (February 1984): 20–22.

Wearne-Hiebert, Diana, and James Hiebert. "Junior High School Students' Understanding of Fractions." *School Science and Mathematics* 83 (February 1983): 96–106.

Wiebe, James. "Discovering Fractions on a 'Fraction Table'." *Arithmetic Teacher* 33 (December 1985): 49–51.

11

Decimals, Ratio, Proportion, Percent, and Rate

Key Question: *What do the concepts of decimal fractions, ratio, proportion, percent, and rate have in common and how can students prepare to use these concepts in everyday life?*

Activities Index

Introduction

The study of ratio, proportion, percent, and rate are largely activities for grade five through grade eight; however, exploratory work with decimals should be included in primary grades. In most textbooks, intensive work with decimals does not begin until students have acquired skill with the manipulation of common fractions. With the increased use of calculators by students, educators may need to reconsider when to start teaching decimals. The NCTM curriculum standards (Commission on Standards for School Mathematics, 1987) call for a greater emphasis to be placed on the development of decimal concepts in grades K–4. Decimals should be related to models and the oral language should develop slowly. Early work with decimals in grades K–4 should relate fractions to decimals. Building a number sense about decimals is an important

329

goal. For example, students should recognize that 0.3 is less than $\frac{1}{2}$ because $\frac{1}{2}$ is the same amount as 0.5.

From decimals, students proceed to the study of ratio and percent, two other focal points in the study of rational numbers. Proportion is an extension of the comparisons developed when studying ratio. Rate is another way of comparing relationships. All five concepts have many real-world applications and frequently become topics of word problems in elementary and middle school problem solving. Performance on items from the National Assessment of Educational Progress fourth mathematics assessment (Kouba et al., 1988) involving decimal concepts indicates a lack of conceptual knowledge. For seventh-grade students, about 60 percent could express simple fractions as decimals, but only 40 percent could express improper fractions as decimals. Less than 10 percent could express a common repeating decimal as a fraction. About one-half of the seventh-grade students were successful comparing decimals. The role of place value is involved with

this skill and it appears that many seventh-grade students do not have a firm understanding of the place value concepts needed to know the difference between 0.02 and 0.002.

A summary of research findings (Hiebert, 1987) shows that many students, including high school students do not realize that decimal fractions are just another way of writing common fractions and do not associate decimals as an extension of the base 10 place-value system. Researchers recommend that students have many introductory experiences with concrete materials before extensive rules are taught. The concrete materials discussed in this chapter will provide a firm background with decimals.

Decimals are as much a part of the rational number system as are common fractions. In fact, decimals are just another name for rational numbers in the base 10 numeration system. Part of the base 10 blocks in Figure 11.1 have already been studied in Chapter 6 on numeration. Chapter 6 concentrated on the build-up of the base 10 whole number system (seen on the left of Fig. 11.1). A new part has been added to the right. It

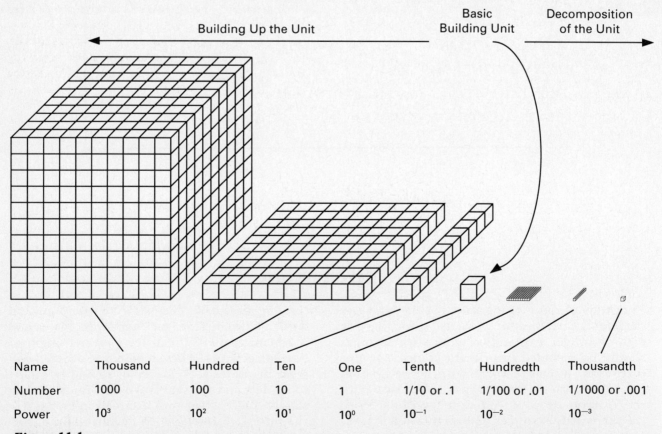

Name	Thousand	Hundred	Ten	One	Tenth	Hundredth	Thousandth
Number	1000	100	10	1	1/10 or .1	1/100 or .01	1/1000 or .001
Power	10^3	10^2	10^1	10^0	10^{-1}	10^{-2}	10^{-3}

Figure 11.1

illustrates the decomposition of the base 10 system. By extending the place-value system into smaller segments of the basic building unit, we obtain the rational numbers. Notice that throughout this section, the discussion will continue to key to the basic building unit just as it was stressed in Chapter 10.

Just as the whole numbers are built up by groups of ten, so the fractional parts of the basic building unit may be partitioned by groups of ten. Every time we move one place to the right, the value of the digit is one-tenth as great. When the first regrouping occurs, its shape is that of a FLAT or one-tenth of the basic building unit. The next regrouping is one-tenth of the small FLAT or one-hundredth of the basic building unit. It looks like a small LONG. The next regrouping is one-tenth of the small LONG or one-thousandth of the basic building unit. It looks like a little BLOCK. Notice that the three new categories of base 10 blocks fit into the same configuration as the whole number system, namely, block-long-flat, block-long-flat (reading from right to left).

Just as the powers of the whole numbers signify which regrouping is taking place in the build-up of the system by tens, so the negative (−) powers tell which regrouping is taking place in the decomposition of the base 10 system. The fractional parts of the base 10 system are known as the *decimal* part of the system. "Deci" is Latin for "part of ten" or "part of a whole broken down by tens." Elementary textbooks stress the point to students by writing a decimal part of a whole with a lead zero like this:

$$0.625 \quad \text{instead of} \quad .625$$

The zero in the above example reminds students that there is no whole number represented there. This text will follow the same principle so that teachers can familiarize themselves with writing decimal representation in this way.

Students should be shown a pictorial model like the one in Figure 11.1 so they can easily see how the total base 10 system works together. Base 10 blocks are only manufactured in units, LONGS, FLATS, and BLOCKS. Upper grade children are encouraged to think of the big BLOCK as the basic building unit, since it is the same configuration as the unit. That way students can use the FLATS as tenths, the LONGS as hundredths, and the small units as thousandths. As shown by the Scope and Sequence

Chart following this introduction, children study the decimal system in the upper grades when they are capable of using an object to represent more than one concept. Hence, changing the representation of the base 10 blocks should present no problems to students, when normal cognitive development occurs (explained in Chapter 2).

Centuries ago in some cultures 01 meant "no whole units and one-tenth". As whole numbers and decimals started to be written together, the "0" in 01 became smaller and smaller to denote where the decimal part of the entire number began. This explanation helps children who sometimes wonder how one-tenth is written as .1 with only one place in its number whereas tens have two places and so on.

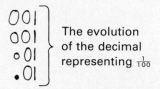

The evolution of the decimal representing $\frac{1}{100}$

There are two kinds of decimal fractions: terminating and repeating decimals. Early and middle grades deal with terminating decimals while repeating decimals are taught in the sixth, seventh, and eighth grades in most mathematics texts. Students should observe the patterns which emerge when exploring terminating and repeating decimals.

Terminating Decimals

Terminating decimals are those that have a definite position or ending point in the base 10 number system. The point can be charted on a number line or its position can be seen as a definite number of base 10 blocks. Common fractions can be seen as terminating decimals in the base 10 number system. An example would be $\frac{1}{2}$ (Fig. 11.2).

One-half ($\frac{1}{2}$) can be seen as $\frac{5}{10}$ (0.5) or $\frac{50}{100}$ (0.50) or $\frac{500}{1000}$ (0.500). It is the same position on the number line. The only thing that is different is whether the basic unit of measure (equaling 1) is subdivided into tenths, hundredths, or thousandths.

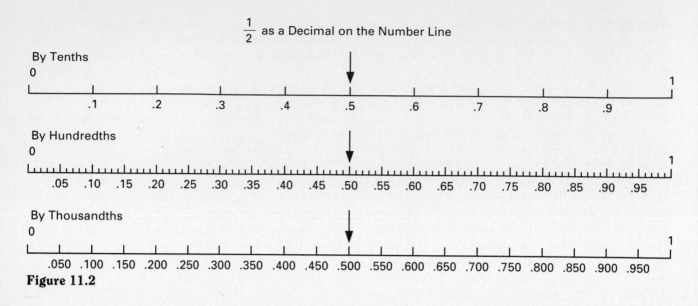

$\frac{1}{2}$ as a Decimal on the Number Line

Figure 11.2

The same fraction, $\frac{1}{2}$, can be seen using base 10 blocks as shown in Figure 11.3. Notice that $\frac{1}{2}$ can be seen in the first regrouping of the basic building unit as $\frac{1}{2}$ of the tenths or five-tenths of the basic unit (written as 0.5). The base 10 blocks also show the markings for the next regrouping (hundredths). If divided by hundredths, there are fifty-hundredths of the basic unit (written as 0.50). If divided by thousandths, (the next and smallest marking on the blocks), there are five hundred-thousandths in $\frac{1}{2}$ of the basic unit (written as 0.500). It represents the same amount whether one talks of tenths, hundredths, or thousandths. There is a definite amount of blocks that can be partitioned to show the termination of the fraction $\frac{1}{2}$; hence, it is a terminating decimal.

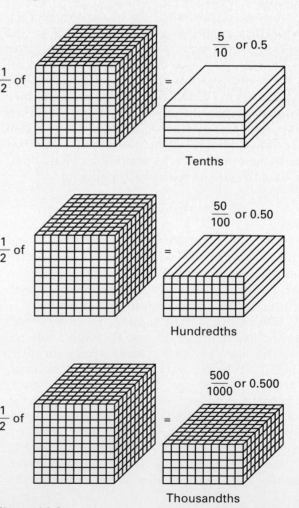

$\frac{1}{2}$ as a Decimal with Base 10 Blocks

Figure 11.3

Activity

MANIPULATIVES

PROBLEM
SOLVING

Terminating Decimals

What will happen if we try to find ¼ as a terminating decimal? Project what will happen before you test your assumptions with physical manipulatives.

1. Set up a base 10 decimal number line like the one in Figure 11.3.

2. Mark where ¼ of the basic building unit would be in tenths, hundredths, and thousandths. What did you discover?

3. Find ¼ using base 10 blocks in tenths, hundredths, and thousandths. What did you discover this time? Does it bear out what you discovered with the number line example? Is that to be expected? What will happen if we apply the same procedure to find ⅛ as a terminating decimal? Follow the same steps in reasoning as you did above. What did you discover?

Repeating Decimals

The existence of repeating decimals is proven in the seventh and eighth grades in most textbook series, although calculators may support an earlier introduction. *A repeating decimal is a nonterminating decimal in which the same digit or block of digits repeats unendingly.* As has been stressed numerous times previously, each new concept should be illustrated with manipulatives. The following exploration can be used with students in middle school as well as adults learning the principle for the first time.

There are some decimal fractions for which we can find no definite point on a base 10 number line and no definite amount of base 10 blocks but we know they are there just the same. Let's use the example of ⅓. Using the same base 10 decimal number line, we can divide it so that there is exactly ⅓ in each segment of the basic building unit (see Fig. 11.4). Remember that the basic building unit equals one whole.

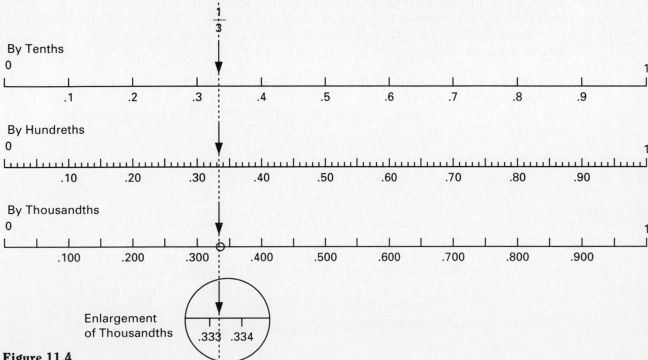

Figure 11.4

Notice that there is no position in tenths, hundredths, or thousandths when $\frac{1}{3}$ touches a definite point; hence it does not terminate. If the number line were divided into ten-thousandths and beyond, $\frac{1}{3}$ would never meet at an exact point in the base 10 decimal system. Children can do some interesting investigations with decimals on the calculator. Let's see what they will discover when they try to divide $\frac{1}{3}$ on a calculator.

Activity

CALCULATORS

Non-terminating, Repeating Decimals

Enter $\frac{1}{3}$ into the calculator.* What is the result?
It could be recorded as:

$\frac{1}{3} = 0.\overline{333333333333}$ The bar over the numbers means that it continues infinitely.

$\frac{1}{3} = 0.\overline{333333333333}$ Now add two more thirds to the original one.

$+ \frac{1}{3} = 0.\overline{333333333333}$

$\frac{1}{3} = 0.\overline{999999999999}$

Mathematicians assert that

$$0.\overline{999999999999}$$

is equal to one whole because of examples like the one above. When a whole unit is divided by thirds nothing is taken away from or added to the basic unit; it remains intact. Therefore:

$$0.\overline{999999999999}$$

is the non-terminating, repeating decimal for the whole number 1. It stands for one complete unit just as $\frac{3}{3}$ stands for one complete unit. This concept is a common question on college aptitude tests because it can be an easily-forgotten mathematical principle.

*If the calculator has a MEMORY PLUS key, you can do the following explorations quickly by entering:
$\frac{1}{3}$ M+ M+ M+ [then press = or MT (for MEMORY TOTAL) or MR (for MEMORY RECALL)]

Activity

CALCULATORS

More Work with Non-terminating, Repeating Decimals

Does every non-terminating, repeating decimal equal:

$$0.\overline{999999999999} \text{ which, in turn, equals 1 ?}$$

Try this assumption with $\frac{1}{7}$. Enter $\frac{1}{7}$ in the calculator. Record the results below.

$$\frac{1}{7} =$$

$$\frac{1}{7} =$$

$$\frac{1}{7} =$$

$$\frac{1}{7} =$$

$$\frac{1}{7} =$$

$$\frac{1}{7} =$$

$$+\frac{1}{7} = \underline{\hspace{3cm}}$$

$$\frac{7}{7} =$$

Find another non-terminating, repeating decimal whose denominator is from 1 to 9. Follow the same procedure as above with the calculator. Record your findings. State in words what you have proven or disproven by your calculations.

If we were to record $\frac{1}{7}$ to eighteen decimal places, this is what we would see:

$$0.\overline{142857}\,\overline{142857}\,\overline{142857}$$

The bars indicate the groups of decimals that are repeated. Because this series goes on infinitely, the decimal is written:

$0.\overline{142857}$ and signifies that this set repeats itself endlessly.

Technically speaking, every decimal can be written as a repeating decimal since

$0.5\overline{0}$
$0.5\overline{00}$
$0.5\overline{000}$
$0.5\overline{0000}$
$0.5\overline{00000}$
$0.5\overline{000000}$

all represent the same place on the number line with an infinite number of zeros annexed. If this approach is followed, the decimal fractions are classified in two categories: (1) repeating, terminating decimals and (2) repeating, non-terminating decimals.

Different Meanings of Rational Numbers

The three different meanings that can be represented by the use of rational numbers were given in Chapter 10. A further explanation is presented here.

Part of a Complete Set and the Complete Set: *A rational number, a/b, is a set of all ordered pairs (a, b) for which a and b are integers (where b ≠ 0).* The denominator of the pair states the number of equal portions into which the whole is divided. The *a* states the portion of *b* as a whole or complete set that is to be considered at any given time. In decimal notation, the denominator is always shown as parts of the base 10 numeration system. Therefore,

$$.a = \frac{a}{10}$$

$$.ab = \frac{ab}{100}$$

The explanations in Chapter 11 to this point are explanations of rational numbers with comparisons to the complete set or unit.

Division: One definition of a rational number is *a number that can be expressed as a quotient of integers:* which means division. Many textbooks introduce this concept after the decimal representation is present because the answers will always be fractional parts of the base 10 number system. The division questions remain the same as those presented in Chapter 7. For the fraction:

$$\frac{1}{4} = 4\overline{)1.00}$$

The question is how many objects are in a set if the basic building unit, the whole number 1 (one), is divided into 4 equal sets? The following activity will show how base 10 blocks can be used with the division of decimals.

Activity

MANIPULATIVES

Division of Decimals with Base 10 Blocks

The Question:

How many base 10 blocks are in a set if the basic building unit (the whole number one) is divided into four equal sets?

The Concrete Manipulations:

The block is partitioned first into tenths and each of the four sets is given an equal number of tenths.

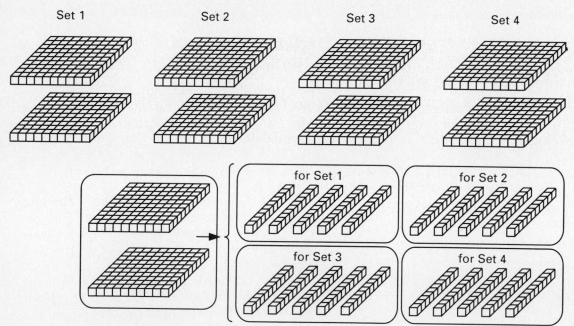

Set 1   Set 2   Set 3   Set 4

for Set 1   for Set 2

for Set 3   for Set 4

There are two-tenths remaining that cannot be divided equally among the four sets unless . . .   the two-tenths are regrouped to the hundredths.

Now there are enough hundredths to divide them equally among the four sets, with each set getting five hundredths.

The answer to the division question is that there are two sets of tenths and five hundredths or twenty-five hundredths in each set.

See how the preceding example is shown at the symbolic level:

$$\begin{array}{r} 0.25 \\ 4\overline{)1.00} \\ \underline{8} \\ 20 \\ \underline{20} \\ \end{array}$$

One whole is going to be divided into 4 equal sets. Eight tenths can be partitioned equally with 0.2 going to each group. That leaves two-tenths to be regrouped to 20 hundredths which are then divided equally so each set gets .05. None remain for another round.

Children need to use base 10 blocks to work many division problems and record step-by-step actions at the symbolic level. Representations should proceed from concrete to connecting (pictorial) to symbolic. This careful sequence helps children to generalize ideas from working with models.

Activity

MANIPULATIVES

PROBLEM
SOLVING

Division of Common Fractions into Decimals

Directions:

1. Study the preceding example and show the actual movements of the blocks for division using the rational number:

$$\frac{2}{4}$$

2. What does $4\overline{)2.00}$ mean in words?

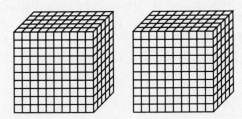

3. If you are working with the whole number 2 you will need to use 2 blocks to start your trades. Write the symbols as you regroup.

4. Now try $\frac{1}{8}$.

Many more examples of common fractions becoming decimal fractions may be tried using the blocks. Other problems to try might be $\frac{1}{2}$, $\frac{1}{5}$, and $\frac{1}{8}$. By problem solving, children can reason that if I find $\frac{1}{5}$, I can also find $\frac{3}{5}$. Children should predict the answer and then check with the blocks to make sure they were correct in their reasoning.

Ratio: The last meaning of rational numbers is that of ratio. *Ratio is defined as the quotient of two numbers or quantities showing the relative sizes of two numbers or quantities.* A ratio may be written one of two ways:

$$\frac{3}{4} \quad \text{or} \quad 3:4$$

This ratio is read 3 to 4 and may be described as 3 elements of one set are present for every 4 elements of another set. Students need to learn how to show comparison of quantities using various forms of ratio. Students should learn the importance of the order of the numbers in the ratio pair.

From the concepts of ratio, students begin to build ideas about proportion, percent, and rate. *A proportion may be defined as the statement of equality of two ratios; an equation whose members are ratios.* Four numbers *a, b, c, d* are in proportion when the ratio of the first pair equals the ratio of the second pair. The numbers *a* and *d* are called the *extremes* of the proportion and *b* and *c* are called the *means* of the proportion.

$$\frac{a}{b} = \frac{c}{d} \quad \text{or} \quad a{:}b \text{ as } c{:}d$$

With numbers

$$\frac{1}{2} = \frac{2}{4} \quad \text{or} \quad 1{:}2 \text{ as } 2{:}4$$

Both forms may be read as "one is to two as two is to four."

A continued proportion is an ordered set of three or more quantities such that the ratio between any two successive ones is the same.

$$\frac{1}{3} = \frac{3}{9} = \frac{9}{27} \quad \text{or} \quad 1{:}3 \text{ as } 3{:}9 \text{ as } 9{:}27$$

Proportional reasoning is important in applications of mathematics. The NCTM curriculum and evaluation standards (Commission on Standards for School Mathematics, 1987) state that students need to have problem situations that can be solved through proportional reasoning. Scale drawings and similar figures involve proportional reasoning.

Percent: *Percent is a ratio, where a quantity is compared to a hundred parts.* The ratio $\frac{1}{2}$ can be represented as a percent by setting up a proportion involving hundredths:

$$\frac{1}{2} = \frac{x}{100} = 50 \text{ percent or } 50\%$$

The word, percent, is short for the Latin, *per centum,* meaning *by the hundred.* Therefore, 50 percent means 50 of the hundred parts.

Activity

MENTAL MATH

CALCULATORS

Percentages

Look at the ratios on the left and figure what proportion of a hundred each would be. That will be the percentage on the right. Try to figure the proportion in your mind without the use of pencil or paper. Some are done for you as examples.

Ratio	Proportion of 100 $\left(\frac{x}{100}\right)$	Percent
$\frac{3}{5}$	(Do this step mentally)	60
$\frac{1}{8}$	?	12.5
$\frac{3}{8}$	?	?
$\frac{4}{8}$	?	?
$\frac{7}{8}$	?	?
$2\frac{7}{8}$	$\frac{200}{100} + \frac{87.5}{100}$	?
$20\frac{7}{8}$	?	?

Check with a calculator to see if you were correct with your mental math. Save your work from this activity. It will be used again in the Teaching Strategies section under Percent to make a computer spreadsheet for students.

Rate: Rate is a term for a ratio when the ratio involves comparisons of two units uniquely different from one another. The definition of rate states that a *certain magnitude, element, or quantity is compared in relation to a unit of something else.* Some examples of rate are:

a rate of eight cents a pound

a rate of fifty-five miles per hour

an interest rate of twelve percent on every dollar

Rate questions apply to real-life situations and are sure to be included in problem-solving activities in elementary and middle school textbooks and standardized tests.

Scope and Sequence

Decimals

Second Grade
- Using decimal notation writing money.
- Adding and subtracting money.

Third Grade
- Adding and subtracting decimals.
- Reading and writing tenths.
- Adding and subtracting money.

Fourth Grade
- Place value through hundredths.
- Writing decimals in word form.
- Rounding to nearest dollar.
- Adding and subtracting decimals and money.
- Estimating sums and differences.
- Operations in horizontal form.
- Ordering and comparing decimals.
- Decimals and decimal fractions.
- Multiplying and dividing with money.

Fifth Grade
- Place value through thousandths.
- Rounding decimals and money.
- Adding and subtracting decimals with different decimal places.
- Multiplying decimals and whole numbers.
- Estimating sums, differences, and products.

- Comparing and ordering decimals.
- Dividing decimals by whole numbers.
- Using decimals with percents and decimal fractions.

Sixth Grade
- All operations with decimals.
- Place value through ten-thousandths.
- Changing fractions to decimals and decimals to fractions.
- Writing, ordering, comparing.
- Rounding decimals and annexing zeros.
- Using decimals with money for all operations.

Seventh Grade
- All operations with decimals.
- Place value through ten-thousandths.
- Changing fractions to decimals and decimals to fractions.
- Writing, ordering, comparing decimals.
- Scientific notation using decimals.
- All operations with decimals in money.
- Fractions, decimals, percents.
- Terminating and repeating decimals.

Eighth Grade
- Review of all previous concepts.

Ratio, Percent, Proportion, Rate

Fifth Grade
- Ratio as fraction and equal ratios.
- Interpret scale drawings with ratios.
- Write fractions as percent.
- Write percent as fraction.
- Write a decimal for a percent.

Sixth Grade
- Ratio as fraction and equal ratios.
- Solve proportions.
- Use cross products.
- Find dimensions using scale drawings.
- Change percent to fractions or decimals.
- Find percent of a number.
- Change decimals to percent.
- Find a number when percent is known.
- Use percents and proportions.

Seventh Grade
- Write ratios.
- Solve proportions.
- Write a fraction as percent and percent as fraction.
- Write decimal as percent or percent as decimal.
- Find percent; find what percent one number is of another; find a number when percent is known.
- Use proportions for scale drawings.

Eighth Grade
- Review all previous concepts.
- Use algebraic expressions to solve equations.

Teaching Strategies

Decimals

Decimals are presented as an extension of the concepts of fractions in most elementary texts. An understanding of place value and common fractions helps students learn decimal notation. This means that a lot of work with fractions such as $\frac{3}{10}$, proves valuable when connecting new symbolic notation (decimals) for fractions. Early exposure to decimal notation occurs in second or third grade when money concepts are developed. Pennies are related to hundredths and dimes are related to tenths. Children generally understand that pennies and dimes are parts of a dollar. Money notation found in textbooks provides experiences to count collections of dimes and pennies and to determine the total amount of money. This concrete, meaningful approach is an important bridge for learning ways to represent decimals. While the cent sign (¢) is used with young children, decimal notation is being used in earlier grades since calculators record money transactions easily and some primary economics units include collections of more than a dollar.

How much money?

Dimes   Pennies

$0.32 in all

As with whole numbers, reading, writing, and interpreting decimals are important skills. Opportunities should be provided for children to take oral dictation of decimals. Refer to activities presented in Numeration, Chapter 6, for ideas about developing these skills. It is important to note again that the word "and" is said for the decimal point. It separates the whole units and the parts of wholes. This is why it is important for students to read large whole numbers *without* the word "and." It can be emphasized by the following example:

Correct: 645.03—six hundred forty-five and three hundredths

Incorrect: 645.03—six hundred and forty-five and three hundredths

When learning about decimals, it is important to require students to read the decimal point with the place-value terms (tenths, hundredths, thousandths) rather than as a "point" and the names of the digits. If this is not stressed, students cannot acquire the place-value ideas as quickly nor will the connection between fractions and decimals be seen as clearly. The following examples may help point out this difference.

Activity

PROBLEM SOLVING

Reading Decimal Numbers

Which reading makes you understand the number better so you can write it, model it, and interpret its value?

The number	What if it were read as
378,294	"three hundred seventy-eight thousand, two hundred ninety-four"
378,294	"three seven eight two nine four"
284.325	"two hundred eighty-four point three two five"
284.325	"two hundred eighty-four and three hundred twenty-five thousandths"

The following activity helps students to read decimals correctly. Notice how the activity can be integrated with mental math. Informal activities of this nature are valuable to build place value awareness.

Activity

MENTAL MATH

Cover Up Game and Add Some More

Directions:
Start with a mixed number:

3 sections cut in
top of flip chart

1. Begin with:

38 . 479

2. Cover up the
   decimal fraction

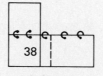

38

Flip paper so
decimal point is
now showing

3. Read "and"

4. Read number following
   decimal point as if it
   were a whole number

479

5. Read place-value label
   of last digit

"thousandths"

Add the Skill of Mental Math . . .

6. As students practice reading various numbers in the sequence suggested in steps 1 to 5, have them add one more tenth, hundredth, or thousandth to the number and read it again. The procedure could continue while students are waiting in line with a few minutes to spare. They get the needed practice breaking down the parts of the mixed fraction and have a chance to practice mental math at the same time.

From students' earlier experiences with fractions, they should know that tenths represent a whole unit partitioned into ten equal parts. Likewise, they need to realize that 10 tenths equal the whole unit ($\frac{10}{10} = 1$). In a similar fashion, they should know that if the whole unit is partitioned into one hundred equal parts, $\frac{100}{100}$, each part is 1 out of the 100 or $\frac{1}{100}$ (one hundredth).

One of the most common and readily available models for decimals are place-value blocks (base 10 blocks). Another effective model is decimal squares made from chart or graph paper. Here, the unit square is divided into 10 equal parts (LONGS) and each one represents a tenth. If the unit square is divided into additional equal parts (each tenth is partitioned into 10 equal parts), the overall effect is to divide the unit square into 100 equal parts (Fig. 11.5).

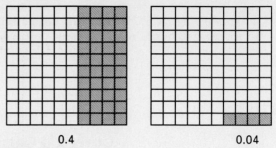

0.4   0.04

Figure 11.5

Provide opportunities for students to practice reading, writing, and representing decimals. An understanding of the place-value interpretation of decimals is a prerequisite to introducing computation with decimals. Students should identify the place-value positions of each digit as well as determine the relationships among the places, such as 10 hundredths equal 1 tenth.

Activity

MANIPULATIVES

Different Ways to Represent Decimals

Directions:

1. Read decimals like these: 0.4   0.7   0.45   0.758

2. Shade decimal squares to represent the decimals read. (See the figure shown directly before this activity.)

3. Model with FLATS, LONGS, and units as the decimals are read. (See introduction to the chapter.)

4. Connect with the pictorial level by drawing decimal squares and base 10 blocks for the examples above.

5. Use place-value grids and record each decimal.

Ones	Tenths	Hundredths	Thousandths

6. Connect with the symbolic level by writing the two ways to represent decimals—$\frac{4}{10}$ and 0.4.

Before handling operations with decimals, students need to explore and understand relationships among decimals by comparing decimals. Again, it is important to provide experiences for modeling the numbers with place-value blocks or with graph paper (decimal squares) as concrete manipulatives. A pictorial comparison can be created by this Logo program.

Activity

COMPUTERS

Decimal Comparisons with Logo Pictures

Directions:

1. This program is saved under the name FLAG.

2. Type LOAD   "FLAG to load the program into the computer.

   (If you have no disk, type in the following program . . .) *Please Note:* Leave a space *before* the : but NOT after it—just as it is shown below:

```
TO FLAG :SIZE
FD :SIZE * 60
REPEAT 3 [FD :SIZE * 60 RT 120]
HT SETPOS [0 0]
END
```

3. Now type FLAG 1 to see the basic building unit of one to which all the decimal flags will be compared.

4. Now type FLAG .5   FLAG .25   FLAG .333 and see what happens. Try other decimals.

5. If you wish to see one flag alone, type CS for clear screen between decimal flag commands.

6. If you wish to see decimals greater than one, change the TO FLAG :SIZE procedure by changing the 60 to 20 in the two places. Appendix B contains the directions for editing a program. Now try decimals such as FLAG 1.5 FLAG 2.93 etc.

After introducing hundredths, the difference between four-tenths and four-hundredths must be called to the students' attention. Modeling helps focus on the size of the number and provides the visual reference needed to correctly record numbers as fractions. Confusion may occur when comparing and interpreting the numbers 0.20 and 0.02. Many opportunities with materials at the concrete and pictorial level are necessary for children to internalize the notation and values of decimals. The following program in Logo helps develop the comparisons using the pictorial level.

Activity

COMPUTERS

Decimal Comparisons with Hundredths in Logo

Important: This program has features that can be used as a lesson model for a variety of Logo learning programs that you, as a teacher, may wish to develop for children.

Directions:

1. This program is saved under the name DECIMALS. Type ERALL and press return.

(continued)

2. Load the program by typing `LOAD "DECIMALS`

   (If you have no disk, type the set of procedures which follow step 6.)
   NOTE: To save space, procedures have been typed side by side, but each
   is to be typed separately when duplicating what is printed here on the
   computer.

3. After the program is loaded, type the word `DECIMALS` to see the whole
   program. Answer some questions incorrectly to see what the program
   will do.

4. The procedures are printed in clusters with a brief explanation of the
   role each cluster plays in the program. It will be helpful to see the procedures printed after step 6 as you gain an appreciation of the work that
   goes into making creative programs.

5. Now type the following names of procedures one at a time to learn what
   part each contributed to the overall program.

```
FACE    SMILE   CRY   TEAR1   TEAR2    NOTEAR1
NOTEAR2   TEARS   LINE   GRID   SOLIDSQ
NEWGRID  EXAMPLE1  EXAMPLE2  EXAMPLE3  EXAMPLE4
DIRECTIONS   LESSON   DECIMALS
```

6. This set of procedures can be used to tell students if their answers are
   correct or not.

```
TO FACE                        TO SMILE
SETH 0                         FACE SETH 180
PU SETPOS [75 -30] PD          PU SETPOS [95 -40] PD
SETPC 3 CIRCLER 30             SETPC 4 ARCL 10 180
PU SETPOS [90 -30] PD          HT WAIT 120
SETPC 5 CIRCLER 2              END
PU SETPOS [120 -30] PD
CIRCLER 2 SETPC 2              TO CRY
PU SETPOS [105 -40] PD         FACE SETH 0 PU
FD 5                           SETPOS [95 -50] PD
END                            SETPC 4 ARCR 10 180
                               TEARS
                               END


TO TEAR1                       TO TEAR2
SETH 180 PU                    SETH 180 PU
SETPOS [90 -30] PD             SETPOS [90 -30] PD
SETPC 1 RT 90 FD 5 BK 5        SETPC 1 RT 30 FD 5 BK 5
LT 30 FD 8 BK 8 LT 30          LT 30 FD 8 BK 8 LT 30
FD 5 BK 5                      FD 5 BK 5
END                            END


TO NOTEAR1                     TO NOTEAR2
SETH 180 PU                    SETH 180 PU
SETPOS [90 -30] PD             SETPOS [120 -30] PD
SETPC 0 RT 30 FD 5 BK 5        SETPC 0 RT 30 FD 5 BK 5
LT 30 FD 8 BK 8 LT 30          LT 30 FD 8 BK 8 LT 30
FD 5 BK 5                      FD 5 BK 5
END                            END


TO TEARS
HT
REPEAT 3 [TEAR1 TEAR2 NOTEAR1 NOTEAR2]
END
```

7. The next set of procedures are used to represent the flat or graph paper.

```
TO LINE                      TO SOLIDSQ
LT 90 BK 10 RT 90            PD REPEAT 10 [FD 10 RT 90
FD 100 BK 100                FD 1 LT 90 BK 10]
END                          END


TO GRID                      TO NEWGRID
REPEAT 4 [FD 100 RT 90]      SETPC 2 SETH 0 PU ST
REPEAT 9 [LINE]              SETPOS [-50 0] PD
LT 90 BK 10                  GRID GRID
END                          END
```

8. The next set of procedures present the learning concepts of the lesson.

```
TO EXAMPLE1
NEWGRID REPEAT 10 [SOLIDSQ]
PR [IS THIS DECIMAL PART .1 OR .01?]
TYPE [TYPE AN ANSWER,] TYPE CHAR 32
PR :NAME
TEST READLIST = [.1]
IFTRUE [SMILE]
IFFALSE [CRY]
WAIT 120
CLEARSCREEN
CLEARTEXT
END


TO EXAMPLE2
NEWGRID SOLIDSQ
PR [IS THIS DECIMAL PART .1 OR .01?]
TYPE [TYPE AN ANSWER,] TYPE CHAR 32
PR :NAME
TEST READLIST = [.01]
IFTRUE [SMILE]
IFFALSE [CRY]
WAIT 120
CLEARTEXT
PR [1 OUT OF 100 PARTS IS .01.]
PR [10 OUT OF 100 PARTS IS .1.]
WAIT 360
CLEARSCREEN
CLEARTEXT
END


TO EXAMPLE3
NEWGRID REPEAT 3 [SOLIDSQ]
PR [IS THIS DECIMAL PART .3 OR .03?]
TYPE [TYPE AN ANSWER,] TYPE CHAR 32
PR :NAME
TEST READLIST = [.03]
IFTRUE [SMILE]
IFFALSE [CRY]
WAIT 120
CLEARSCREEN
CLEARTEXT
END
```

(continued)

```
TO EXAMPLE4
NEWGRID
REPEAT 3 [REPEAT 10 [SOLIDSQ] RT 90 BK 100 LT 90 FD 10]
PR [IS THIS DECIMAL PART .3 OR .03?]
TYPE [TYPE AN ANSWER,] TYPE CHAR 32
PR :NAME
TEST READLIST = [.3]
IFTRUE [SMILE]
IFFALSE [CRY]
WAIT 120
CLEARSCREEN
CLEARTEXT
PR [3 OUT OF 100 PARTS IS .03.]
PR [30 OUT OF 100 PARTS IS .3.]
WAIT 360
CLEARSCREEN
CLEARTEXT
END
```

9. The next set of procedures can be used as a general format to put a lesson together.

```
TO DIRECTIONS
PR [I WILL SHOW YOU A GRID THAT]
PR [REPRESENTS THE WHOLE NUMBER, 1.]
PR []
NEWGRID WAIT 120
PR [SOME PART WILL BE SHADED.]
PR [TELL WHAT DECIMAL IT REPRESENTS.]
TYPE [HERE WE GO,] TYPE CHAR 32
PR :NAME
WAIT 460
CLEARSCREEN
CLEARTEXT
END

TO LESSON
EXAMPLE1
EXAMPLE2
EXAMPLE3
EXAMPLE4
PR :NAME
SMILE
PR [THANKS FOR WORKING WITH ME.]
END

TO DECIMALS
SMILE
PR [HI! WE ARE GOING TO LEARN ABOUT]
PR [DECIMALS.]
PR [TYPE IN YOUR NAME.]
MAKE "NAME READLIST
PR []
PR [THANK YOU,] PR :NAME
WAIT 100
DIRECTIONS
LESSON
END
```

10. Students love the personalization that PR :NAME allows, and it can become a motivating factor when children of all ages are doing a program.

11. Notice that Logo automatically drops the 0 in the whole number place value in examples like 0.3 and 0.03. This may be a disadvantage of the program if a teacher is trying to stress the way most textbooks begin teaching decimals with the 0 in the whole number place value. Some Logo versions will code a decimal like .03 as 1.N2 when the procedure is stored in the program, but students who type in .03 will be counted as correct at the appropriate times.

For those who like a challenge:
Create two more examples following the formats given above. Call them EXAMPLE5 and EXAMPLE6. Place them in the LESSON procedure and run the program.

A quick game may be used on a pictorial level. Using a deck of cards, the students take turns drawing a card. Each student models the number on the card. Each person must read the number to the group, explain its value, and the group compares the numbers within the group to determine greatest to least in value. It is important to include equivalent forms (0.6 and 0.60).

In teaching the four basic operations with decimals, it is important to build upon prior understanding of the place value of decimals and the algorithms used with whole numbers. Since the algorithms are the same, modeling them with a variety of materials provides greater understanding and faster generalization.

Addition: The following procedure should be followed when working with addition of decimals (See Fig. 11.6).

1. Represent the decimals with a concrete material.
2. Estimate whether the answer will be more than one or less than one. The same principle applies as discussed in Chapter 10 with common fractions.
3. Combine the concrete material making exchanges when necessary.
4. Record the actions and the answer in symbols.

When introducing the symbolic level, students must begin with the smallest unit (the unit farthest to the right) and combine these units first as in whole number addition. An important task to master and remember when adding or subtracting decimal numbers is to align the decimal points so the place values are kept in mind. This is especially necessary when problems are presented in a horizontal form. Some textbooks stress the alignment of the decimal point while placing the zero in the appropriate decimal place values. The base 10 blocks seen in the introduction to this chapter show why 0.8 and 0.80 are the same amount. Working with the manipulatives makes this apparent to students. Have one student work with the manipulative materials while the other student does the symbolic recording of the answer.

Subtraction: The same basic principles apply to subtraction as apply to addition since one is the inverse of the other (See Fig. 11.7).

According to the seventh-grade students' performance on the NAEP items (Kouba et al., 1988) involving decimal computation, students have learned to compute with decimals without internalizing the related conceptual understanding. About 40 to 60 percent could perform simple computations. Students were more successful adding (59 percent) or multiplying (62 percent) with decimals than subtracting (48 percent) or dividing (36 percent) with decimals.

The Problem   1.75 + 0.80

1 and 75 hundredths of the next whole

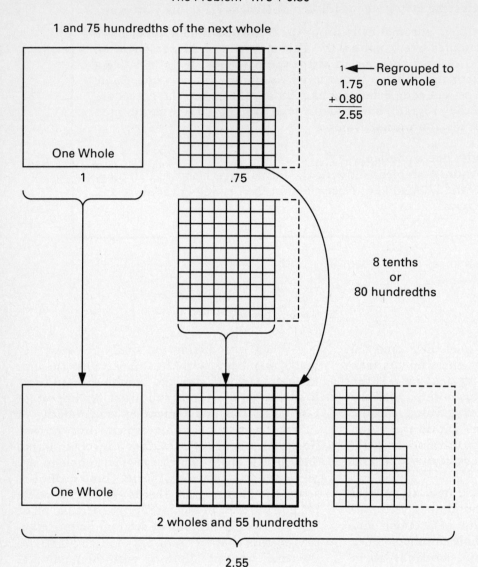

1.75
+ 0.80
‾‾‾‾‾
2.55

8 tenths
or
80 hundredths

2 wholes and 55 hundredths

2.55

Figure 11.6

1. Represent the first decimal with a concrete material.
2. Estimate if the answer will be greater or less than one.
3. Working with the smallest unit first, subtract the appropriate decimal making exchanges when necessary.
4. Record the actions and the answer in symbols.

The steps in mental computation include: estimating the answer, rewriting the problem with decimal points aligned, annexing zeros when appropriate, and following the procedure given above to check the answer.

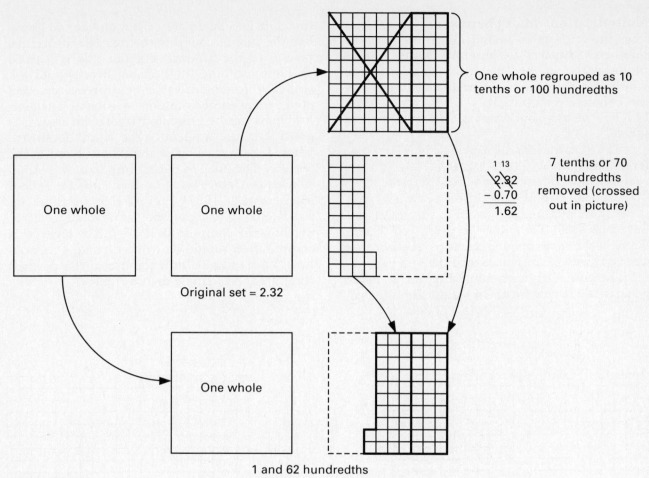

One whole regrouped as 10 tenths or 100 hundredths

$$\begin{array}{r} {\scriptstyle 1\ 13}\\ \cancel{2.}\cancel{3}2\\ -\ 0.70\\ \hline 1.62 \end{array}$$

7 tenths or 70 hundredths removed (crossed out in picture)

One whole

One whole

Original set = 2.32

One whole

1 and 62 hundredths

Figure 11.7

Activity

ESTIMATION

MENTAL MATH

Rounding with Decimals

Directions: Look at the problems on the left and estimate how many whole numbers will be in the answer, if any. Fill in the rest of the chart. The first two are completed as a model.

Problem	Mental Rounding	Estimated Answer
3.4 − 2.9	Think: 3 − 3	Less than 1
7.85 − 4.2	Think: 8 − 4	Around 4
11.34 − 0.895		
1.11 − 0.999		
345.25 − 245.5		
0.927 − 0.398		

Check how close you came in the above activity by computing the answer.

Multiplication: Multiplication of decimals follows the same rules as multiplication of whole numbers (Chapter 8). Therefore, multiplication of decimals can be modeled using the same words and patterns with the base 10 blocks as the concrete manipulatives. In Figure 11.8 the FLAT is the basic building unit representing one whole.

The illustrations of a partial FLAT are enlarged in Figure 11.9 and Figure 11.10 to the actual size of the base 10 blocks. Therefore, they may appear out of proportion to you as you study these pictures compared to the others you have just seen. They may be presented to students in the same manner as strips of paper that can be folded or measured into the 100 parts.

Encourage students to find patterns and to generalize the rule for the multiplication of dec-

imals. It may seem easier and quicker to establish the rule of counting the number of decimal places for the product, but the rule is learned without meaning. Rather, teachers should ask students for place-value interpretations and guide them in deciding on reasonable answers. Problems can be presented where the answer is given but the students task is to determine where the decimal point should be placed in the answer. The NCTM curriculum and evaluation standards (Commission on Standards for School Mathematics, 1987) suggest that valuable instructional time should not be devoted to learning how to compute 0.31×0.588 since this computation should be worked using a calculator. The standards limit the use of the decimal computation to single-digit numbers.

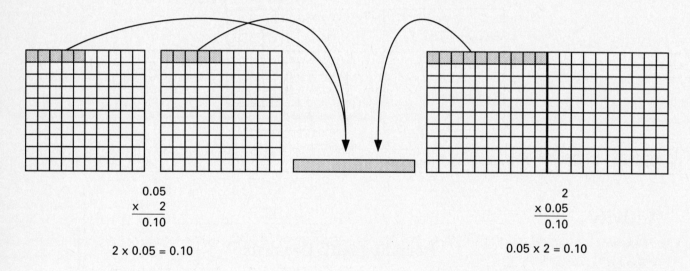

$$
\begin{array}{r}
0.05 \\
\times\ \ \ 2 \\
\hline
0.10
\end{array}
\qquad
\begin{array}{r}
2 \\
\times\ 0.05 \\
\hline
0.10
\end{array}
$$

2 x 0.05 = 0.10   0.05 x 2 = 0.10

Two sets of five-hundredths

Five-hundredths of the whole number two (two has to be divided into 100 parts and then 5 of the 100 is found)

Figure 11.8   The answer is one-tenth.

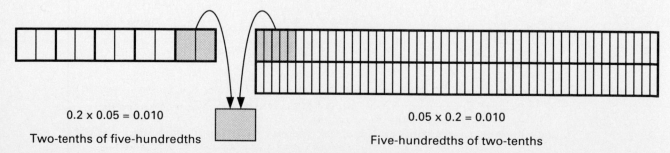

0.2 x 0.05 = 0.010   0.05 x 0.2 = 0.010

Two-tenths of five-hundredths   Five-hundredths of two-tenths

The answer is one-hundredth.

Figure 11.9

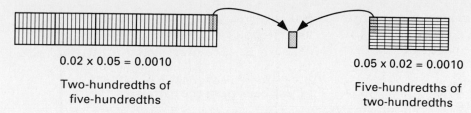

0.02 x 0.05 = 0.0010   0.05 x 0.02 = 0.0010

Two-hundredths of   Five-hundredths of
five-hundredths   two-hundredths

The answer is one-thousandths of the entire flat.

Figure 11.10

Activity

PROBLEM
SOLVING

Observations of Multiplication with Base 10 Blocks

Directions:

1. Look at the pictorial and symbolic representations of multiplication shown in Figures 11.8, 11.9, and 11.10.

2. Fill in the chart. The first three are done as an example.

Problem	Answer	Number of Decimal Places in the Problem	Answer
2 × 0.05	0.10	2	2
0.05 × 2	0.10	2	2
0.2 × 0.05	0.010	3	3

3. Look for a pattern when working with decimals.

4. Generalize a rule for multiplying decimals using the pattern for clues.

Activity

COMPUTERS

A Spreadsheet for Multiplication of Decimals

MENTAL MATH

Directions:

1. Set up the spreadsheet as shown on page 352. Type the top row and left column first.

2. You may type the decimals with the zero as the place holder for the whole number (i.e., 0.01), but the computer program will automatically record it *without* the zero (i.e., .01).

3. Now set up the formula for multiplication in each cell, and watch the numbers that result as the multiplication is done. What patterns do you see?

(continued)

	A	B	C	D	E	F	G
1	X	1	.1	.01	.001	.0001	.00001
2	.05						
3	.1						
4	7						
5	8.74						
6	16.8						
7	24.91						
8							

Mentally compute what will happen in these cells from the pattern.

For cell B2, type the formula +a2*b1
For cell C2, type the formula +a2*c1
For cell B3, type the formula +a3*b1
(Complete the other cells, following the same pattern.)

4. Now change the numerals in the left column to other decimal combinations with and without whole numbers in various place values. What happens instantly?
5. Change the numerals in the top row to other decimal notations, keeping the pattern . . . (i.e., 3 .3 .03 .003). What happens instantly?
6. Change other top row and left column numerals until you can estimate what will happen without changing any more numerals.
7. Generalize a rule for the multiplication of decimals based on what you have discovered.
8. Save this spreadsheet to be used with division in the next section.

If you do not have a computer spreadsheet program available, you can do the same thing by constructing a chart and using a calculator to see the answers for each cell. Record each answer on the chart as the calculator gives it to you. The calculator will not allow you to see what will happen instantaneously as the spreadsheet program will do.

Provide experiences where students decide where the decimal point should go in a given answer. Students can use estimation to justify their answers. There should be time spent on simple calculations without a calculator; however, valuable instructional time should not be spent on tedious paper-and-pencil calculations. This can be done more readily using a calculator.

Division: Division of decimals follows the same rules as division of common fractions (Chapter 10). Therefore, division of decimals may be modeled with the same words and patterns as have been used before. For example, the base 10 blocks seen in the multiplication of decimals can be used if the question is changed to represent division. An example of partitive division was found earlier in this chapter on page 336. Table 11.1 shows models in measurement division so you can become familiar with both kinds of division as they are used with decimals. (If needed, review Chapter 7 for discussion of the two kinds of division.) As the decimal place values extend into the thousandths and ten thousandths, it becomes virtually impossible to do the entire problem with concrete manipulatives. The answers must be reasoned from the experience children have had in the past. Estimation should be used in problem-solving situations with decimals. Students should learn to estimate answers and evaluate the reasonableness of answers. Calculators can be used to perform routine computations and should be the appropriate tool for solving division of multidigit numbers.

$$0.001\overline{)1000}$$

Division still asks the question, "*How many sets of one-thousandths (0.001) are there in 1000?*" Manipulatives can show the beginning of the thought process. If a cube represents one whole, then a tiny cube will be one-thousandth (0.001) of the whole (Fig. 11.11).

Now imagine that there are 1000 cubes like the one on the left representing the whole number of 1000. Then the same one-thousandth pictured here would make up a very, very tiny portion of the new number. So the question, "How

Table 11.1 Measurement Division

Figure	In Symbols	The Questions in Division Are
11.9	$0.05\overline{)0.10}$ with quotient 2	How many sets of five-hundredths are there in 0.10? THERE ARE TWO SETS OF 0.05 IN 0.10.
	$2\overline{)0.10}$ with quotient 0.05	How many sets of two are there in 0.10? THERE ARE 5 OF THE HUNDREDTHS (0.05) IN 0.10.
11.10	$0.05\overline{)0.010}$ with quotient 0.2	How many sets of five-hundredths are there in 0.01? THERE ARE TWO-TENTHS OF 0.05 IN 0.01. (The 0.2 can be seen as $\frac{1}{5}$ of 0.05 with the base 10 blocks but since $\frac{1}{5}$ is NOT a power of ten, 0.2 is the answer.)
	$0.2\overline{)0.010}$ with quotient 0.05	How many sets of two-tenths are there in 0.01? THERE ARE FIVE OF THE HUNDREDTHS (0.05) IN 0.01.
11.11	$0.05\overline{)0.0010}$ with quotient 0.02	How many sets of five-hundredths are there in 0.001? THERE ARE TWO-HUNDREDTHS OF 0.05 IN 0.001.
	$0.02\overline{)0.0010}$ with quotient 0.05	How many sets of two-hundredths are there in 0.001? THERE ARE FIVE-HUNDREDTHS OF TWO-HUNDREDTHS IN 0.001.

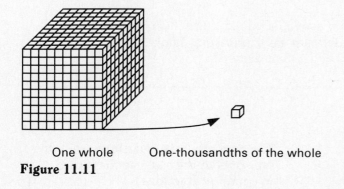

One whole   One-thousandths of the whole

Figure 11.11

many sets of one-thousandth (0.001) are there in 1000?," would result in a very large number.

By going over the division question, the students will become adept at predicting whether answers to such problems will be large or small numbers. Teachers can show problems and children can state if the answer will be a large or small amount. This is an excellent activity to use with the every pupil response cards, EPRs, (reviewed in Chapter 2). These activities help children visualize decimals to build number sense.

Activity

CALCULATORS

PROBLEM
SOLVING

Division of Decimals

The following activity promotes problem solving as well as calculator use. Find the correct pathway by dividing the first number by the second one as you trace your way to the ending numeral.

Start

1000	10	0.1	100
1000	1	0.001	0.01
0.001	1	0.01	10
100	100	10	100
			End 1

Activity

COMPUTERS

Division of Decimals with Spreadsheets

Directions:

1. Using the spreadsheet you made in the multiplication section, change the following commands to make the spreadsheet into division:

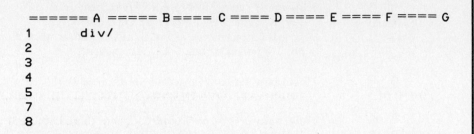

```
====== A ===== B ==== C ==== D ==== E ==== F ==== G
1        div/
2
3
4
5
7
8
```

The cell formula will change to:
For cell B2,   +a2/b1   to show division instead of multiplication. Change the other cells following this example.

2. Follow steps 3 through 7 to see patterns and generalize a rule for the division of decimals.

3. *If you have no spreadsheet program available:* Make a "discovery chart" like the last problem-solving activity for multiplication. Look for patterns and generalize a rule for the division of decimals.

After working with division of decimals, it is easy to see why many mathematics educators believe that this concept must wait to be taught until the upper elementary grades when children can visualize intellectually what is happening with minute quantities. If upper grade students are at Piaget's Concrete Operational Stage, there may be limited comprehension.

Ratio

The concept of ratio is an extension from common fractions and may be written using the same symbolic notation:

$\frac{2}{5}$ The concrete manipulations are represented the same as they were in common fractions.

Two ways to express the meaning of ratio in words are:

two to five

or

2:5   For every two in one set, there are five in another set.

Pattern blocks and Cuisenaire rods are two of the materials children can use to express ratio (Fig. 11.12).

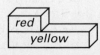
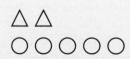

Cuisenaire Rods   Pattern Blocks
Figure 11.12

Activity

COMPUTERS

Ratios in Logo

Directions:

1. This program is saved under the name RATIO.

2. Load the program by typing the words `LOAD "RATIO`

   (If you have no disk, type in the following procedures.)

```
TO 1PART                        TO 2PART
PU SETPOS [-133 50] PD          PU SETPOS [-133 0] PD
END                             END

TO MOVE                         TO PENT :N
RT 90 PU FD 35 LT 90 PD         REPEAT :N [REPEAT 5 [FD 20
END                             RT 72] MOVE]
                                END

TO STAR :N                      TO TRI :N
LT 90 PU BK 5 RT 90 PD          REPEAT :N [REPEAT 3 [FD 30
REPEAT :N [REPEAT 5             RT 120] MOVE]
[FD 30 RT 144] MOVE]            END
END

TO HEX :N                       TO OCT :N
REPEAT :N [REPEAT 6             REPEAT :N [REPEAT 8 [FD 10
[FD 15 RT 60] MOVE]             RT 45] MOVE]
END                             END
```

3. Now type the following commands just as they are printed here and see what happens:

   ```
   1PART TRI 4 2PART PENT 6
   ```

4. What ratio did you produce? What commands did you type that would tell you?

5. Use CS to clear the screen between ratio examples.

6. Here are some more to try: What's the ratio?

   ```
   1PART STAR 5 2PART HEX 7
   1PART OCT 5 2PART TRI 2
   1PART PENT 7 2PART STAR 3     Yes, it can be this way.
   ```

7. Make up some of your own ratios using the procedures given here.

8. If you have a color monitor, add the SETPC command to make colored shapes. The following color chart shows what to type to get the desired color. Place the SETPC command in the desired color at the beginning of a shape procedure. It is set automatically to white unless told to do another color.

Command	To Obtain
SETPC 1	white
SETPC 2	green
SETPC 3	violet
SETPC 4	orange
SETPC 5	blue

9. What ratios produced these Logo printouts? Type in the commands that make these pictures.

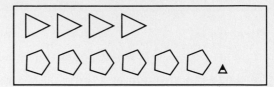

Ratios are used in much the same way as equivalent fractions. The previous Logo program created the following pictures. One can say that the ratios are:

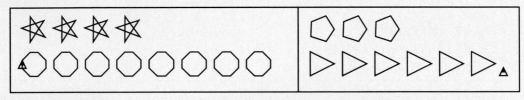

4 : 8   3 : 6

or

in the same ratio of 1 : 2 because for every star or pentagon, there are two octagons or triangles respectively.

Activity

MANIPULATIVES

Equivalent Ratios

Directions:

1. Using pattern blocks, make equivalent ratios (ratios that have the same relationship). For children, one student can start and another student can make an equivalent ratio with different pattern blocks.
Which of these represent the same ratio?

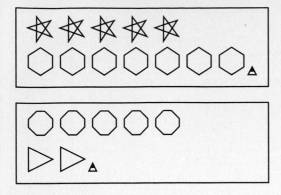

2. Continue the activity using Cuisenaire rods.
3. Think of other manipulatives that could show ratios.

Proportion

Equivalent ratios are really proportions. A 12:8 ratio and a 3:2 ratio may be read as a proportion statement, "12 is to 8 as 3 is to 2." Proportions are checked the same way equivalent fractions are:

$$\frac{a}{b} = \frac{c}{d} \quad \text{if} \quad ad = bc$$

$$\frac{12}{8} = \frac{3}{2} \quad \text{if} \quad \begin{array}{c} 12 \times 2 = 8 \times 3 \\ 24 = 24 \end{array}$$

Students can see the connection between proportions and equivalent fractions when they are encouraged to use the BASIC and Logo programs originally introduced in Chapter 10 as a check for equivalency of fractions.

Activity

COMPUTERS

BASIC and Logo Check for True Proportions

Directions:

1. Use the BASIC and Logo programs appearing in the Introduction of Chapter 10 and saved under the names:
   EQUIV FRACTIONS (BASIC)
   EQUIVFRACTIONS (Logo)

2. Try these proportions with the BASIC and Logo programs. See which ones work and which ones do not work. On a sheet of paper, keep track of the ones that are true proportions. Use the NOT EQUAL sign ($\neq$) to show when the examples are not true proportions.

   $$\frac{10}{55} = \frac{16}{88} \qquad \frac{47}{63} = \frac{139}{189} \qquad \frac{13}{269} = \frac{12}{144}$$

   $$\frac{12}{16} = \frac{76}{95} \qquad \frac{39}{54} = \frac{13}{18} \qquad \frac{2002}{2040} = \frac{1001}{1020}$$

3. Make up true proportions and non-proportions on your own. You can learn a lot by creating non-examples and comparing them with true examples.

Activity

MATH ACROSS THE CURRICULUM

Proportions in English

Directions:

1. Proportions must maintain a definite relationship among four magnitudes such that the first is in relation to the second in the same way as the third is in relation to the fourth.

2. *Analogies* are really proportions with words as the magnitudes. For example:

   Oar is to rowboat as wheel is to car.

   Some standardized tests even use the mathematical symbols

   oar : rowboat = wheel : car

3. Think of other word analogies and write them using the mathematical symbols.

Proportions are the essential elements of many problem-solving activities that appear as word problems in many elementary textbooks. Typically, one ratio is given and the next ratio is only partially finished. The problem solver must "see" the two ratios as a proportion to figure out the missing segment.

Activity

PROBLEM SOLVING

Proportion Problems

Directions: Read these word problems and set up the proportion required. The first is done for you.

1. If it takes 12 seconds to go 60 miles in a rocket ship, how many seconds will it take to go 55 miles?

$$\frac{12}{60} = \frac{?}{55}$$

Think : Reduction

$$\frac{12 \div 12}{60 \div 12} = \frac{1}{5}$$

so

$$\frac{n \div 1}{55 \div 5}$$

so

n = 11 seconds

Think : Cross Multiplication

$$12 \times 55 = 60 \times n$$

$$\frac{660}{60} = \frac{60 \times n}{60}$$

$$11 = n$$

11 seconds

2. Labels for the proportions are important. In the preceding example, the following relationship patterns would produce valid proportions and arrive at the same answers.

Relationship $\uparrow\downarrow \left[\dfrac{12 \text{ sec.}}{60 \text{ mi.}} = \dfrac{N \text{ sec.}}{55 \text{ mi.}} \right]$ $\left[\dfrac{60 \text{ mi.}}{55 \text{ mi.}} = \dfrac{12 \text{ sec.}}{N \text{ sec.}} \right]$

Relationship $\longleftrightarrow$

The arrows show the pertinent relationship. The example on the right is a good one to use when introducing percent word problems.

3. If it costs $1.50 for 3 candy bars, how much will 27 candy bars cost? Set the proportion up in more than one way using the example in step 1.

4. Lee can swim across a 60 foot pool in 3 minutes. If the pool is half the size, how many minutes will it take Lee to swim across the pool?

There are many real-world problems that involve proportion. Using proportions in solving problems is a part of applications of mathematics. Scale models are one example involving proportional reasoning. The following can be done as a homework assignment by cutting out prices from the newspaper or by writing down prices when shopping.

Activity

ESTIMATION

CALCULATORS

MATH
ACROSS THE
CURRICULUM

Proportions in Economics

Directions:

1. List the two prices you found in the paper or by shopping and figure which is the better buy. Use mental math to estimate by rounding the prices, because in a real shopping spree, you may not have the time to stop and use paper or a calculator.

2. Check by setting up the proportions with the calculator. See how close you came to making the better buy. The first one is done for you.

Prices	Estimate	* = Better Buy	Calculator Check
4 @ $1.19 3 @ $0.79	4 @ $1.20 3 @ $0.80	*	4/$1.19 : 3/$.79 $0.29 ≠ $0.26
2 @ $1.59 3 @ $2.39			
32 oz @ $1.79 24 oz @ $1.59			
50 oz @ $1.99 32 oz @ $1.69			

3. Add some more to the list.

Percent

Data from the fourth mathematics assessment of the NAEP (Kouba et al., 1988) on items involving percents indicate many seventh-grade students appear to lack understanding of the underlying concepts of percents. Only 30 percent of the students could express 0.9 as a percent and 8 percent as a decimal. Students were more successful when working with common percents for which they knew fractional equivalents. Again students were more successful calculating percent of a number than in solving other forms of percent problems. About 70 percent of the seventh-grade students responded that when a quantity is partitioned into parts represented by percentages, the sum of the percentages total 100 percent.

Percent, when standing alone as in 32 percent, may be considered a ratio—32 parts out of 100. All percents are compared to a hundred parts or hundredths. To find a percent when a common fraction or decimal fraction is given requires a proportion. Frequently, finding percent is the first proportion situation introduced to students.

$$\frac{2}{5} = \frac{?}{100} = \frac{40}{100} = 40\%$$

$$0.175 = \frac{175}{1000} = \frac{?}{100} = \frac{175/10}{1000/10}$$

$$= \frac{17.5}{100} \text{ or } 17.5\%$$

$$22.6 = 22\frac{6}{10} = 22\frac{?}{100} = 22\frac{6 \times 10}{10 \times 10}$$

$$= \frac{60}{100} = 22.60\%$$

Percent is usually taught right after decimals and fractions because of the close relationship between the two. Students can generalize the rule for finding percents by the use of charts and spreadsheets. The % symbol is taught from the beginning as a quick way to make sure that a person is talking about percent without writing the word. One of the first activities using percent is usually the shading of figures or graph paper once students have learned how percent is related to fractions.

Activity

COMPUTERS

Spreadsheet

Directions:

1. Use the chart you saved from the last activity in the Introduction section of this chapter.

2. From what you have learned about spreadsheets, construct one using the information from the chart. (Use the % symbol when constructing the chart rather than writing out the word since columns and rows are narrow.)

3. Sketch what your spreadsheet program looks like or save it on a disk to show your instructor.

4. How could you change the spreadsheet to show more examples of going from decimals to percents and back again?

5. From the patterns you have observed in the preceding examples, generalize a rule for finding:
   a. percents when the decimal fraction is given.
   b. percents when the common fraction is given.
   c. decimals when the percent is given.

Activity

MANIPULATIVES

Percent with Pattern Blocks or Graph Paper

Directions:

1. Using the actual pattern blocks as models, trace around them and shade the percent of the figure indicated below each one. Be ready to use graph paper also.

Shade 75%

Shade $33\frac{1}{3}$%

Shade 60%

Shade 20%

2. Trace around four other pattern blocks and shade a percent of the figure. Write the percent you chose to shade or color underneath each figure.

3. Do the same activity as in step 2, using graph paper this time.

4. Check each figure to see if it is the correct percentage by finding the common fractional equivalent. For example in step 1:
   If 75 percent is shaded then $\frac{3}{4}$ of the figure (or 3 equal sections out of the 4 equal sections) should be shaded.

5. Write the fractional equivalent under each percentage to show you have checked each figure.

The computer allows students to transfer the knowledge of percent to pictorial approaches while using problem solving at the same time.

Activity

COMPUTERS

Logo Shading of Fractions to Percents

Directions:

1. Look at the first Logo program in the Teaching Strategies section of Chapter 10 saved under the name of TRUEFRACTIONS.

2. Recall what you learned from doing the program and think about how you could use the program to shade in figures to show percents. What would you have to adjust in the program, if anything? Justify your answer.

3. Test your prediction by loading the program and making the necessary changes.

Activity

COMPUTERS

Percentage Pictures in Logo

Directions:

1. This program is saved under the name PERCENTPICTURES. Type LOAD "PERCENTPICTURES to load the program on the computer.

PROBLEM SOLVING

(If you have no disk, type in the following procedures.)

```
TO 33.3%CIRCLE
HT
REPEAT 36 [FD 1 RT 10 * .333]
END

TO 75%CIRCLE
HT
REPEAT 36 [FD 1 RT 10 * .75]
END

TO 25%CIRCLE
HT
REPEAT 36 [FD 1 RT 10 * .25]
END
```

2. Run each program by typing only the name (not the TO). For example, 25%CIRCLE

3. Think of a way you could test to see if you really have drawn 25 percent of the circle. *HINT:* If you know how many 25 percents make a whole, typing that many should give you a whole figure. Try it and see.

4. Do step 3 for all the procedures.

5. You may have been tempted to try 25% instead of .25 within the Logo procedure itself. If you did, you received a message telling you that Logo does not understand 25% in the procedure statement. That's why knowing decimal conversions is important. It can be used as a motivating factor for children to learn the conversions in order to do the Logo.

(continued)

6. Look for the common pattern in each procedure. If you wanted to draw 72 percent of the circle, what would you have to change? Write a new procedure for each of the following percentages and test out your procedures.

$66\frac{2}{3}$% of the circle

20% of the circle

59% of the circle

1.57% of the circle

Here's where knowing decimal conversions really becomes important:

5% of the circle

0.05% of the circle

1.7% of the circle

7. Think of other possibilities and try them. Write down your procedures so others can try them too. Use Appendix B to see how to edit your procedures.

Percent is one of the mathematical concepts that occurs frequently throughout a person's life. Therefore, most elementary textbooks move into real-world problem solving as soon as possible. Here are just a few of the activities that can be done.

Activity

CALCULATORS

MATH
ACROSS THE
CURRICULUM

Percentages in Sports

Directions: Use calculator to solve.

Marksmen	Total Shots	Shots Hitting Target	Percent Hitting
1	600	475	____
2	500	375	____
3	200	105	____
4	150	75	____
5	50	31	____
6	75	52	____
7	30	20	____
8	960	777	____
9	280	213	____
10	25	5	____
11	5	2	____
12	140	78	____
13	750	702	____
14	650	602	____
15	750	435	____

a) The winner____best percent is ____
b) The second place marksman is ____
c) The last place marksman is ____

The most common use of percent in the real world is on sale items. Students need to know how to figure percents when they are buying reduced items so they have a clear understanding of the discounted price. Often a store will have a sign over a counter announcing, "20% OFF THE TICKET PRICE." It is up to the consumer to figure out how much the item is reduced before deciding if it is affordable at the sale price. Computer programs like the ones below can help students analyze what steps a person must take to find percentage reductions. Additional practice can then be done on the calculator.

Activity

Consumer Math

Directions:

1. Find several grocery store or business ads in the newspaper.

2. Use the calculator to check advertised discounts.

3. Use the calculator to compute percent decrease in the sale price to original cost.

4. Use the calculator to find the percent increase in price from the sale price to original cost.

5. Write the procedure for using your calculator and the % key to compute the following:

discounts	percent increase
sale price	percent decrease
commissions	selling price

Activity

Sale Prices in BASIC and Logo

Directions:

1. To load each program, type:

```
LOAD SALE PRICE    (BASIC)
LOAD "SALEPRICE    (Logo)
```

(If you have no disk, type in the following programs. Do them separately, following steps 2 to 4 for each program.)

BASIC

```
10 REM FIND THE SALE PRICE
20 PRINT "WHAT IS THE ORIGINAL PRICE?"
30 INPUT P
40 PRINT "WHAT PERCENT IS IT REDUCED?"
50 INPUT X
60 LET X = X * .01
70 LET R = P * X
80 LET S = P - R
90 PRINT "YOUR SALE PRICE IS ";S;"."
100 END
```

(continued)

```
Logo
TO SALEPRICE
CLEARTEXT
TEXTSCREEN
PR [WHAT IS THE ORIGINAL PRICE?]
MAKE "P FIRST READLIST
PR [WHAT PERCENT IS IT REDUCED?]
MAKE "X FIRST READLIST
MAKE "R :P * :X * .01
MAKE "S :P - :R
PR [YOUR SALE PRICE IS]
PR :S
END
```

2. Here is a list of original prices with the percent reduction listed. Use these in both programs to find the sale prices. Type *only* the number (not the %). Type RUN (BASIC)   Type SALE PRICE (Logo)

20% off	40% off	10% off	30% off
34.99	15.99	1.69	14.99

3. Show on a calculator how you would do the essential steps seen in the computer programs above. List the steps.

4. *For those who like a challenge:*
How could you change the computer programs if you needed to know what percent was saved on a product if you knew the sale price and the original price? Make the adjustments in the programs and run each with· the answers to the problems in 2 as the sale prices. If the percent comes out to the same (or will be the same if rounded up or down from the thousandths place) as shown in 2, then you know you have done the steps correctly in the computer program.

Note: Children enjoy finding pictures of sale items in the circulars that come with the newspaper. They can make task cards with the original price and the sale price and then find what percent was saved on each item.

Rate

Rate is a ratio where the first term is compared in relation to a unit of something else in the second term. For example, 55 miles per hour is the ratio of miles to hour.

Many textbooks include problems where the ratios are expressed as rates. When the rate involves a proportion, the problem becomes more difficult. These are found frequently on standardized tests from the upper elementary grades through college entrance examinations.

Activity

MENTAL MATH

Finding the Ratio in the Rate

Directions:

1. Think of the ratio that is implied in each of these rates:

Rate	Ratio
55 miles per hour	55:60 minutes
8 cans for a dollar	8:$1.00
12 eye blinks a minute	12:60 seconds

2. Think of others you can add to the list.

Activity

PROBLEM
SOLVING

Rates and Proportions

Directions: Answer the following problems. Set up each problem as a proportion using equivalent fractions or tables to show the proportions. The first is done for you.
Example:

1. A train travels at the rate of 120 miles per hour on a 600-mile trip. The train leaves at 2 p.m. How long will the trip take and at what time will the train arrive?
The proportion is:

$$\frac{\text{hours}}{\text{miles}} = \frac{1}{120} = \frac{n}{600}$$

Solved by:

Equivalent Fractions

$$\frac{1}{120} = \frac{?}{600}$$

$$120n = 600$$

$$\frac{120n}{120} = \frac{600}{120}$$

$$n = \frac{600}{120}$$

A Proportion Table

Hour	Miles covered
1	120
2	240
3	360
4	480
5	600

Divide 600 by 120 to find the proportionate difference between them

$$600 \div 120 = 5 \text{ so---}$$

$$\frac{1}{120} \quad \frac{\times 5}{\times 5} = \frac{5}{600}$$

The trip takes 5 hours, so a 2 p.m. leaving time + 5 more hours makes the arrival time at 7 p.m.

2. If train 1 travels at a rate of 120 miles per hour on a 600-mile trip and train 2 travels at a rate of 105 miles per hour on a 450-mile trip to reach the same destination, which one arrives first if they both leave at 3 p.m.?

3. If a stereo costs $250 with 12 percent interest over the payment time of one year in store A and the same stereo costs $290 with 4 percent interest over the same time period in store B, which is the better buy? How much money will you save by purchasing the better buy?

The teaching strategies in this chapter require much practice to understand fully how they apply in a variety of real-world experiences. A teacher must avoid rushing from one principle to another without adequate exploration with manipulative models. Otherwise, students will have difficulties solving problems like the preceding three, which require an integration of all the mathematical principles studied in this chapter. The NCTM curriculum standards (Commission on Standards for School Mathematics, 1987) state that mathematics curriculum in grades 5–8 should investigate relationships between fractions, decimals, and percents. Students should understand the various representations of the same number, such as $\frac{15}{100}$, $\frac{3}{20}$, 0.15, 15%, and 15:100.

Diagnosis and Remediation

Successive vs. Simultaneous Thought Processing

Both thought processing styles will be compared side by side to demonstrate how different students work together in cooperative learning situations.

Shaded Regions

Successive processors have more success finding their own fractional part from the percent given and then shading it. They should do the activity and then give it to a simultaneous processor to figure out the percent from a list of made-up answers with distractor items.

Simultaneous processors like to see the whole example and figure out the percent from there. They should explain how they get the answer for the successive processor because most standardized tests use this form.

Shade 33 1/3%

What % = Shaded
Choose one:
33 1/3%
75%
12.5%

Shade 12.5%

What % = Shaded
Choose one:
33 1/3%
75%
12.5%

Spreadsheets

Observe the students as they work to see which of the actions described below are occurring and pair students with opposite processing styles together for computer work.

A successive processor enjoys building the table, working from cell to cell. Creating the formulas such as +a2*b1 does not seem to be a problem. They often stop after each formula and study what has happened as the number appears in each cell. They may have trouble telling all the parts of the pattern if the table is shown to them fully constructed.

Simultaneous processors are bothered with what they perceive as long, tedious work building the formula. They can often "spot" patterns more quickly and in more depth than the successive thinker once the table is constructed.

BASIC and Logo

While it is too early to tell if the next observations are generally true from research, there is some indication that . . .

Successive processors like the BASIC and Logo programs equally well and do not mind typing in directions one step at a time, analyzing each step as they go.

Simultaneous processors enjoy Logo more than BASIC and want to see what a procedure will do graphically before analyzing the procedure's steps.

Use successive processors to type procedures in programs but allow simultaneous processors to explore what the graphics will do in the program. Perhaps you noticed the same frustrations while working through the computer programs in this text.

Correcting Common Errors with Decimals

Non-alignment of the Decimal Points: Students are used to writing numerals from left to the right and some students proceed to write the decimals in the same manner as in the example on the left below. Students need practice transferring decimals written in the horizontal form to the vertical form. A teaching method to mark the decimal point is as follows: Add: 235.06 + 41.25 + 9.345 = ?

```
      235.06        ↓
       41.25        .
    +  9.345      + .
    _____       ___
```

Ask children to place an arrow above the decimal points, marking the points before writing any numerals.

Annexing Zeros: Some students see 0.500 as greater than 0.50, and 0.50 as greater than 0.5. They do not understand that annexing zeros to the right of a decimal does NOT change the value of the decimal. Students must go back to work with the concrete manipulatives as shown in the Teaching Strategies section. Some students will need many experiences representing decimals to see numerical relationships, while other students see the pattern after a few times. Let students use the concrete materials as long as needed, using a place-value chart like figure 11.13 to record the amounts.

Attention to Decimal Point as Place Holder: Some students, when faced with a number like .529, totally disregard the decimal point, treating .529 as if it were a whole number. We have already stressed the importance of writing such decimals as 0.529 to emphasize that 0.529 is a part of a whole. Calculators will also reinforce this approach because every decimal without a whole number is automatically recorded with the zero as a place holder whether the child enters it that way or not. Unfortunately, most computer languages do not maintain that consistency and will not show an answer with zero as the place holder. Perhaps students with such a problem should be encouraged to use the calculator rather than the computer if the problem persists.

Name Value Confused with Place Value: Just as young children become confused with the name, forty-seven, and write the numeral as 407 (seen in Chapter 6), when students learning decimals hear such name values as "eight hundredths" or "fifty-two thousandths," they write:

0.800 or 0.80 (if they remember that hundredths has only two decimal places)

0.5200 or 0.520

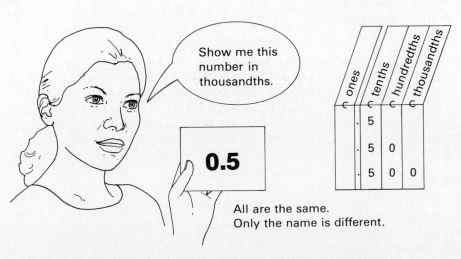

All are the same.
Only the name is different.

Children use Every Pupil Response Cards or individual blackboards *so everyone has the chance to answer.*

Figure 11.13

Writing the numeral on a place-value chart helps if students have a folder with the rule written out to the side as in the following example.

Rule:
1. Find the place that corresponds with the word you are saying.
2. Start writing the number there, moving from right to left.
3. Fill in any 0's you need to get to the decimal point.

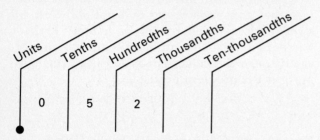

Example: "fifty-two thousandths"

Poor Estimation Skills in Multiplication and Division: Some students have difficulty computing with decimals because they lack a sense of place-value and estimation. Therefore these following error patterns occur.

$$0.3 \times 0.3 = 0.9 \qquad 0.52\overline{)104}^{\,2}$$

They need to model such problems with manipulatives where possible, and at all times, they should accompany the modeling with the oral language as shown in the Teaching Strategies part of the chapter. This procedure helps students determine whether their answers are reasonable or not.

Correcting Common Errors with Percent

Writing the Decimal with the Percent: It is common for children to hear 52 percent and write .52%, remembering that the number with two decimal places is "per hundredth." Reading the percent aloud with place-value labels is often all that is needed for the student to see the mistake. Fifty-two hundredths of a percent sounds much less than 52 percent.

If the problem persists, spreadsheet programs where children can change the numbers quickly and see the results are often a good way for them to distinguish the difference because they can compare what .52% and 52% will give them. The use of a calculator to check percents is useful if the calculator has a % key.

Follow these steps for use with most calculators:

1. Enter 1.00 × (the percent you wish to check) = _____
   (This fixes the number in the calculator as a variable.)

   Example: 1.00 × 0.52 = 0.52 (calculator answers)

2. Press the % key. Now the number on the calculator will be treated as a percent and the resulting number is the decimal equivalent.

   Example: [press % key] results in 0.0052

3. On some calculators, a quicker way is possible. Entering 0.52 and pressing the % key will give the student a good way to see that the answer is 0.0052 instead of 0.52

Percent of a Given Value: There are three common mistakes when figuring percent of a given value.

• **Forgetting the meaning of the term "of":**
  Students know that a percent of an original number means it will be less than the original quantity so they divide rather than multiply to find the percent. They reason that multiplication yields a greater number rather than a lesser number. Again, the word, "of" is crucial to the problem. If this point is stressed throughout the study of multiplication from basic facts to percents, students will recognize that "of" means multiplication and that the answer is smaller than the original number, which does make sense. For example:

29% off of 52

Common Error	With OF as Multiplication
$0.29\overline{)52.00}^{\,179.31}$	$\begin{array}{r} 52 \\ \times\ .29 \\ \hline 15.08 \end{array}$

- **Stopping After Percent is Found Without Further Reduction:**

Frequently students find the percent correctly, but forget that the problem is not finished because the original price must be reduced by that percent, meaning subtraction must be used. Role-playing many shopping examples with things they enjoy buying seems to help in two-step problem-solving situations.

- **Thinking of Percent as Money to be Reduced:**

When converting percents to decimals, the decimal reminds the student of the money that is involved in the answer. Therefore, 0.29 becomes $0.29 and it is reduced from the original price by subtraction. This is where estimation becomes valuable as a tool. If students have had much practice finding 10 percent of amounts, reasonable answers can be figured:

Faulty Reasoning

$$\begin{array}{r} \$52.00 \\ - \quad .29 \\ \hline \$51.71 \end{array}$$

Reasoning by Estimation

10 percent of 52 = 5.20
30 percent of 52 = 5.20 × 3 = 15.60
29 percent is slightly less than 30 percent so percent has to be a little less than 15.60 and 51.71 is nowhere close. What will bring me closer to 15.60?
Answer: Multiplication NOT subtraction

Average and poor students will not be able to solve problems like the one above unless they have had effective teachers who help them structure their thoughts.

Summary

Base 10 blocks were used to illustrate decimal fractions. Several kinds of decimal fractions, including terminating and repeating decimals, were introduced with activities that upper grade and middle school students can do. Money was used to introduce decimal notation. A Logo program explored decimal comparison with hundredths while multiplication and division of decimals were demonstrated with a computer spreadsheet. The teaching strategies for ratio, proportion, percent, and rate involved pattern blocks and Cuisenaire rods. Calculator and computer activities were used to develop the concepts in more detail. The Diagnosis and Reme-

diation section concentrated on common errors of annexing zeros, confusing place value with name value, and forgetting that the decimal point is a place holder.

Chapter 12 will develop the concepts of graphing, statistics, and probability. The chapter emphasizes experiments for probability and utilizes comparison statistics with newspaper data. The ideas developed in this chapter are used as an integrated part of the following chapter when working statistical problems, averaging trends for graphs, etc.—all of which require a working knowledge of decimals and the specialized topics of this chapter.

Exercises

Directions: Read all questions *before* answering any one exercise. Frequently the last question in one category leads to the first question in the next category.

A. Memorization and Comprehension Exercises
Low-Level Thought Activities

1. Find all the base 10 blocks that represent the decimal equivalent for $\frac{1}{4}$, $\frac{6}{10}$, and $\frac{21}{100}$.

2. Show the difference between 0.08 and 0.80 with base 10 blocks and with graph paper.

3. What do $0.\overline{33333}$ and 0.33333 represent? Which one is more? What teaching materials could you use to prove your answer?

4. List the repeating decimals found between $\frac{1}{2}$ and $\frac{1}{10}$. Generalize a rule for finding non-terminating, repeating decimals in relation to the whole unit 1.

5. Show how you could check the answers in A4 with a calculator.

6. Practice mental computation with estimation in the following examples:

Problem	Number of Decimal Places	Answer Starting at Right to Left
0.2 × 0.2	2	0.04
0.02 × 0.2		
0.02 × 0.02		

Try other combinations.

7. Search the newspaper and circulars to find real-life percent problems. Set up the problem mathematically and solve.

B. **Application and Analysis Exercises**
Middle-Level Thought Activities

1. Read these proportions as one would say them in English to emphasize the mathematical meaning. Give an answer to make each proportion equivalent.
Birds:Migration = Bears: ?
Snow:Snowflake = Rain: ?
Furnace:Heat = Air Conditioner: ?

2. Using the textbook evaluation forms in Appendix F, review an elementary and middle school series to see how it introduces and extends the learning of decimals. Analyze its strengths and weaknesses based on the research and teaching strategies covered in this chapter.

3. Find and label the following points on a number line. As you do this exercise, make a list of the similarities and differences that you notice about these rational representations.

1.) 0.8   2.) 0.82   3.) 0.825
4.) 0.330   5.) $\frac{7}{10}$ 6.) $\frac{7}{100}$
7.) $\frac{7}{1000}$ 8.) 0.03   9.) 0.$\overline{1111}$
10.) 0.111

4. Analyze the following two lines. What are they asking the computer to do? This formula helps complete a mathematical definition mentioned in this chapter. Which one? (*CLUE:* These are the lines omitted from the BASIC and Logo programs found in the introduction to Chapter 10, but they apply to a mathematical principle covered in this chapter as well.)

In BASIC
`110 IF A * D = B * C THEN 150`
In Logo
`IF :A * :D = :B * :C [PR [THE FRACTIONS ARE EQUIVALENT.]]`

5. Look through elementary textbooks for word problems involving ratio, proportion, percent, and rate. What are the similarities and differences between textbook series?

6. Look at one of the mathematics textbook series for the elementary and middle schools. Analyze how many different approaches to decimals, ratio, proportion, percent, and rate are presented within one grade level. How many pages are devoted to concrete and/or pictorial models? What "extra" materials might you as a teacher need to use as supplementary work for your students?

7. Search professional journals for current articles on research findings and teaching ideas with decimals, ratio, proportion, and percent. You may wish to look at the first activity under Synthesis and Evaluation Activities below and coordinate activities. Follow the form below to report on the pertinent findings:

Journal Reviewed: _____
Publication Date: _____ Grade Level: _____
Subject Area: _____
Author(s): _____
Major Finding:

Study or Teaching Procedure Outlined:

Reviewed by: _____

Some journals of note are listed below:
Journal for Research in Mathematics Education
Arithmetic Teacher
Mathematics Teacher
School Science and Mathematics

8. Computer programs and computer reviews for classroom use are found in the following journals. Their articles are by no means the only ones. Search the periodical section of the library for more. You may wish to look at the first activity under Synthesis and Evaluation Exercises to coordinate activities. Using the evaluation form found in Appendix A, select commercial software for any three topics from this chapter. In the selection, include drill and practice, simulation, tutorial, and problem-solving software. For each software program selected, describe how you would integrate it in a lesson plan.
Teaching and Computers
Classroom Computer Learning
Electronic Learning
Computers in the Schools
The Journal of Computers in
  Mathematics and Science Teaching
The Computing Teacher

9. To reinforce math across the curriculum:
   a. List five mathematical questions/problems that come from popular storybooks for children. Show how they are related to the topics of this chapter.
   b. List five ways to use the topics of this chapter when teaching lessons in reading, science, social studies, health, music, art, physical education, or language arts (writing, English grammar, poetry, etc.).

10. What do the concepts of decimal fractions, ratio, proportion, percent, and rate have in common and how can students prepare to use these concepts in everyday life?

C. Synthesis and Evaluation Exercises
High-Level Thought Activities

1. Create your own lesson plan for teaching one concept using decimals, ratio, proportions, or percents. Follow the following steps:
   a. Look at the Scope and Sequence Chart. Find a grade level and a concept within that grade level. Trace the previous knowledge of the concept taught in earlier grades so you will know where to begin your instruction.

b. Use current articles from professional journals to help plan the direction of the lesson. Document your sources.
   c. Include at least one computer software program as a part of the lesson. Show how it will be integrated with the rest of the lesson.
   d. Include behavioral objectives, concrete and pictorial materials, procedures, and evaluation of student mastery.
   e. Use the research-based lesson plan format of i) review, ii) development, iii) controlled practice, and iv) seatwork.

2. Here is the work of an actual student. Create a remediation plan to help the student overcome the misconceptions in the representation of decimals. Your plan should include:
   a. An analysis of the error pattern.
   b. An evaluation of the thought processing required in the examples and the thought processing displayed by the student.
   c. The sequential development of concrete, pictorial and symbolic models to reteach the concept.

One student's written response:

Seventy-two hundredths = .720
Three tenths = .30
Eight hundredths = .800
Twenty-four thousandths = .2400

3. Design your own Logo picture to be used as a decimal comparison like the FLAG and the PERCENTPICTURES program.

4. Using the Logo lesson program found in the strategies section, create a lesson to teach percent or ratio. What would you need to change in the directions and in the examples? Try the new creation with 4 examples. Show your program to others. Compare the different ways your colleagues changed the program to meet their chosen concept.

5. Prepare a student lesson that has the student create a computer spreadsheet on the percent of time spent on daily activities.

The spreadsheet could follow this format:

```
===== A ===== B ===== C        D ===== E ===== F
 1|Time Spent on Activities
 2|
 3|         Weekday Weekend Week Tot. Mon. Tot. % Spent
 4|
 5|School      6        0      30      129     17.9%
 6|Sleep       8        8      56      240.8   33.3%
 7|Hobbies     2        4      18      77.4    10.7%
 8|Exercise    1        2      9       38.7    5.4%
 9|Study       1        2      9       38.7    5.4%
10|Meals       2        2      14      60.2    8.3%
11|TV          4        6      32      137.6   19.0%
12|
13|Totals     24       24     168      722.4   100.0%
14|
15|
16|
17|
18|
```

The lesson should include cooperative activities, objectives, higher-order thinking questions, and an evaluation plan.

Bibliography

Allinger, Glenn D. "Percent, Calculators, and General Mathematics." *School Science and Mathematics* 85 (November 1985): 567–673.

Clason, Robert G. "How Our Decimal Money Began." *Arithmetic Teacher* 33 (January 1986): 30–33.

Commission on Standards for School Mathematics of the National Council of Teachers of Mathematics. *Curriculum and Evaluation Standards for School Mathematics.* Working Draft. Reston, Va.: The Council, 1987.

Grossnickle, Foster D., and Leland M. Perry. "Division with Common Fractions and Decimal Divisors." *School Science and Mathematics* 84 (November 1984): 556–566.

Hiebert, James. "Decimal Fractions." *Arithmetic Teacher* 34 (March 1987): 22–23.

Kouba, Vicky L., Catherine A. Brown, Thomas P. Carpenter, Mary M. Linquist, Edward A. Silver, and Jane O. Swafford. "Results of the Fourth NAEP Assessment of Mathematics: Number, Operations, and Word Problems." *Arithmetic Teacher* 35 (April 1988): 14–19.

Meyer, Ruth A., and James E. Riley. "Investigating Decimal Fractions with the Hand Calculator." *School Science and Mathematics* 84 (November 1984): 556–565.

Paull, Sandra. "Balancing a Checkbook: Children Using Mathematics Skills." *Arithmetic Teacher* 33 (March 1986): 32–33.

12

Graphing, Statistics, and Probability

Key Question: *How can the subjects of statistics and probability, which sound complicated and obscure, be taught to elementary and middle school students? How does graphing play a part in developing these concepts?*

Activities Index

Introduction

The NCTM curriculum standards (Commission on Standards for School Mathematics, 1987) list statistics and probability as a part of the curriculum beginning in primary grades. The investigation of statistical and probabilistic concepts illustrates the role of mathematics in organizing and presenting information. Students must work with data, namely, collect it, organize and display it, analyze and interpret it, and understand how decisions are made on the basis of collected evidence. The standards also call for studying probability in real-world settings. Students need to investigate the notion of fairness and chances of winning. Since many predictions are based on probabilities, time should be al-

lowed to make predictions based on experimental results or mathematical probability. Some items on the National Assessment of Educational Progress (Kouba et al., 1988) assessed organization and interpretation of data presented in graphs and tables. It also included items about finding and using measures of central tendency. As expected, students did better on making direct reading from graphs and tables than at deciding relationships among data. Only about 50 percent of the seventh-grade students could calculate an average, whereas about 40 percent could calculate a mean or median.

This chapter will emphasize the different methods of representing information as well as producing the results. *Graphing* is a common method of displaying information pictorially. *Statistics* is used to discuss data numerically.

Probability is the mathematics used to predict outcomes.

Scope and Sequence

Kindergarten
- Make and interpret real graphs.
- Make and interpret bar graphs.

First Grade
- Make and interpret real graphs.
- Make and interpret bar graphs.
- Use picture graphs.
- Collect and organize data for graphs.

Second Grade
- Read and interpret picture graphs.
- Collect and organize data for graphs.
- Make and interpret bar graphs.
- Record outcomes for probability.

Third Grade
- Make and interpret bar graphs.
- Interpret picture graphs.
- Read charts and tables.
- Collect and organize data.
- Use tallies to make a table or graph.
- Intuitive probability—random drawing.

Fourth Grade
- Make and interpret bar and picture graphs.
- Make and read charts and tables.
- Interpret line graphs.
- Graph ordered pairs.
- Collect and organize data.
- Read a fraction circle graph.
- Conduct probability experiments and record outcomes.

Fifth Grade
- Make and interpret charts and tables.
- Make and interpret bar and picture graphs.
- Read double bar graphs.
- Make and interpret line graphs.
- Collect, organize, interpret data.

- Interpret circle graphs.
- Record fraction of outcomes and find probabilities.
- Use tree diagrams for probability.
- Determine mean (average) and range.

Sixth Grade
- Make and interpret bar, line, circle graphs.
- Collect, organize, interpret data.
- Make and interpret charts and tables.
- Interpret double bar graphs.
- Interpret double line graphs.
- Use frequency tables.
- Interpret points on a grid.
- Use tree diagrams and find probabilties.
- Use outcomes of probability experiments to predict.
- Find range, mean, median, mode.

Seventh Grade
- Review all concepts with graphs and tables.
- Review all probability concepts.
- Probability of zero and one.
- Use possible combinations and empirical probability.
- Determine probability of independent and dependent events.
- Permutations and combinations.
- Use factorials in probability.
- Venn diagrams.
- Find range, mean, median, mode.
- Use samples to predict.

Eighth Grade
- Review concepts with graphs and tables.
- Histograms and scattergrams.
- Graph equations and inequalities.
- Graph slope and y-intercept.
- Review all probability concepts.
- Permutations and combinations.
- Review all statistics concepts.

Teaching Strategies

Graphing

Graphing presents data in a concise and visual way that allows relationships in the data to be seen more easily. Students need to learn how to tally information, arrange data in a table, and display the information visually using a graph. Students can be introduced to graphing as a means of representing or organizing data early in the elementary years. Young children can make their own graphs and will benefit from collecting and organizing the information. Baratta-Lorton (1976) offers a suggested developmental sequence for presenting graphs:

1. Real graphs: actual objects to compare two or three groups.
2. Picture graphs: pictures or models to represent real things to compare two or three groups.
3. Real graphs: actual objects to compare four groups.
4. Picture graphs: pictures or models to compare four groups.
5. Symbolic graphs: most abstract using only symbols.

Ask children questions about the graph to help focus the discussion of information available for the graph. Questions should include comparisons such as which has the most or the least, how many more or how many less, and how many in all. Children enjoy taking a survey, collecting information, and constructing opinion graphs about their classmates or school. These graphs can be displayed in the hallway and create student interest and attention. Graphing should be at least a weekly activity as it is a part of daily exposure in newspapers, magazines, and other forms of media. Five kinds of graphs will be discussed: real, picture, bar, line, circle, and scatter.

Creating Real Graphs

Children should begin experiences with graphing like everything else in mathematics—concretely. They can form actual graphs themselves such as their hair color, sex, short pants or long pants, shoes with laces or without, children who eat bag lunch or hot lunch. The possibilities for topics to graph are endless and should be determined by the interests of the class. A graphing plastic can be made from oil cloth with colored masking tape for the sections. Children can stand in the sections, or place their favorite fruit (actual object), or their shoes (laces or without) and the sections allow for easier counting and comparing (Fig. 12.1).

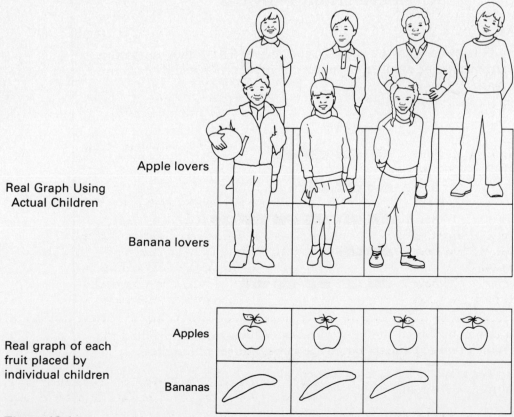

Real Graph Using Actual Children

Apple lovers

Banana lovers

Real graph of each fruit placed by individual children

Apples

Bananas

Figure 12.1

Creating Picture Graphs

A picture graph represents data in the form of pictures, objects, or parts of objects. Cut pictures of real things, draw pictures to represent real objects, or use some models to stand for the objects in picture graphs. For example, have children draw or cut from magazines pictures of their favorite pet, season of the year, sport, toy, or cookie. Be sure not to extend the categories to four until many experiences have been provided for comparing two and three groups. Another example would be to count the number of girls in each row of the class. The graph may look like Figure 12.2.

The graph shows row 1 with 4 girls, row 2 with 3 girls, row 3 with 7 girls, row 4 has no girls, and row 5 has 5 girls. Questions can be asked to determine which row has the least number of girls, or how many girls are in row 1.

Older children can be taught to interpret picture graphs in which a symbol represents multiple units (i.e., a picture of one milk carton = 5 people who choose milk for lunch). This is a more difficult concept and can be done when multiplication is being studied in third or fourth grade.

Children In Class

Represent each girl with the symbol ♟

Row 1. ♟ ♟ ♟ ♟
Row 2. ♟ ♟ ♟
Row 3. ♟ ♟ ♟ ♟ ♟ ♟ ♟
Row 4.
Row 5. ♟ ♟ ♟ ♟ ♟

Figure 12.2

Activity

MATH
ACROSS THE
CURRICULUM

Picture Graph with School-Community Activities

Directions:

1. A class is helping with the school paper drive. The following picture graph represents the number of bundles of papers collected by each row in the class.

Paper Drive

Row 1. ▰ ▰ ▰ ▰

Row 2. ▰ ▰ ▰

Row 3. ▰ ▰ ▰ ▰ ▰ ▰

Row 4. ▰ ▰

Row 5. ▰ ▰ ▰ ▰ ▰

2. Which row contributed the most bundles? Which row contributed the least? Which two rows contributed the same number of bundles?

3. Convert the graph to let each picture equal three bundles. (i.e., one bundle pictured = 3 bundles.)

Activity

Picture Graph

Directions:

Draw the picture graph.

Children In Class

Let ♟ = 2 students

Row 1 ♟ ♟ ♟ ♟ ♟
Row 2 ♟
Row 3 ♟ ♟ ♟
Row 4 ♟ ♟ ♟

Ask questions such as: How many children are in each row? Which row has the least, most, etc?

Directions:

1. Make a new graph with the following information:

Let 👤 = 10 students

Class 1—10 students
Class 2—20 students
Class 3—30 students
Class 4—25 students
Class 5—50 students

2. Create questions similar to the ones listed previously.

Creating Bar Graphs

A bar graph uses discrete (separate, distinct) data on each axis. Bars represent the information by the x and y axis. Bar graphs can begin with concrete materials such as colored cubes or connecting cubes. Have each child select a cube from four color choices to represent a favorite color. Later, have a color represent a choice of favorite ice cream (i.e., brown for chocolate, white for vanilla, and red for strawberry), a favorite fast food chain, or a favorite Saturday night television program. In this way, children can observe how the graph is created into the bar graph. They can take graph paper and transfer that data into a more abstract form that will hold meaning for them.

Creating bar graphs can begin with a grid. Using the example of girls in each row of the class, have students divide each row into spaces and place the name of each girl in the space next to the row she is in (Fig. 12.3).

This is the beginning of a bar graph. The students could now shade in the squares with girls' names to give a horizontal bar graph as in Figure 12.4.

The same activity could be done with the paper drive data to give a horizontal bar graph like the one in Figure 12.5.

Girls in Classroom Rows

Row 1	Lynn	Lauren	Mary	Marilyn			
Row 2	Violet	Alice	Maude				
Row 3	Emma	Betty	Carolynn	Blanche	Lois	Sue	Helen
Row 4							
Row 5	Ruth	Sarah	Paula	DeAnna	Pamela		

Figure 12.3

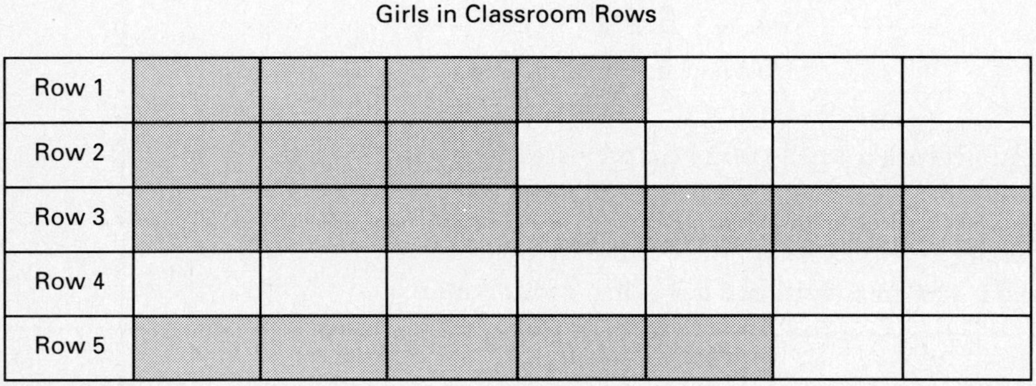

Girls in Classroom Rows

Figure 12.4

Paper Drive Participation by Row

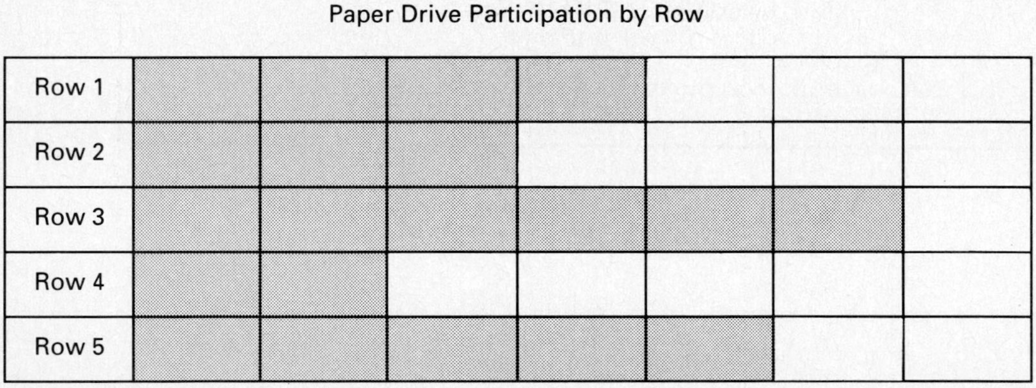

Figure 12.5

Bar graphs can have vertical or horizontal bars. Students need experiences constructing graphs in both formats. They also need activities where they decide the labels for the two axes. Increments should be in consecutive whole numbers and in later grades, bar graphs can have other increments such as counting by twos, fives, tens. Be sure to limit the list of choices to four or five or comparisons become more difficult.

Activity

Favorite Academic Subject

Directions:

1. Look at the following picture:

Favorite Academic Subject

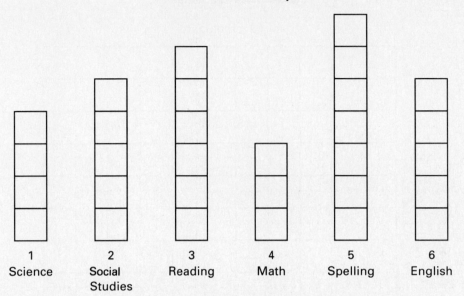

| 1 Science | 2 Social Studies | 3 Reading | 4 Math | 5 Spelling | 6 English |

2. Now, transfer this to a piece of graph paper by using each square to represent those students preferring each academic area and shade in the appropriate number of squares.

Favorite Academic Subject

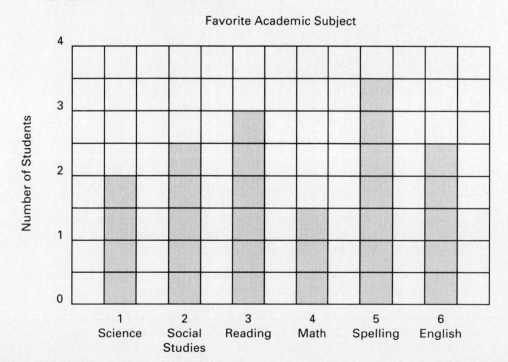

Activity

PROBLEM
SOLVING

Bar Graphs

Directions:

1. Make a survey of the class to collect data on favorite sports.

2. Plot it on a bar graph. It could look like this:

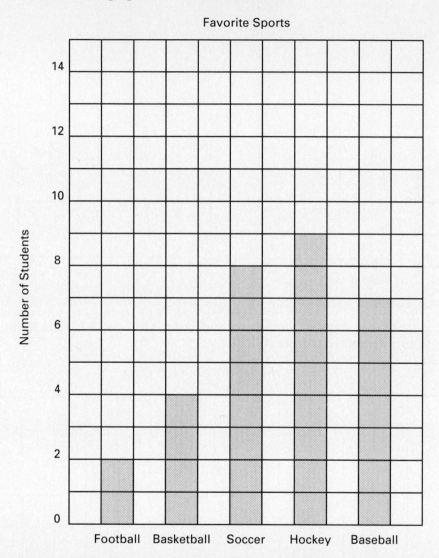

Favorite Sports

3. The bar graph shows 2 students in the class preferring football, 4 students preferring basketball, 8 students each preferring soccer, 9 students preferring hockey, and 7 students preferring baseball.

4. A variation: Graph the opinions of the class regarding their favorite pet, favorite school lunch menu, or any topic they select.

Creating Line Graphs

The authors' experiences indicate that upper elementary students have more problems interpreting information on a line graph than on a bar graph. It may be related to the visual interference from all lines seen simultaneously—horizontal, vertical, and broken diagonal lines. Perhaps the child does not know where to center attention. When given identical information in a bar graph and a line graph, an increased number of errors are made by children when interpreting the line graph. Creating a line graph can be done by connecting the midpoint of the top of the bars in the bar graph.

Start with simple line graphs, like the following that tell a message. What information can be obtained from this line graph?

The graph expresses the relationship between Greg's and Travis's height and weight. Travis is taller while Greg is heavier.

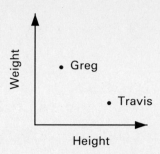

Line graphs are often used to illustrate a continuous activity. Tell what each graph illustrates.

Activity

PROBLEM SOLVING

Line Graphs

Directions:

1. Write the message for each of the following line graphs:

A

• Joy

• Judy

Height / Weight

Message:

B

Annette   Sherian

Weight / Height

Message:

D

• J.C.

• Bob

Weight / Height

Message:

E

• Karen   • Tammy

Age / Weight

Message:

C

• Krista

• Jackie

Height / Weight

Message:

F

• Paul   • Lyman

• Frank

Age / Height

Message:

(continued)

Example:

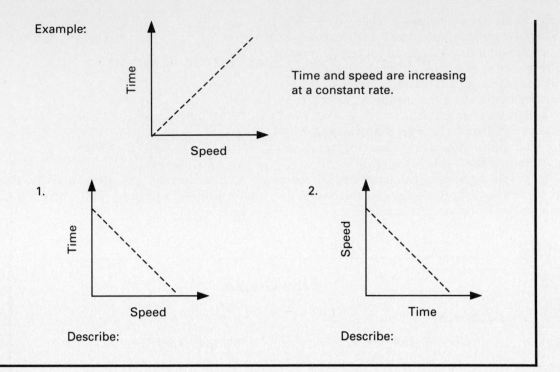

Time and speed are increasing at a constant rate.

1.

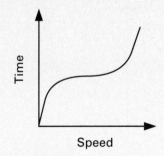

Describe:

2.

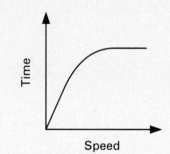

Describe:

Activity

MATH
ACROSS THE
CURRICULUM

PROBLEM
SOLVING

Continuous Line Graphs in Science

Directions: These graphs show a *continuous* activity. What is the message in each graph?

A

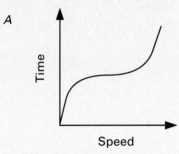

Conclusion:

B

Conclusion:

C

Conclusion:

D

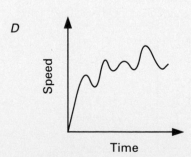

Conclusion:

Activity

MATH
ACROSS THE
CURRICULUM

PROBLEM
SOLVING

Plotting Continuous Line Graphs in Science

Directions:

1. How could you label and plot these graphs? Be sure they give the message intended. An example is presented first.

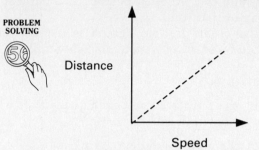

Distance

Speed

This is one way to graph an accelerating car.

Car approaching a stop sign.

Running a long distance.

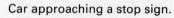

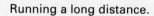

2. Label and plot these graphs to fit the situation given for each:

A horse race.

Gasoline level and a running automoblie.

Exercise and heart beats.

Growth of person in height.

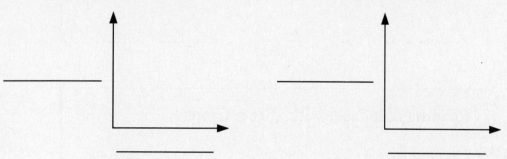

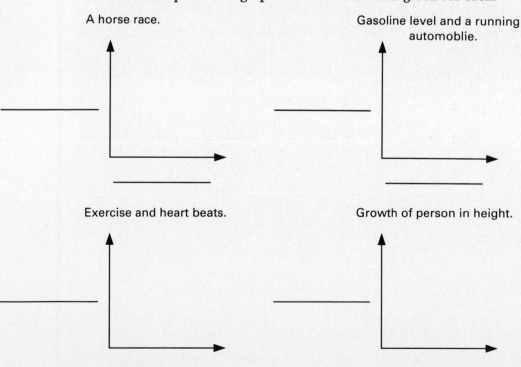

The data points are plotted and lines are drawn connecting the data points. The scale of a graph is very important. The scale used should be uniform and appropriate for the quantities to be displayed. The scale should be large enough to adequately represent the quantity being plotted. It is often necessary to use multiple units to keep the graph in reasonable proportions. Line graphs are often used because they provide a continuous representation of data. Therefore, the scale is important to adequately reflect the data.

Creating Circle Graphs

The circle or "pie" graph is used to show the relationship between a whole and its parts. A good example of the use of circle graphs is to illustrate a family's budget in relation to its income (Fig. 12.6). Circle graphs often show percents so they should not be introduced until children have an understanding of percent. This kind of graph is generally encountered in the fifth or sixth grades.

The circle represents the total income of the family and each item is specified as a percent or portion of that total.

The circle or "pie" graph is usually constructed by starting with a table showing:

(a) facts,

(b) the fractional part each quantity is of the whole,

(c) the number of degrees in each fractional part. (Total degrees in a circle is 360 degrees.)

Activity

COMPUTERS

MATH
ACROSS THE
CURRICULUM

Reading/Alphabet with Circle Graphs

Directions:

1. Which letters in our alphabet are used most often?

2. Predict the frequency ranking of the letters on the following chart.

3. Give the ranking of "1" to the letter you feel is most frequently used, a "9" to the least.

4. Select short paragraphs from each of several books and tally the frequency of each letter.

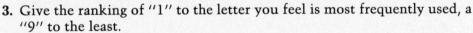
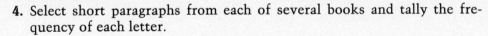

	Your Ranking	Sample 1	Sample 2	Final Ranking
A				
E				
I				
O				
U				
T				
Z				
B				
S				

5. Rank this information and compare it to your predictions.

6. Enter the information into a computer graphics program and have the computer draw a circle graph of each sample.

Family Budget

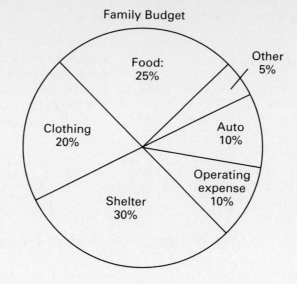

Food: 25%
Clothing: 20%
Shelter: 30%
Operating expense: 10%
Auto: 10%
Other: 5%

Figure 12.6

Activity

COMPUTERS

Circle Graphs

Directions:

1. Take a survey of the nationalities represented in a school. The table of data from this survey might look like:

Number of People	Fractional Parts	Percent	Degrees	Cumulative Total
Caucasians—60	$\frac{60}{160}$	37.5% × 360 = 135		135
Blacks—20	$\frac{20}{160}$	12.5% × 360 = 45		180
Asians—48	$\frac{48}{160}$	30% × 360 = 108		288
Hispanics—32	$\frac{32}{160}$	20% × 360 = 72		360
Total: 160	$\frac{160}{160}$	100% × 360 = 360		360

The graph would be as follows:

Population Distribution in a School

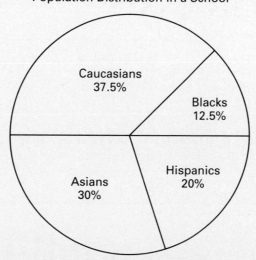

2. Use a computer graphing program or Logo to develop this circle graph.
3. Change the numbers to illustrate the population of your school.

385

Creating Scatter Graphs

A scatter graph is the plotting of pairs of values and observing the "scatter" form they take on. Often real-world data is not "very uniform" and it is difficult to construct an accurate picture, line, bar, or circle graph. Students can plot the pairs of points and attempt to find some trend or tendency from the scatter pattern. For example, the following scatter graphs can be used to help students increase their problem-solving skills.

Activity

PROBLEM
SOLVING

Story Writing with Scatter Graphs

Directions:

1. Graphs I, II, and III all use the same data.

2. Label the X values and Y values with identifiers.

MATH
ACROSS THE
CURRICULUM

3. Write a brief newspaper story using the data.

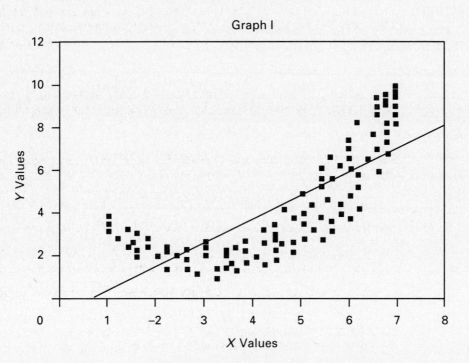

Graph I

4. What has changed in Graph II? Does your newspaper story change from Graph I?

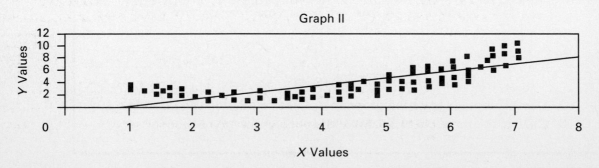

Graph II

5. What has changed in Graph III? Does your newspaper story change from Graph I or II?

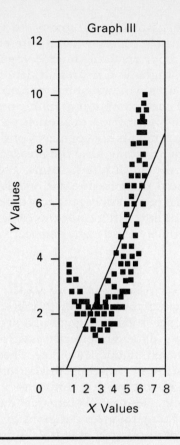

Graph III

The same information can be seen by the use of different graphs. Each graph produces a different visual presentation and may convey a different perspective for the same data. Students should be encouraged to start with the graphs that seem easiest for them. Through the use of computer programs, they can see how the same information can be represented in other more difficult graphs. Computer software can greatly enhance the work of organizing and collecting data. Database programs make it easy to sort data by different categories and organize it in a variety of ways. The choice of a scale to use in the graph is another important aspect of constructing a graph. Scale changes can be done easily with the computer to compare different pictures of the same data. Students can make inferences, analyze the data from various perspectives, and formulate key questions.

Activity

COMPUTERS

Comparing Different Types of Graphs

Directions:

1. Collect height data in the classroom; the findings are: 2 students measure 63 inches, 4 measure 65 inches, 3 measure 64 inches, and 5 measure 68 inches.

2. Make a bar graph of this information using a computer graphing program.

3. Set the scale of each division. Try different scales to represent the data.

4. Using the same data, represent the information in a different type graph.

5. Compare the two graphs.

As you can see, graphs can tell many stories and the same data can be represented in different ways. Newspaper reporters using microcomputers and graphing programs can present data in many views. Pick up any newspaper presenting statistical data and use a computer graphing program to display the information in another way. It is important that students are cognizant of the importance of understanding how the presentation of data can take on different forms. Each gives a unique visual presentation and yields a different impact. To be an intelligent consumer, we must be aware of how different forms of data presentation can convey a different perspective.

Statistics

The study of statistics is important in the elementary grades since society frequently organizes and expresses data numerically (statistically). Since early childhood, we have been exposed to statistics in weather predictions, newspaper ads, and test grades, and with advances in technology, additional statistical data is available. Daily living requires decisions to be made based upon processing of statistical information continuously.

Students need to be aware of how statistics can be manipulated to say whatever one desires. Media experts can use statistics to mislead consumers. Reports of medical information or opinion surveys must be examined carefully before drawing conclusions. Students should be aware of how statistics shape our lives and influence our decisions. There are many misconceptions about data because key questions have not been asked: how many people were in the sample, how was the population sampled, who conducted the survey, how was the data summarized. We are inundated with statistical information and we must learn how to process this information accurately and effectively to function as knowledgeable citizens in society.

One of the first concepts of statistics encountered in daily life is that of average. The doctor tells a child "your height is about average for a 6-year-old girl" or "you are a little over the average weight of an 8-year-old boy." The teacher says "the class average on the last test was 68" or "you scored above the average of the class on the last exam." *The average is a measure of central tendency of data, or an attempt to describe what is "typical" in a set of data.* There are three measures that are commonly used to describe data: mean, median, and the mode.

Mean

The mean, or arithmetic mean, is most commonly used to describe the "average" of a set of data. The mean is computed by dividing the sum of the numbers by the number of members in the set. For example, if we want to know the average weight of 4 boys who weigh 105, 99, 110, and 90 pounds, we add $105 + 99 + 110 + 90 = 404$ and divide by 4. Thus, the average weight of the four boys is 101 pounds. Suppose we want to find the average weight of 6 bags of apples if the bags weigh 2 pounds, 2 and $\frac{1}{2}$ pounds, 3 pounds, 1 and $\frac{1}{2}$ pounds, 3 and $\frac{1}{2}$ pounds, and 2 and $\frac{1}{2}$ pounds. Add the weights: $2 + 2\frac{1}{2} + 3 + 1\frac{1}{2} + 3\frac{1}{2} + 2\frac{1}{2} = 15$ pounds and divide by 6 (the number of bags) to find an average weight of $2\frac{1}{2}$ pounds. Students should be encouraged to perform simple activities to develop an understanding of the arithmetic mean. Have students work with objects until they understand the algorithm for determining the average (arithmetic mean) of a set of data.

Activity

MANIPULATIVES

Mean

Directions:

1. The weight of the bags of apples can be represented by objects, such as Cuisenaire rods. If the purple rod is one pound and the red rod is $\frac{1}{2}$ pound, the six bags of apples could be represented by:

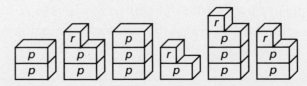

2. Rearrange the rods so there are six stacks of equal heights:

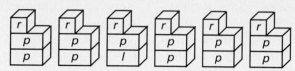

3. This is a method to discover the arithmetic mean or average of 6 bags of apples weighing 2 pounds, $2\frac{1}{2}$ pounds, 3 pounds, $1\frac{1}{2}$ pounds, $3\frac{1}{2}$ pounds, and $2\frac{1}{2}$ pounds is $2\frac{1}{2}$ pounds.

4. Another approach using Cuisenaire rods is to have the students find the average height of 4 boys in a class. Their heights are 63 inches, 65 inches, 64 inches, and 68 inches.

5. Use Cuisenaire rods to represent the heights:

10	10	10	10	10	10	1 1 1	
10	10	10	10	10	10	5	
10	10	10	10	10	10	1 1 1 1	
10	10	10	10	10	10	5	1 1 1

6. Rearrange the rods to make each one the same length:

10	10	10	10	10	10	5
10	10	10	10	10	10	5
10	10	10	10	10	10	5
10	10	10	10	10	10	5

The average height of these four boys is 65 inches.

Activity

PROBLEM SOLVING

Mean

Directions: Judy has a bag with 4 apples, Sally has a bag with 6 apples. Carmen has a bag with 2 apples. What is the average number of apples in each bag? Bob places all the apples in a pile and then begins to divide them into 3 separate piles.

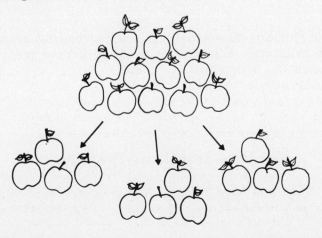

(continued)

He finds that there are 4 apples in each pile—showing an average (arithmetic mean) of 4 apples in each bag. This type of activity can lead to the discovery of the algorithm for finding the arithmetic mean. Take the total sum of the elements and divide by the number of members of the set. Add the apples: $6 + 4 + 2 = 12$ and divide by the number of members of the set (3 bags), $12 \div 3 = 4$ and the average is 4 apples in each bag.

Activity

CALCULATORS

Finding the Mean

Directions: Using a calculator, determine if the mean (average) in each square yields a value that allows the square to be a "magic square." A magic square is a square array of numbers where the sum of any row, column, and diagonal equals the same number.

159.35 217.98 149.73 230.38	16.93 36.32 18.71 19.15 27.24	123.63 217.14 219.67 91.68 133.37 66.63
78.60 52.93 45.12 71.25 68.39 88.88 91.90	69.75 78.96 45.69 147.32 250.03	123.45 189.79 183.83
78.69 121.23 81.56 99.99 87.78 98.83	258.63 187.86 203.29 202.34	45.21 51.58 44.36 39.78 63.11 45.79 51.09 37.80

Provide experiences in calculating means in practical problems using a calculator. A good source of practical problems is the daily newspaper. Class discussions could include "average rainfull," "batting average," "average speed," "stock market average," and "average points per game."

Activity

COMPUTERS

BASIC—Test Score Program

Directions: When you have entered all your scores, type 9999 and the computer will give you the average. Type LOAD TEST SCORES and run the program.

```
10 LET NUMBER = 0
20 LET T = 0
30 INPUT "YOUR TEST SCORE: ";  TEST
40 IF TEST = 9999 THEN 80
50 LET NUMBER = NUMBER + 1
60 LET T = T + TEST
70 GOTO 30
80 PRINT "YOUR AVERAGE SCORE IS ";T/NUMBER
90 END
```

Mode

The second measure of central tendency is the mode. Since an extremely high number or an extremely low number can distort the picture of the tendency shown by the arithmetic mean, the mode is frequently used. *The mode of a set of numbers is that number that occurs most frequently in the set.* For example, the following scores were recorded for 10 pupils taking a test: 8, 7, 6, 8, 9, 6, 8, 5, 7, 8. If we put this data into a frequency table, the mode becomes more readily apparent.

Score	Number of Pupils
9	/
8	////
7	//
6	//
5	/

We can see that the most frequent score was 8, hence the mode of this set of number is 8.

Activity

MATH
ACROSS THE
CURRICULUM

Mode

Directions:

Suppose we ask a group of 12 girls on the playground to tell us their ages. They report:
8, 10, 7, 10, 11, 8, 9, 10, 12, 10, 7, 6
The frequency table follows:

Ages	Number of Students
6	/
7	//
8	//
9	/
10	////
11	/
12	/

The mode of this set of data is 10.
NOTE: The mean is 9, which is different from the mode.

Median

A third measure of central tendency is the median. *The median of a set of numbers is the middle number of the set.* For example, take a group of test scores such as: 7, 9, 3, 1, 6, 4, 10 and arrange them in order from greatest to least: 10, 9, 7, 6, 4, 3, 1. There are seven scores. Now determine the position of the middle score. It must be the fourth score from either end; therefore, 6 is the median of this set of test scores.

Here is another example. To find the median of the following ages of a class:

12, 6, 7, 8, 7, 9, 10, 6, 7, 11, 12, 11, first, arrange the ages in order: 6, 6, 7, 7, 7, 8, 9, 10, 11, 11, 12, 12. How many ages are there? Twelve. Now determine the position of the middle age. It must be between the age of 8 and the age of 9. The median age is $8\frac{1}{2}$ since half of the ages are below $8\frac{1}{2}$ and half of the ages are above $8\frac{1}{2}$. Since there are an even number of ages, the median age lies halfway between the sixth and seventh age, $8\frac{1}{2}$.

Range

Another common statistical measure is the range. *The range of a set of numbers is the difference between the highest and lowest numbers in a set.* The range tells us the spread of a set of data.

Looking at one of our earlier examples of scores from a test:

Score	Number of Pupils
9	/
8	////
7	//
6	//
5	/

The range here (or spread of scores) is $9 - 5$ or 4.

In another example with ages:

Ages	Number of Students
6	/
7	//
8	//
9	/
10	////
11	/
12	/

The range of ages here is $12 - 6$ or 6.

Range, like the mean, is sometimes not a good measure of a distribution, because it includes the extremes and often gives a distorted picture of the data. For example, look at these two sets of numbers:

1, 1, 7, 7	mean 4; range 6
1, 4, 4, 7	mean 4; range 6

The above two sets of numbers have the same mean and range, but have a very different "scatter." The mean and range in the example do not give us an accurate picture of what is happening. We do not know that the points in the second set are grouped near the mean and those in the first set are located in the extremes of the range.

Activity

CALCULATORS

Finding the Average

The following National Hockey League (NHL) standings were published:

Campbell Conference

Patrick Division

Team	W	L	T	PTS	GF	GA
Philadelphia	27	2	10	64	168	120
NY Rangers	19	17	7	45	164	156
NY Islanders	16	17	6	38	133	130
New Jersey	16	18	5	37	131	139
Pittsburgh	11	23	6	28	125	151

Smythe Division

Team	W	L	T	PTS	GF	GA
Calgary	15	14	12	42	116	121
Los Angeles	16	19	6	38	127	139
Vancouver	15	21	7	37	135	145
Winnipeg	12	25	5	29	111	163
Edmonton	9	22	9	27	133	174

Whales Conference

Adams Division

Team	W	L	T	PTS	GF	GA
Buffalo	28	11	3	59	159	112
Hartford	21	9	8	50	162	113
Boston	21	12	6	48	144	115
Quebec	17	18	6	40	130	137
Montreal	17	19	4	38	144	154

Norris Division

Team	W	L	T	PTS	GF	GA
Minnesota	20	16	6	46	157	142
Toronto	19	13	8	46	172	151
St. Louis	17	13	11	45	144	141
Detroit	14	18	7	35	129	132
Chicago	9	20	10	28	122	148

Terms to Remember

W = Win   T = Tie   GA = Goals Against
L = Loss   GF = Goals For   PTS = Points

Directions:

1. Using a calculator, find the averages for GF and GA per game for each team in the Patrick Division.

2. Using a calculator, find the averages for GF and GA per game for each team in the Smythe Division.

3. Using a calculator, find the averages for GF and GA per game for each team in the Adams Division.

4. Using a calculator, find the averages for GF and GA per game for each team in the Norris Division.

5. Make a bar graph of the above information.

Activity

COMPUTERS

PROBLEM
SOLVING

Spreadsheet

Directions:

1. The following recap of a basketball game appeared in the sports page.

2. Build a computer spreadsheet with this information:

Griffons

Player	MIN	FG	FT	R	A	F	T
Horner	39	10–14	3–6	6	4	4	23
Sisco	39	10–17	3–5	5	2	4	23
Losh	36	6–7	0–0	3	5	2	12
LaFave	40	11–15	1–2	1	7	3	23
Klein	27	6–12	0–0	2	3	3	12
Miles	20	5–11	0–0	0	2	2	10
Ratliff	12	3–6	1–1	1	0	2	7
Edwards	12	0–0	0–0	2	0	0	0
Warren	9	0–2	2–2	0	1	1	2
Ford	6	0–1	0–0	4	0	0	0

(continued)

Wildcats

Player	MIN	FG	FT	R	A	F	T
Reinholz	36	3–9	0–0	3	0	3	6
Salmon	37	0–1	2–4	10	4	3	2
Bitter	45	8–14	1–2	16	2	2	17
Crowder	40	11–14	1–1	2	5	4	26
Hatfield	40	16–27	7–9	6	3	3	39
Spiers	26	3–5	4–4	1	2	3	10
Boydston	16	7–13	0–0	2	1	2	16

Terms to Remember

Min = Minutes Played	T = Totals
R = Rebounds	A = Assists
FG = Field Goals	F = Foul
FT = Free Throw	

3. Create a spreadsheet following the sample format and find the averages for Min, R, A, and T for each team.

Griffons

	Team Total	Average/Player
Min		
R		
A		
T		

Wildcats

	Team Total	Average/Player
Min		
R		
A		
T		

4. Use the spreadsheet to find the averages for both teams combined.

5. Prepare a line graph of each player's FG percentage by team.

Probability

Probability is the branch of mathematics concerned with analyzing the chance that a particular event will occur. The basic purpose of probability theory is to attempt to predict the likelihood that something will or will not happen. Probability is computed on the basis of observing the number of actual outcomes and the number of possible outcomes.

Probability of an event =

$$\frac{\text{Number of actual outcomes}}{\text{Total number of possible outcomes}}$$

Studying probability will help children develop critical thinking skills and to interpret the probability that surrounds us daily. The study of chance needs to begin in an informal way in the early grades. Many students and teachers have misconceptions about the outcome of real events in life. For example, if the Watson family has three girls and is expecting a fourth child, is there an equally likely chance of having a boy or girl? Many people feel the next child "is bound to be a boy" since there are already three girls in the family. They base their predictions on their opinions about what they feel should happen (a biased reaction) rather than on factual data. Statements involving probability abound in everyday situations and it is important to make students aware of them. Students should be introduced to probability through activities with measuring uncertainty. Choice devices can be used to provide initial experiences for the study of chance—tossing a coin, rolling a die, drawing a card from a deck or a colored cube from a bag.

Students should begin by doing experiments, predicting the outcome, and recording the results. In most cases, a sample must be selected. *A sample is a small number that represents a large group called a population.* In most cases, it is impossible to test the total population so a sample or samples are chosen.

Activity

MANIPULATIVES

Coin Toss

Directions:

1. A good experiment is to toss a coin 30 times and record how many heads and how many tails you get from a set number of tosses.

2. Represent the occurrences as a fraction. For example, if the outcome was 16 heads, the fraction $\frac{16}{30}$ would be the ratio of heads to the number of tosses and $\frac{14}{30}$ would be the ratio of tails to the number of tosses.

   Heads $\frac{16}{30}$

   Tails $\frac{14}{30}$

3. Record your results as a ratio.

From experience like the coin toss, students can begin to predict the probability of certain events occurring. If a coin is tossed, there are two probable outcomes: heads or tails. There is an equally likely chance of it landing as a head or as a tail. If a coin is tossed 12 times, the theoretical probability is that it will land heads 6 times and tails 6 times. If one were actually to do this act, the experimental probability (actual outcome) may be quite different. Children need experiences with experimental probability. Another study of chance is tossing a die. Toss a die 100 times and record how many times each number appears. The results will indicate on the average that each number occurs about $\frac{1}{6}$ of the time. The theoretical probability of the number 4 appearing on the toss of a die is $\frac{1}{6}$. Likewise, the probability of the number 2 is $\frac{1}{6}$. The following activity will let students experiment themselves.

Probability of 4 = $\frac{1}{6}$

Activity

Die Toss

Directions:

1. Take one die and roll it 100 times. Keep a record of the number of times each number appears.

 _____ times   _____ times   _____ times

 _____ times   _____ times   _____ times

2. Divide the number of times you threw the die into the number of times your number appeared. This ratio tells you the chances of your number appearing in a certain number of trials.

3. Compare your results in the die activity to the theoretical probabilities.

4. From your record, what was the probability of a number less than 3 showing? What is the theoretical probability of this event occurring?

5. What is the theoretical probability of an odd number (1, 3, 5) showing?

6. What is the theoretical probability of a prime number (2, 3, 5) showing? A multiple of 3 (3, 6) showing? A multiple of 2 (2, 4, 6) showing?

Students should engage in many experiments to explore probability problems such as tossing a coin, tossing a die, or flipping two-colored counters until they become skillful in assigning probabilities. From experiments of this type with one coin or one die, the teacher should progress to experiments with two coins, colored balls, and checkers. Discussions should include notions of fairness. Have students work through these activities as they continually ask, "Is this a fair situation?"

Activity

Money Toss

Directions:

1. Take two coins—a nickel and a penny or a penny and a dime—and predict the outcome of tossing the two coins.

2. The class will probably predict an H–H, H–T, T–T outcome with the probability of $\frac{1}{3}, \frac{1}{3}, \frac{1}{3}$. Yet when the experiment is performed, the H–T combination seems to occur about twice as often as predicted.

3. Look at all possible outcomes:

Penny	Dime	Ordered Pair
Heads	Heads	H–H
Heads	Tails	H–T
Tails	Heads	T–H
Tails	Tails	T–T

4. We see there are four possible outcomes rather than the three we thought originally. The penny can be heads and the dime tails or the penny tails and the dime heads, giving two ordered pairs (H–T) and (T–H). The probability becomes:

Event	Penny	Dime	Probability
1	H	H	$\frac{1}{4}$
2	H	T	$\frac{1}{4}$
3	T	H	$\frac{1}{4}$
4	T	T	$\frac{1}{4}$

5. What is the probability both coins are heads?

6. What is the probability one coin is heads and one coin is tails?

7. What is the probability at least one coin falls tails?

Activity

Creating Tables with Dice Toss

Directions:

1. Throw a pair of dice 25 times and give the sum of the two faces. Record the results. Each die should be a different color from the other to make the record keeping in step 3 easier.
   Occurrences:
   Sum of

   2 _____   5 _____   8 _____   11 _____
   3 _____   6 _____   9 _____   12 _____
   4 _____   7 _____   10 _____

2. Which sum occurred the most often? _____ least? _____

3. Complete the table to determine how often the sums will occur.

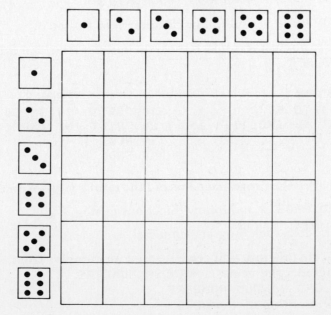

All Possibilities with Two Dice

(continued)

4. Mathematically, which sum occurs the most often? _____ least? _____

5. Did your experimental answer agree with the mathematical outcome as shown on the table?

6. If not, explain why there would be different answers.

Activity

MANIPULATIVES

Probabilities with Cards

Directions:

1. Place 5 cards with the numbers 0 through 4 on them in a cardboard box.

2. What is the probability of drawing a card with a number less than 5 on it? There are five possible outcomes and they are all favorable; therefore, the probability of drawing a card with a number less than 5 on it is $\frac{5}{5} = 1$.

3. What is the probability of drawing a card with a number greater than 4 on it? There are five possible outcomes and none of them are favorable—the probability of this occurring is $\frac{0}{5}$ or 0.

4. If an event is sure to happen, the probability is one. If there is no favorable outcome and an event is sure not to happen, the probability is 0.

Activity

COMPUTERS

PROBLEM SOLVING

BASIC

Directions:

1. The two programs are saved on the BASIC disk under the names:
   PROB 1 and PROB 2

2. Type LOAD PROB 1 and follow the directions in step 3.

   (If you have no disk, type each program one at a time and follow directions in step 3.)

```
        PROB 1                        PROB 2
10 FOR I = 1 TO 50           10 FOR I = 1 TO 50
20 PRINT INT (6*RND(1) + 1)  20 PRINT INT (RND(1) + .5)
30 NEXT I                    30 NEXT I
40 END                       40 END
```

3. Type RUN after the program and record the results you see on the screen.

4. Type LOAD PROB 2 and follow directions in step 3.

5. What event does each program simulate?

6. The following program will generate random numbers from −50 to 50. It is saved under the name RANDOM NUMBERS
   Type LOAD RANDOM NUMBERS
   Type RUN RANDOM NUMBERS

(If you have no disk, type the following:)

```
100 REM RANDOM NUMBERS FROM -50 TO 50
110 NUMBER = INT (RND(1) * 101) - 50
120 PRINT NUMBER
130 GOTO 110
140 END
```

Press the control key and S key at the same time to stop the scrolling on the screen. Press return to resume.
Press the control key and C key together to end.

7. What is the value of a random number generator?

In most experiments, we have more than one set of outcomes associated with the experiment. For example, in our previous example of tossing two coins, our probable outcomes were H–H, H–T, T–H, T–T. A pictorial way of illustrating all possibilities is with a *tree diagram.* To illustrate, use a bag containing 5 red and 3 black marbles. Draw twice with replacement (return the marble to the bag before making the second draw). What are the possible outcomes of this experiment?

You could draw a red marble first then a black marble, a red marble first and then another red marble, a black and then a red or a black and a black. We can abbreviate this as: rb, rr, br, bb.

We can represent this visually as follows.

On the first draw, we can obtain either a red marble or a black marble:

Regardless of what occurred on the first draw, there are still 2 possibilities for the second draw. The tree diagram is very useful in determining possible outcomes of an experiment. We can extend our diagram to include the second draw:

1st	2nd	Outcomes
	r	rr
r	b	rb
	r	br
b	b	bb

Activity

Tree Diagrams

Directions:

1. Draw a tree diagram illustrating the outcomes of draws from a bag containing 12 red marbles and 4 black marbles.

2. Marbles are drawn from the bag with replacement. Three draws are made. What are the possible outcomes?

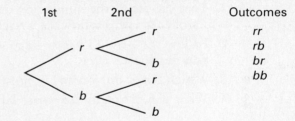

Outcomes

rrr
rrb
rbr
rbb
brr
brb
bbr
bbb

Assigning Probabilities

Middle school children need to approach problems in which they represent the probability of the outcome of a certain experiment in a mathematical way. For example, use a bag with 12 red marbles and 4 black marbles, and represent the probability that a red marble is drawn, or that a black marble is drawn. Since the red marbles make up $\frac{12}{16} = \frac{3}{4}$ of the total number of marbles, $\frac{3}{4}$ can be used to represent the probability of drawing a red marble. The black marbles represent $\frac{4}{16} = \frac{1}{4}$ of the total number of marbles. Therefore, $\frac{1}{4}$ can be used to represent the probability of drawing a black marble.

In another situation, if the same bag contained 4 black, 3 green, 3 red marbles, the probability of drawing a black marble would be $\frac{4}{10}$ (0.4), a green $\frac{3}{10}$ (0.3), and a red $\frac{3}{10}$ (0.3).

The probability of each outcome will be a number between 0 and 1, and the sum of all the probabilities in the experiment must equal 1. What is the probability that in the toss of a coin, "tails" will appear? $\frac{1}{2}$. What is the probability of rolling a 3 when tossing a die? $\frac{1}{6}$. Students should be allowed to perform experiments and record the results while learning to assign probabilities. Toss a die and ask how often does a 3 occur? About $\frac{1}{6}$ of the time? How often does a 2 occur? About $\frac{1}{6}$ of the time?

Experiments can be developed using die, coins, spinners, urns, cards, or whatever to give students experience assigning probabilities. Assigning probabilities (making selections) without bias is essential for computing the probability of events occurring. The next step is to identify different types of events.

Activity

PROBLEM SOLVING

Probability Path

Directions: Follow the probability path. Tossing a die 5 times will get you from the START position to one of the lettered boxes.

1. Toss the die. If the number on the die is odd, follow the odd path. If the number is even, follow the even path.

2. Follow the path 25 times. Keep a tally mark record of the box in which you finish each time.

3. Which boxes do you end in most often? Least often?

4. What percent of the 25 tosses lands in each box?

Start

Toss 1

even odd

Toss 2

even odd even odd

Toss 3

even odd even odd even odd

Toss 4

even odd even odd even odd even odd

Toss 5

even odd even odd even odd even odd even odd

| A | B | C | D | E | F |

Two events are said to be mutually exclusive if they have no outcomes in common. For example, if X and Y are mutually exclusive events with a finite (Z) set of outcomes, we could represent these sets as:

If the intersection (represented by ∩) of the two sets is the empty set, then

$$X \cap Y = 0 \qquad \text{and } P(X \cap Y) = 0$$
(no outcomes in common)   (where P = the probability)

The addition property of two mutually exclusive events is the sum of their probabilities, which is $P(X \cup Y) = P(X) + P(Y)$. For example, if a marble is drawn at random from a bag containing 4 red, 5 black, and 6 green marbles, what is the probability of drawing:

a) a green marble?
$P(G) = \frac{6}{15}$

b) a green or red marble?   which can be written $P(G \cup R)$
$P(G) = \frac{6}{15}$
$P(R) = \frac{4}{15}$ (The symbol for "or" is the ∪ symbol)

and, since these two events are mutually exclusive, the combined probability is

$$P(G \cup R) = \tfrac{6}{15} + \tfrac{4}{15} = \tfrac{10}{15} = \tfrac{2}{5}$$

To check our answer, compare this answer with the probability of drawing some color other than black.

$$P = 1 - \tfrac{5}{15} = \tfrac{10}{15} = \tfrac{2}{5}$$

Two dice are thrown. What is the probability that the sum is 5 or the sum is 7?

The possible outcomes 5 = (1,4) (2,3) (3,2) (4,1)

The possible outcomes 7 = (1,6) (6,1) (2,5) (5,2) (3,4) (4,3)

The probability of either event happening:

$$P(5 \cup 7) = \tfrac{4}{36} + \tfrac{6}{36} = \tfrac{10}{36} = \tfrac{5}{18}$$

Practice is recommended to understand the concept of mutually exclusive events. However, if students are continually reminded that mutually exclusive events are two events that have no common outcomes, they will have less difficulty with the topic.

Activity

MATH
ACROSS THE
CURRICULUM

Science

Directions:

1. Get a pencil and measure the length of it.

2. Draw parallel lines on the paper that are 1.5 times as far apart as the length of the pencil.

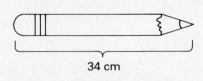

34 cm

34 cm × 1.5 (distance between parallel lines)

3. Drop the pencil on to the paper from a height of about 1 meter.

(continued)

4. Record on a chart, such as the one shown here, how many times the pencil touches a line or does not touch a line. Don't count it if the pencil doesn't land on the paper.

	Tally	Total
Touches a line		
Doesn't touch		

5. After a certain number of trials, such as 50, figure the probability that the pencil will touch a line when dropped. Does it make a difference if your pencil is longer or shorter?

In probability experiments we are often restricted by certain conditions imposed on the outcomes. How does the outcome change when a "known" is imposed on the experiment?

Suppose we throw two dice, one red and one green. If the red die shows a number that is divisible by 4, what is the probability that the total of the two dice is equal to or greater than 7? This condition that has been imposed on the experiment (the red die has a number divisible by 4) cuts down the sample or outcome.

The original sample of 36 possibilities is shown in Figure 12.7. When the condition that the number on the red die has a number divisible by 4 is imposed, the sample reduces to 6 possibilities, and the probability of having a total greater than or equal to 7 is $\frac{4}{6} = \frac{2}{3}$ (Fig. 12.8).

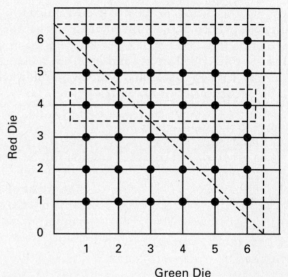

Figure 12.8

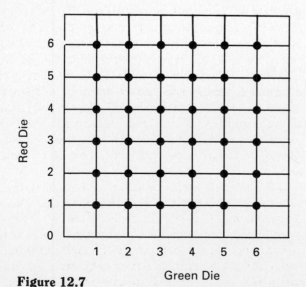

Figure 12.7

Probability that is dependent upon an outcome is referred to and is represented as P(A|B). This means the probability of A happening once B has occurred.

In the previous example, A equals the set of points representing a total dice sum of 7 or over. B equals the set of points for which the number on the red die is divisible by 4.

A ∩ B are the 4 points common to A and B (7, 8, 9, 10), and the probability of B, P(B) = $\frac{6}{36}$ = $\frac{1}{6}$. So, the probability of A occurring after B has occurred is the intersection of A and B divided by the probability of B.

The formula is:

If P(B) ≠ 0, then the conditional probability that A will occur once B has occurred is:

$$P(A|B) = \frac{P(A \cap B)}{P(B)}$$

Example:

P (sum > 7 | Red die divisible by 4)

$$= \frac{\frac{4}{36}}{\frac{6}{36}} = \frac{4}{6} = \frac{2}{3}$$

There are certain events that are not dependent on what occurred prior to their occurrence. These events are called *independent events*. For example, if a coin is tossed, the probability of it being tails on the second toss is not dependent on what it was on the first toss. The probability of it being a tail on the second toss is $\frac{1}{2}$ regardless of whether it was a head or a tail on the first toss.

The definition for the *multiplicative property for independent events is: the probability of independent events x and y to occur is the product of their probability such that:*

$$P(X \cap Y) = P(X) \times P(Y)$$

Previous bag problems, where marbles were being drawn from a bag with replacement after each draw, were an example of an independent event. Our original problem had a bag containing 5 red and 3 black marbles. Two marbles are drawn one after another, with replacement after each draw.

The probability of red on the first draw is $\frac{5}{8}$. Since the bag is restored to its original state after the drawing, the probability of drawing a red marble on the second drawing is $\frac{5}{8}$ also. A tree diagram of this experiment would look like Figure 12.9.

Repeat the same experiment without replacing the marbles. Again, make two draws, but this time do not replace the first marble drawn. The probability of a black marble being drawn on the first draw is $\frac{3}{8}$. If the marble is drawn and not replaced, there are now 7 marbles in the bag. The probability of a black marble being drawn on the second draw is $\frac{2}{7}$. A tree diagram of this experiment looks like Figure 12.10.

The probability for the second draw changes because the marbles are not replaced and the content of the bag changes.

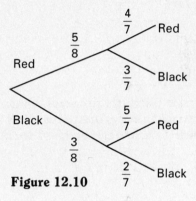

Figure 12.10

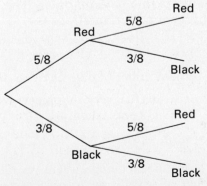

P (Red on 2nd|Red on 1st) = 5/8

Figure 12.9

Cards, colored die, and coins are good manipulative aids for probability experiments. Drawing diagrams and recording data can be used to develop skills in determining if events are independent or dependent. Probability is rich with interesting problems to experiment and stimulate. It provides a setting to engage middle school students in making hypotheses, testing conjectures, and justifying their conclusions. Once students have experimented, a computer can generate hundreds of simulated results.

Probability includes the study of permutations and combinations. These probability applications have to do with mathematically predicting the results of games and real-life situations. Before discussing permutations and combinations, the term factorial needs to be defined. *Factorial is a series of multiplications of consecu-*

tive integers. For example, 6! means 6 × 5 × 4 × 3 × 2 × 1 = 720 and is read "six factorial."

Factorial is used in evaluating combinations and permutations.

Zero Factorial is defined to be equal to one, i.e., 0! = 1

There are many problems involving permutations. Here is one example: Suppose we wish to seat four students (Al, Beth, Cathy, and Diego) in four seats in the front of the room. How many different ways are there to seat these students?

The problem can be solved by experimenting with the actual arrangements and constructing a tree diagram as follows.

How many different ways are there to fill seat 1? It can be filled with:

1. Al
2. Beth
3. Cathy
4. Diego

Seat 1 can be filled in four ways. When seat 1 is filled, how many ways are there to fill seat 2? If Al is in seat 1, seat 2 can be filled with:

1. Beth
2. Cathy
3. Diego

There are 3 ways of filling seat 2. Suppose Beth is in the first seat, how many ways are there of filling seat 2?

1. Al
2. Cathy
3. Diego

There are three ways of filling seat 2 with seat 1 being occupied by Beth. The same is true when the first seat is filled by Cathy and when it is filled by Diego.

So there are 3 + 3 + 3 + 3 = 12 different ways of filling seats 1 and 2. They are as follows:

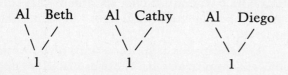

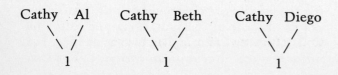

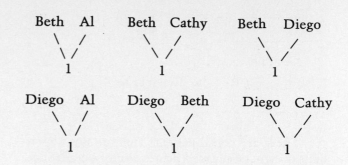

There are twelve ways of filling the first two seats. There now remain two different ways of filling the remaining two seats. For each way of filling the first two seats, there are a total of twenty-four ways to seat the four children in four seats. The tree diagram of this seating would look like Figure 12.11.

This experiment and others like it should actually be done cooperatively with students so they can visualize the concept of ordered arrangements. Placing n objects in a particular order, gives the following result:

n = 4: n(n − 1)(n − 2)(n − 3) ways

n = n: or n! = n(n − 1)(n − 2)(n − 3) . . . 3,2,1

The generalization of an ordered selection of x of the n objects can be done in n(n − 1) (n − 2) . . . (n − x) ways, which is equal to n!/(n − x)! The generalized formula can be applied to the previous example as follows:

Example: 4 students
4 chairs

$$\frac{4!}{(4-4)!} = \frac{4 \times 3 \times 2 \times 1}{0!} = \frac{24}{1} = 24$$

A permutation of a number of objects is an ordered arrangement of those objects—i.e., placing the objects in a definite order. The formula for the number of permutations is:

$_nP_x$

x = the number of objects to be ordered

n = the original set of n objects from which x was taken.

$_nP_x = n!/(n - x)!$

Let's do another experiment to determine the number of ways a set of objects can be ordered and then check the answer with the permutation formula.

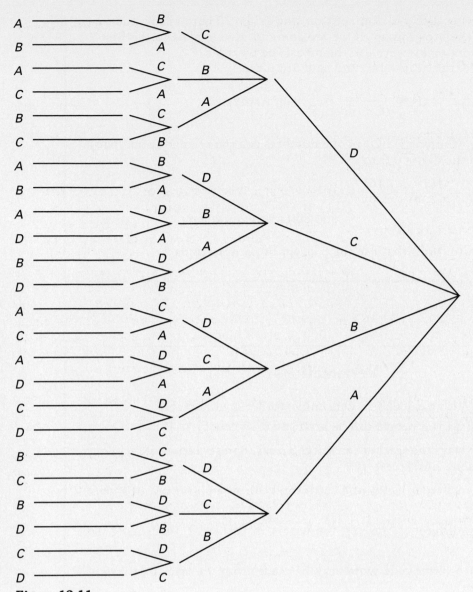

Figure 12.11

Activity

Permutations with Letters

Directions:

1. How many three-letter arrangements can be made from the letters in the word, *favorite?*
   Here the original set n contains 8 letters, and we want to order 3 of the letters, so x = 3.

$$_8P_3 = \frac{8!}{(8-3)!} = \frac{8!}{5!} = 8 \times 7 \times 6 = 336$$

So, there are 336 ways of combining 3 letters out of the original set, *favorite.*

(continued)

2. We are having a school competition this Friday. There are 8 students entered in the broad jump. If we are going to give medals to the first 3 students, how many ways can the medals be awarded?
For example: one way—1st, 2nd, and 3rd place

$$_8P_3 = \frac{8!}{(8-3)!} = \frac{8!}{5!} = 336 \text{ ways}$$

3. How many different 7-digit phone numbers can the telephone company make from the digits 0 through 9?

$$_{10}P_7 = \frac{10!}{(10-7)!} = \frac{10!}{3!} = 10 \times 9 \times 8 \times 7 \times 6 \times 5 \times 4$$
$$= 604,800 \text{ phone numbers}$$

4. List other situations that use the concept of permutations.

Activity

Permutations

Directions: With a calculator, determine: (order is important)

1. How many ways 3 people can be arranged on a bench?

2. How many ways the teacher can select 3 class representatives (President, Vice President, and Secretary)?

3. How many different hands of 4 cards can be dealt a player out of a deck of 52 cards?

4. How many different ways can the letters in the word, *measuring*, be arranged?

5. How many 4-letter code words can be made from the last 10 letters in the alphabet?

As can be seen, permutation situations require a specific arrangement and order. *Events that ignore order are called combinations.* A solution is shown by exhausting all possibilities.

Three people form a line at the post office counter. How many different combinations are possible to be the first two in line?

Three people	The possibilities
Larry	Larry, Julie
Julie	Julie, Jeff
Jeff	Jeff, Larry

NOTICE:
ORDER IS NOT IMPORTANT
The total is 3.

When we want to take x number of objects from a set of n objects *without regard to order,* we are dealing with combinations instead of permutations.

The formula for combinations is as follows:

$$_nC_x = \frac{_nP_x}{x!} = \frac{n!}{x!(n-x)!}$$

Applying the formula for 3 objects taken two at a time, gives the following results:

$$\text{EXAMPLE: } _3C_2 = \frac{3!}{2!} = \frac{3!}{2!(3-2)!}$$
$$= \frac{3 \times 2 \times 1}{2 \times 1 \times 1} = 3$$

Activity

MANIPULATIVES

Exploring Combinations

Directions:

1. Place 8 books on a table in the classroom. How many pairs of books may be selected? (Order doesn't matter.)

2. Keep a tally of the different pairs of books that are selected. Then use the formula

$$_nC_x = \frac{n!}{x!\,(n-x)!}$$

to see if you get the same answer as the experiment gave.

$$_8C_2 = \frac{8!}{2!6!} = \frac{8 \times 7}{2} = \frac{56}{2} = 28 \text{ pairs}$$

Activity

MANIPULATIVES

Boxes of Candy

Directions:

1. Suppose boxes are to be filled with 3 different kinds of candy. If 6 different kinds of candy are available, how many ways can the boxes be filled? (Order is not important.)

2. Collect data and then match the experimental answer with:

$$_6C_3 = \frac{6!}{3!3!} = \frac{6 \times 5 \times 4}{3 \times 2 \times 1} = \frac{120}{6} = 20$$

3. How many committees of 2 can be formed from a group of 12 people?

$$_{12}C_2 = \frac{12!}{2!\,(10)!} = \frac{12 \times 11}{2} = 66$$

Activity

CALCULATORS

Combinations with Cards

Directions: With a calculator, determine the following: (Order is not important.)

1. How many different hands of 4 cards can be dealt to a player out of a deck of 52 cards?

2. How many different basketball teams can be formed from 8 players?

3. The library has 16 books on the "sports" shelf. How many pairs can be selected?

4. Given a set of 14 elements, how many 4 member subsets are there?

5. How many ways can 4 letters be combined from the letters in the word, *problem*?

Summary

There are many interesting activities to teach about probability. Understanding probability is critical to being an informed adult. Real-world situations can be modeled to engage children beginning in primary grades with simulations about probability. As the NCTM's curriculum and evaluation standards (Commission on Standards for School Mathematics, 1987) advocate, the exploration of probability encourages a systematic and logical approach to problem solving.

This chapter describes the six different kinds of graphs taught in elementary and middle school: real, picture, bar, line, pie or circle, and scatter graphs. Graphs can be used to illustrate information and enhance problem solving as they are applied throughout the curriculum. The statistical terms—mean, median, and mode—are used to interpret data and students are exposed to their applications in the scoring of homework assignments. Probability is an integral part of many real-life applications, including insurance, gambling, and various consumer predictions.

Graphing, statistics, and probability are important topics in mathematics. All three play a significant role in a child's education, from kindergarten through grade eight. Experimentation is essential to understanding the applications and concepts of probability and statistics. Many situations in real life are based upon the mathematics of this chapter. Numerous professions and businesses are dependent upon the basics of graphing, probability, and statistics. Technology is making the exploration and application of these topics more an integral part of our life through newspaper graphics, tables, and charts. Therefore, the math curriculum needs to emphasize these topics, throughout the curriculum and especially across the curriculum.

Exercises

Directions: Read all questions *before* answering any one exercise. Frequently the last question in one category leads to the first question in the next category.

A. Memorization and Comprehension Exercises
Low-Level Thought Activities

1. Make a circle graph illustrating the following data table:

Pupils in Class	Fraction of Population	% of Population	Degrees of Circle
Grade 1—160	$\frac{160}{800}$	20%	72
Grade 2—216	$\frac{216}{800}$	27%	97.2
Grade 3—184	$\frac{184}{800}$	23%	82.8
Grade 4—240	$\frac{240}{800}$	30%	108
Total:	800	100%	360

2. Using a microcomputer graphing program, make a circle graph from the following information:

Exam Scores	Fractional Part	Percent	Degrees
160 students passed	$\frac{160}{320}$	50%	180
100 students failed	$\frac{100}{320}$	31.25%	112.5
60 students didn't take the test	$\frac{60}{320}$	18.75%	67.5
Total: 320 students		100%	360

3. The top ten golfers in the Bob Hope Desert Golf Classic scored as follows for the first two rounds:

Bob Hope Golf Scores

Name	Round 1	Round 2
Heerlein	70	65
Harris	69	67
Winston	70	67
Fields	69	68
Aschermann	70	67
Marion	70	68
Hansen	70	68
Minnis-Day	69	69
Haynes	74	65
Arnett-Kump	69	67

Find the mean and mode of the above data for Round 1 and Round 2.

4. The Sunset Travel Agency advertises the following Spring Vacation Tours:

Hawaii—10 days—$468/person
Mexico City—7 days—$298/person
Hong Kong—15 days—$1238/person
Las Vegas—3 days—$150/person
Bahamas—14 days—$964/person

Several groups of teachers wish to take advantage of these special prices.

Complete the chart and graph the results:

Destination	No. in Group	Cost
Hawaii	68	
Mexico City	223	
Hong Kong	149	
Las Vegas	36	
Bahamas	420	

5. The American Council of Athletics sent young people out to sell tickets to Track and Field Qualification Trials to benefit the U.S. Olympic Team. Below is a summary of their activity for two hours one evening.

Salesperson	No. of Tickets Sold	No. of Contacts Made
1 Wilkerson	320	400
2 Mahaffy	64	120
3 Gunn	128	200
4 Coyne	156	180
5 Nunez	204	240
6 Roever	260	276
7 Murphy	150	230

What is the average number of tickets sold per person?
Which salesperson sold the most tickets? How many?
Who sold the least number of tickets? How many?
What was the highest average sold per contact made? Who sold them?
What was the lowest average sold per contact made? Who sold them?
What was the mean number of tickets sold per hour? per minute?

6. Describe an activity that will help students differentiate between mean, median, and mode.

7. Describe an activity that will help students understand the probability of a specific outcome.

8. Describe an activity that illustrates mutually exclusive events.

9. In a 30-problem mathematics test, a fifth grader earned these scores:
23, 20, 30, 26, 22, 25, 28, 19, 21, 29, 28, 24, 27, 24, 23, 24, 27, 20, 28, 26, 22, 26, 29, 24, 30.
Make a frequency distribution of the scores.
Find the range.
Find the mean.
Find the median.
Find the mode.

10. Examine some contemporary elementary school textbooks and determine the extent to which they include topics of probability and statistics; do the same for some older texts. Compare the two textbooks and comment on your feelings about the appropriateness of this topic.

11. Explore a computerized grade book and summarize the statistics the program can supply to the teacher. How would these statistics be helpful in your teaching and parent–teacher conferences?

B. **Application and Analysis Exercises**
Middle-Level Thought Activities

1. Conduct a class discussion about the difference between bar and circle graphs and when it would be appropriate to use one instead of the other.

2. Make a collection of five activity cards patterned after the topic of probability. The activity cards should be sufficiently self-explanatory so that they can be used independently by children.

3. Obtain a school supply catalog. Make a list of at least three concrete models you could use for teaching three different concepts of probability. Obtain one of the models, describe it briefly to a peer, and tell how you would teach the selected concept.

4. In the daily newspaper, locate five instances in which the idea of statistics must be understood before the reader can interpret the material.

5. A parent asks at a conference with you: "Why should my child have to study probability?" Write an answer for the parent.

6. Select a topic or activity from this chapter that is new to you. Find out all you can about it, and then try teaching it to a child. Write your experiences and lesson plans for this activity.

7. Use a computer data bank and write several statistics and/or probability questions that students could answer using the data from the data bank.

8. List five computer software programs that could be used to teach any of the three main topics covered in this chapter. Evaluate each program using the computer software evaluation form in Appendix A. Find a variety of programs of different types—i.e., problem solving, drill and practice, tutorial, simulation, etc.

9. To reinforce math across the curriculum:
   a) List five mathematical questions/problems that come from popular storybooks for children. Show how they are related to the topics of this chapter.
   b) List five ways to use the topics of this chapter when teaching lessons in reading, science, social studies, health, music, art, physical education, or language arts (writing, English grammar, poetry, etc.).

C. Synthesis and Evaluation Exercises
High-Level Thought Activities

1. Develop a lesson plan illustrating the different uses of real, picture, bar, line, circle, and scatter graphs. Include the collection of data from the class. Use a microcomputer graphing program to illustrate each type of graph using the same data. Include questions to explore labels and results.

2. Develop an activity that will help students develop skills in constructing bar and line graphs.

3. Develop an activity that will help students develop skills in constructing circle and scatter graphs.

4. Develop a lesson plan illustrating that the mean and range can give a distorted picture of data sets.

5. Develop a small group cooperative learning activity to help students determine the mode of a set of data.

6. Develop a lesson plan illustrating the use of tree diagrams.

7. Develop a small group cooperative learning activity where students perform an experiment and illustrate the results using a tree diagram.

8. Develop an activity where students get practice in assigning probabilities.

9. Develop a lesson plan illustrating the difference between dependent and independent events.

10. Develop an activity that will reinforce the concept of conditional probability.

11. Develop a lesson plan that illustrates the difference between permutations and combinations.

12. Outline a lesson plan for the introduction of a pictograph to a grade level you select. Include a computer graphing segment to create the graph, label each axis, and give it a title. Prepare several questions and follow up questions to be used for discussion.

13. Design a classroom activity that can be used to introduce probability and/or statistics to children. Specify the grade level for which the activity is intended.

14. Develop five experiments for children patterned after those suggested in the chapter; write the instructions on 5 x 8 cards designed for children to use independently.

15. Use the Logo program called PROTRACTOR (found in Chapter 3) as a starting point to create a Logo program that can be used to make circle graphs.

16. Develop a computer spreadsheet gradebook to use in your classroom. The general template could be as shown in Table 12.1. Include all the information you feel is important and convenient for you to keep accurate records. Other information could include student numbers, letter grades, standard deviation, and attendance records. Provide a printout of the final template including the results of ten student records.

Table 12.1 Computer Spreadsheet Gradebook

A		B	C	D	E	F	G	H
1)	Gradebook							
2)								
3)	Names	#1	#2	#3	#4	#5	Total	Ave.
4)								
5)	Andrews, R.	82	99	78	84	98	441	88.2
6)	Barker, M.	76	79	99	76	65	395	79.0
7)	Fisher, M.	88	100	98	76	100	461	92.2
8)	Johnson, D.	96	95	99	96	100	486	97.2
9)	Michaels, D.	56	77	99	88	79	399	79.8
10)	Parker, T.	89	97	79	76	99	440	88.0
11)	Robertson, J.	94	85	92	100	88	459	91.8
12)								
13)	Test Averages	83	90	92	85	90	440	88.0
14)								
15)								
16)								
17)								
18)								

17. Evaluate how the subjects of statistics and probability can be taught to elementary and middle school students. Evaluate the role of graphing as a part in developing these concepts.

Bibliography

Baratta-Lorton, Mary. *Mathematics Their Way.* Menlo Park, Calif.: Addison-Wesley, 1976.

Beattie, Ian D. "Building Understanding With Blocks." *Arithmetic Teacher* 34 (October 1986): 5–11.

Billstein, Rick. "A Fun Way to Introduce Probability." *Arithmetic Teacher* 24 (January 1977): 39–42.

Bruni, James V., and Helene J. Silverman. "Developing Concepts in Probability and Statistics—and Much More." *Arithmetic Teacher* 33 (February 1986): 34–37.

Burns, Marilyn. *The Math Solution: Teaching for Mastery through Problem Solving.* Sausalito: Calif.: Marilyn Burns Education Associates, 1984.

———. "Put Some Probability in Your Classroom." *Arithmetic Teacher* 30 (March 1983): 21–22.

Christopher, Leonora. "Graphs Can Jazz Up the Mathematics Curriculum." *Arithmetic Teacher* 30 (September 1982): 28–30.

Commission on Standards for School Mathematics of the National Council of Teachers of Mathematics. *Curriculum and Evaluation Standards for School Mathematics.* Working Draft. Reston, Va.: The Council, 1987.

Dossey, John A., Ina V. S. Mullis, Mary M. Lindquist, and Donald L. Chambers. *The Mathematics Report Card, Are We Measuring Up? Trends and Achievement Based on the 1986 National Assessment.* Princeton, N.J.: Educational Testing Service, June 1988.

Dreyfus, Tommy. "A Graphical Approach to Solving Inequalities." *School Science and Mathematics* 85 (December 1985): 651–662.

Fennell, Francis. "Ya Gotta Play to Win: A Probability and Statistics Unit for the Middle Grades." *Arithmetic Teacher* 31 (March 1984): 26–30.

Kelly, Margaret. "Elementary School Activity: Graphing the Stock Market." *Arithmetic Teacher* 33 (March 1986): 17–20.

Kouba, Vicky L., Catherine A. Brown, Thomas P. Carpenter, Mary M. Lindquist, Edward A. Silver, and Jane O. Swafford. "Results of the Fourth NAEP Assessment of Mathematics: Measurement, Geometry Data Interpretation, Attitudes, and Other Topics." *Arithmetic Teacher* 35 (May 1988) 10–16.

Landwehr, James, and Ann E. Watkins. *Exploring Data.* Palo Alto, Calif.: Dale Seymour Publications, 1987.

Lappan, Glenda, William Fitzgerald, Elizabeth Phillips, Janet Shroyer, and Mary Jean Winter. *Mid-*

dle Grades Mathematics Project. Menlo Park, Calif.: Addison-Wesley Publishing, 1986.

National Council of Teachers of Mathematics. *Teaching Statistics and Probability.* 1981 Yearbook of the National Council of Teachers of Mathematics. Reston, Va.: The Council, 1981.

Newman, Claire M., Thomas E. Obremski, and Richard L. Scheaffer. *Exploring Probability.* Palo Alto, Calif.: Dale Seymour Publications, 1987.

Nibbelink, William. "Graphing for Any Grade." *Arithmetic Teacher* 30 (November 1982): 28–31.

Padilla, Michael J., and Danny L. McKenzie. "An Examination of the Line Graphing Ability of Students in Grades Seven Through Twelve." *School Science and Mathematics* 86 (January 1986): 20–26.

Saltinski, Ronald. "Graphs and Microcomputers: A Middle School Program." *Arithmetic Teacher* 31 (October 1983): 17–20.

Shulte, Albert P. "Research Report: Learning Probability Concepts in Elementary School Mathematics." *Arithmetic Teacher* 34 (January 1987): 32–33.

Appendix A

Software Evaluation Form

Program Name: _____   Reviewer: _____

Subject Area: _____   School: _____

Specific Topic: _____   District: _____

Area of Specialization: _____ Date: _____

Comments: _____

_____

OVERALL EVALUATION—Circle one:

5   Excellent program . . . Recommended for purchase without hesitation

4   Very good program . . . Recommended for purchase

3   Good program . . . Consider purchase

2   Fair . . . Might want to wait for something better

1   Not useful for this application/grade/etc. . . . Do not recommend for purchase

PROGRAM

low   high

1 2 3 4 5   Easily used by teacher in teaching mathematics.

1 2 3 4 5   Content is accurate.

1 2 3 4 5   Content has educational value.

1 2 3 4 5   Appropriate use of computer capabilities.

1 2 3 4 5   Content is user friendly.

1 2 3 4 5   Content is clear and logical.

1 2 3 4 5   Instructions well-organized, useful, and easy to understand.

1 2 3 4 5   Flexible application.

1 2 3 4 5   Exhibits freedom from need for teacher intervention or assistance.

1 2 3 4 5   Free of bias: racial, sexual, or political.

1 2 3 4 5   Graphics and color.

1 2 3 4 5   Sound.

1 2 3 4 5   Grade level appropriate.

(continued)

413

1 2 3 4 5   Quality of screen formats.
1 2 3 4 5   No need for external information.
1 2 3 4 5   Freedom from program errors.
1 2 3 4 5   Simplicity of user response.
1 2 3 4 5   Provides for self-pacing.
1 2 3 4 5   Appropriate and immediate feedback.
1 2 3 4 5   Branching occurs through student control.
1 2 3 4 5   Summary of student performance.
1 2 3 4 5   Degree of student involvement.
1 2 3 4 5   Degree of teacher support materials

Program Publisher: _____

Copyright date: _____

Cost: _____

Documentation available: yes   no

Lesson plans provided: yes   no

Available for which microcomputers: _____

Any memory requirements: _____

Any special hardware requirements: _____

Additional comments:

Appendix B

Logo Commands and Control Keys

Apple Logo Graphic Command List

Command	Short Form	Definition
SHOWTURTLE	ST	Makes turtle visible.
HIDETURTLE	HT	Makes turtle invisible.
FORWARD	FD	Moves turtle a designated number of steps forward (e.g., FD 15).
BACKWARD	BK	Moves turtle in the opposite direction from where he is facing (e.g., BK 50).
RIGHT	RT	Turns turtle ____degrees right: clockwise (e.g., RT 45).
LEFT	LT	Turns turtle ____degrees left: counterclockwise (e.g., LT 45).
HOME		Moves turtle to center (0,0) and sets heading to 0.
CLEARSCREEN	CS	Erases screen, moves turtle to home (middle of screen).
RIGHT ARROW		Moves the cursor to the right without erasing.
LEFT ARROW		Moves the cursor to the left erasing as it moves.
PENUP	PU	Puts pen in up position. Turtle stops leaving a trail.
PENDOWN	PD	Puts pen down. Turtle leaves trail.
PENERASE	PE	Puts eraser down. Turtle erases trail.
PENREVERSE	PX	Will do the opposite of what has been done. E.g., if line was there, it will erase when moved over; if no line was there, will draw one.
CLEAN		Erases graphics screen without moving turtle.
CLEARTEXT		Clears text screen.
FULLSCREEN	CTRL-L	Devotes entire screen to graphics.
SPLITSCREEN	CTRL-S	Splits screen; top for graphics, bottom for text.
TEXTSCREEN	CTRL-T	Devotes entire screen to text.
REPEAT		Repeats _____ times. E.g., REPEAT 4[FD 5 RT 90]
WRAP		Makes turtle field wrap around edges of screen.
CIRCLER		Makes a clockwise circle a designated radius size (e.g., CIRCLER 20).
CIRCLEL		Makes a counterclockwise circle a designated radius size (e.g., CIRCLEL 25)

(continued)

Command	Short Form	Definition
ARCRIGHT	ARCR	Makes an arc to the right. Need to designate radius and angle (e.g., ARCR 50 90).
ARCLEFT	ARCL	Makes an arc to the left. Need to designate radius and angle (e.g., ARCL 50 90).
STOP		Stops procedure.
SETPC		Sets the pen color you choose: 0 Black 1 White 2 Green 3 Purple 4 Orange (e.g., SETPC 4) 5 Blue
SETBG		Sets the background color (e.g., SETBG 5)
SETPOSITION	SETPOS	Moves turtle to a position (e.g., SETPOS [0 0])
SETX		Moves turtle horizontally so that X-coordinate is ____. SETX ____
SETY		Moves turtle vertically so that Y-coordinate is ____. SETY ____
SETHEADING	SETH	A command that sets the turtle's angle in degrees ranging from 0 to 360.
SETHEADING TOWARDS		Command that requests turtle to set his HEADING towards a position on the screen.
RANDOM		Command that directs the computer to select the input (e.g., RANDOM 10).
WAIT		Command that tells the turtle to wait for n/60ths of a second (e.g., WAIT 60 = 1 second).
MAKE		Primitive which assigns a value to a variable.

Apple Logo Control Keys

CTRL A moves the cursor to the beginning of the current line without erasing it.

CTRL B moves the cursor left (back) one character position.

* CTRL C is the standard way to exit from the editor.

CTRL D erases the character at the cursor position.

CTRL E moves cursor to end of current line.

CTRL F moves cursor one space forward.

CTRL G interrupts running procedure, cancelling it.

CTRL K erases everything on current line to right of cursor.

* CTRL L scrolls screen to put current line at center (in editor).

CTRL L devotes entire screen to graphics (outside of editor).

CTRL M same as RETURN.

* CTRL N moves cursor down to next line in editor.

* CTRL O opens new line at position of cursor in editor.

* CTRL P moves cursor up to previous line in editor.

CTRL S splits screen: top for graphics, bottom for text.

CTRL T devotes entire screen to text.

CTRL U same as CTRL-F.

* CTRL V scrolls screen to next page in editor.

CTRL W makes Logo stop until another character is typed.

CTRL Y recalls the previous line.

CTRL Z interrupts running procedure, making it pause. Do not use unless running a procedure.

* An asterisk indicates a command that will work in and out of the editor. These CTRL keys are used in the EDIT mode to revise procedures.

Apple Logo
Editing and Management Commands

Command	Short Form	Definition
EDIT "(name)	ED "(name)	Enters the Editor.
TO		Special word which tells Logo you are defining a procedure.
END		Special word to tell Logo you are finished defining a procedure. It *must* be on a line by itself.
POTS		Stands for "Print Out Titles." Displays all the names of procedures in the workspace.
POPS		Stands for "Print Out Procedures". Displays all the procedures in the workspace.
ERASE "(name)	ER "(name)	Erases procedures from workspace.
ERASEFILE "(name)		Erases a file from the workspace.
ERALL		Clears the workspace completely of all procedures.
SAVE "(name)		Creates a file and saves in it all procedures in the resident memory.
LOAD "(name)		Loads the contents of the named file into the workspace from the disk.
CATALOG		Prints on the screen the names of all the files on the disk.

Appendix C

Paper Models and Patterns

Attribute Pieces

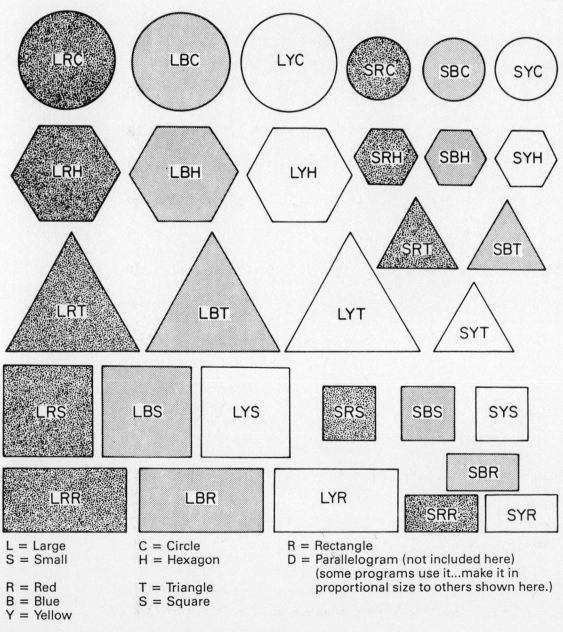

L = Large
S = Small

C = Circle
H = Hexagon

R = Rectangle
D = Parallelogram (not included here)
(some programs use it...make it in
proportional size to others shown here.)

R = Red
B = Blue
Y = Yellow

T = Triangle
S = Square

From *Mathematics: An Activity Approach*, by Albert B. Bennett, Jr., and Leonard T. Nelson.
© Wm C. Brown Publishers. Reprinted by permission.

Rectangular Geoboard Template

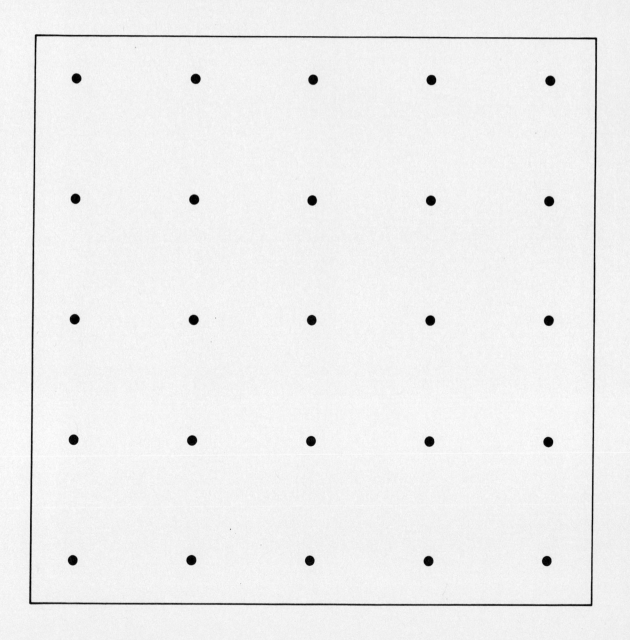

Metric Ruler, Protractor, and Compass

Compass

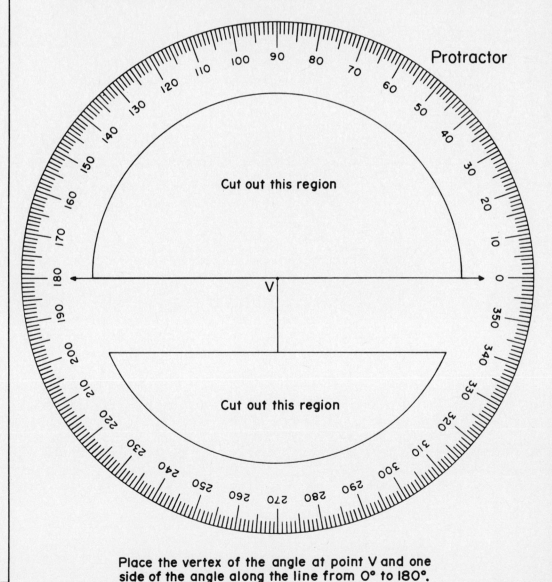

Punch out holes C, 1, 2, 3, ···, 10. Hold point C fixed for the center of a circle and place a pencil at hole 8 to draw a circle of radius 8 centimeters.

Place the vertex of the angle at point V and one side of the angle along the line from 0° to 180°.

Regular Polyhedra

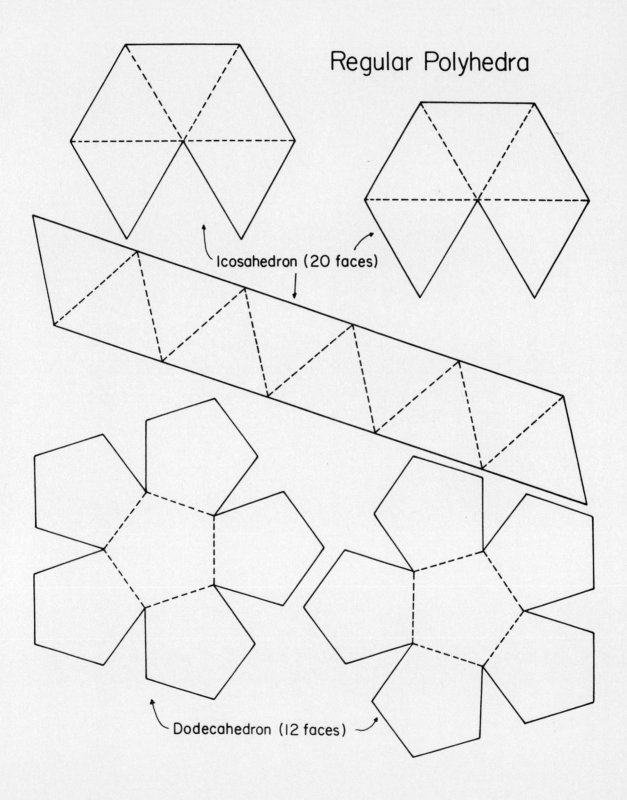

Icosahedron (20 faces)

Dodecahedron (12 faces)

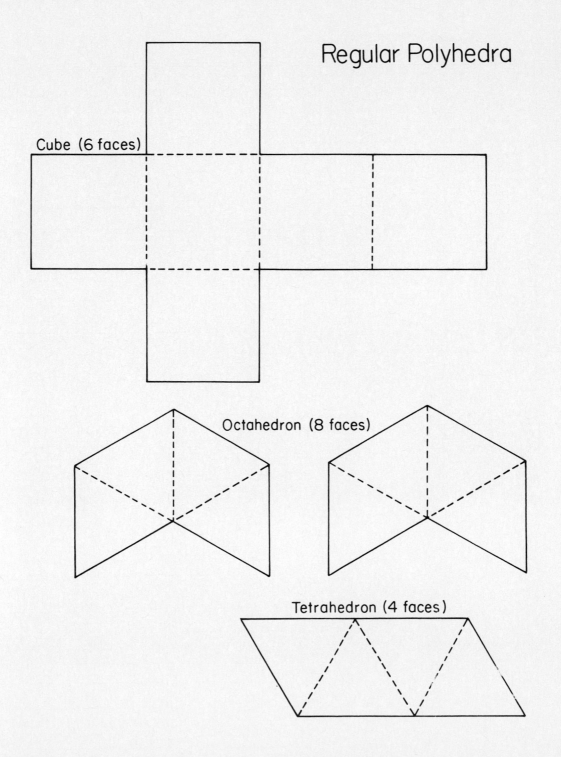

Regular Polyhedra

Cube (6 faces)

Octahedron (8 faces)

Tetrahedron (4 faces)

Prism, Pyramid, and Cylinder

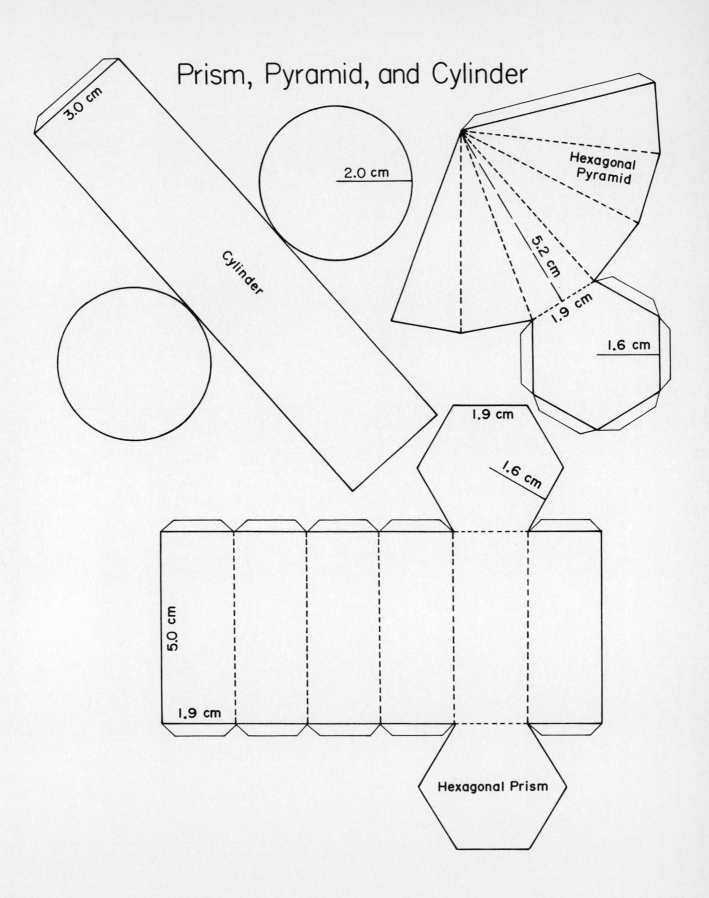

3.0 cm

Cylinder

2.0 cm

Hexagonal Pyramid

5.2 cm

1.9 cm

1.6 cm

1.9 cm

1.6 cm

5.0 cm

1.9 cm

Hexagonal Prism

Cuisenaire Rods

red	red

red	red

red	red	red	red	red	red	red	

green	green	green	green	green	

green	green		purple	yellow

green	green		purple	yellow

green	green		purple	yellow

green	green		purple	yellow

black	purple	yellow

black	purple	yellow

black	purple	yellow

black	purple	yellow

brown	red	dark green

brown	red	dark green

brown	red	dark green

blue		dark green

blue		dark green

orange	dark green

orange	dark green

Chip Trading Chips

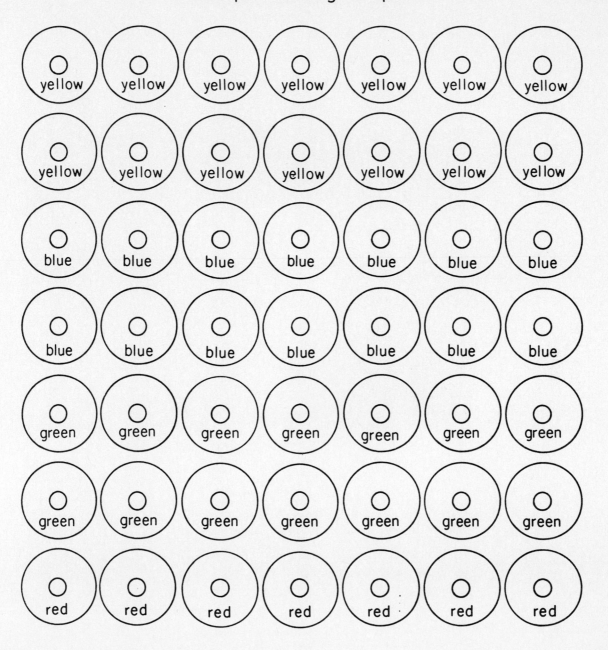

Chip Trading Mat

Red	Green	Blue	Yellow

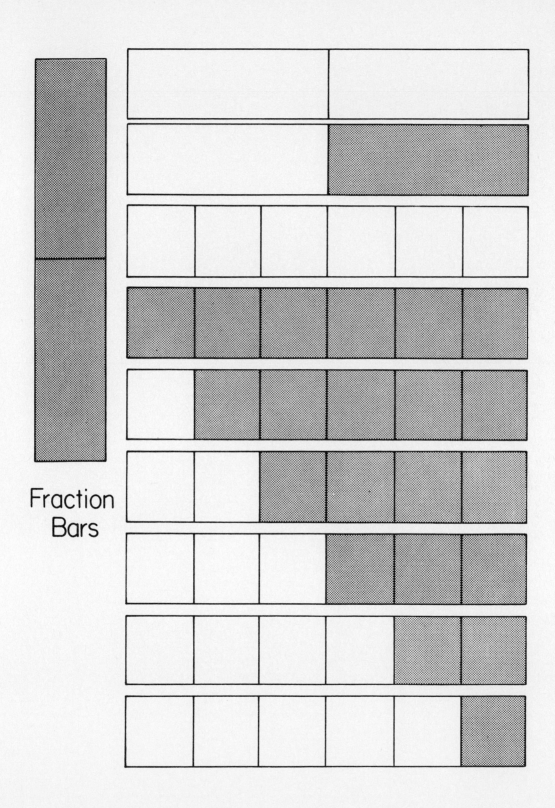

Fraction
Bars

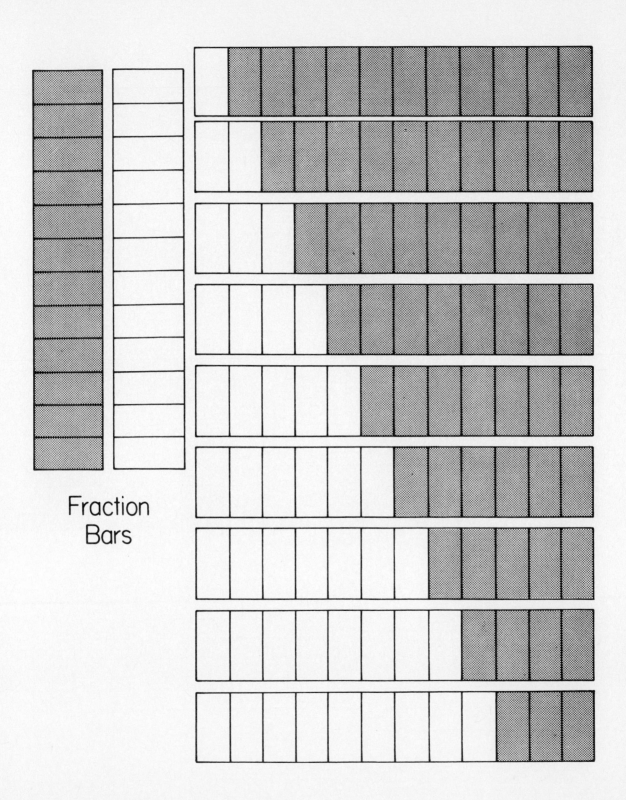

Fraction
Bars

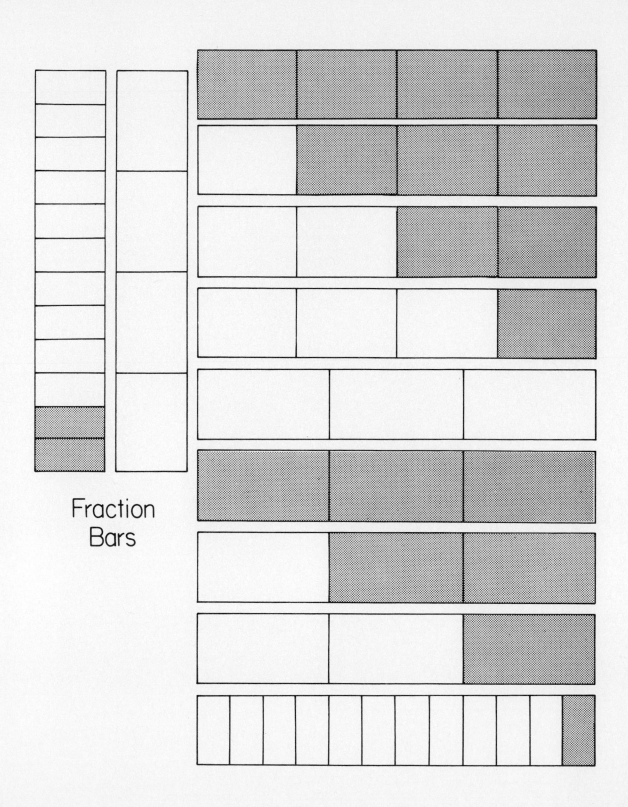

Fraction
Bars

Appendix D

Example of Student Work with Logo

The Logo program below is an example of what upper middle school students are able to create if they have had exposure to Logo through the school years from the early primary grades. This program may be used as a history assignment, studying early cultures and their accomplishments. This activity supports the idea of math across the curriculum.

1. This program is saved on the Logo disk under the name BAB.
2. Load the program and follow the directions below to see how it works.

Computer Exploration with Logo: Creating Early Babylonian Numerals

If you have no disk, type in the following procedures one at a time:

```
TO ONE :F
HT LT 30 REPEAT 3 [FD :F RT 120]
RT 30 BK :F+5 PU LT 90 FD :F
RT 90 PD
END

TO TEN :B
HT LT 30 FD :B+5 RT 90
FD : B BK :B/2 LT 60
BK :B PU LT 90 FD :B
RT 90 PD
END

TO BAB
MAKE "RK READCHAR
IF :RK = "O [ONE 10]
IF :RK = "T [TEN 15]
IF :RK = "S [ONE 15]
IF :RK = "H [TEN 20]
IF :RK = "J [ONE 20]
IF :RK = "K [TEN 25]
IF :RK = "Y [ONE 25]
IF :RK = "Z [CS]
IF :RK = "N [PU SETPOS [120 0] PD]
IF :RK = "E [STOP]
IF :RK = "P [COPY]
BAB
END
```

If you have an Apple dot matrix printer, the following copy procedure will allow you to print the Babylonian numerals you create:

```
TO COPY
>.PRINTER 1
(TYPE CHAR 9 "G CHAR 13)
>.PRINTER 0
END
```

Directions to Start the Activity:
- Type BAB [press return].
- Just press the appropriate keys below (without a return needed) and the Babylonian numerals will appear.
- Press z to clear the screen.
- Press n to start a new numeral.
- Press e to end the program.
- Press p to print the numerals you created (if you have an Apple dot-matrix printer).

Complete the chart below to see the numerals:

Key to Press		Number Amount Represented	Draw Babylonian Numeral Discovered
o	=	1	
t	=	10	
s	=	60	
h	=	600	
j	=	3600	
k	=	36000	
y	=	216000	

Represent the following numbers:

600 + 30 + 1   3600 + 660 + 50 + 4
  Press h   ttt o   Press __ _____ ____

Since position does not matter, you may enter the numeral any way you wish.

Make up your own numbers: **Draw the Babylonian numeral:**

443

Appendix E

Guide Sheet for Working Spreadsheet Programs

Using AppleWorks as the Model

1. You need two disks; they are called . . .
   a. AppleWorks Startup Disk
   b. AppleWorks PROGRAM Disk

2. Boot (start) AppleWorks Startup Disk.
   a. A message will appear at the bottom of the monitor . . .

   ```
   Place the AppleWorks PROGRAM
   disk in Drive 1 and press
   return.
   ```

3. Boot (start) the AppleWorks PROGRAM disk.

4. Follow the instructions by typing the day's date as the program requests.
   Press return after each entry during the entire creation of the table.

5. When the main menu appears, choose 1 . . .

   ```
   1. Add files to the desktop.
   ```

6. When the "add files" part comes on the monitor, move the down arrow key to step 5 . . .

   ```
   5. Spreadsheet
   ```

7. When the "Spreadsheet" part comes on the monitor, choose step 1 . . .

   ```
   1. From scratch
   ```

8. Choose a name of your choice for the file key (any name will do). It is a good idea to indicate the chapter on which you are working as a part of the filename. It should have no spaces in it. Example: HalCh6

9. Type the categories seen in this text along the top row and the left column, using the arrow keys to move around the table. AppleWorks accepts either lower or upper case typing. The labels will be typed on the bottom of the monitor, but it will affect the box that is highlighted on the chart. Press an arrow key and the typed material will appear in the desired cell of the table.

10. If you make a mistake in typing, press the escape key and it will restore the table to the way it was before your last entry.

11. To type the formulas for the values in the table (appearing under the top row and the left column), follow the directions in step 9.

12. When you are finished, move to a part of the monitor screen that is outside the categories of the table. Press escape and the program will return to a direction "page" in the program. Press "6. Quit" to finish the program. If you do NOT want to quit, press an arrow key to reach step 2 . . .

    ```
    2. Work with one of the files
    ```

    and keep pressing return even when it tells you that you have no files on the desktop.

13. You do not need to save the material you have typed unless you want to do so for your own benefit. If you wish to save your table, you will need a new disk formatted in AppleWorks.

Appendix F

Criteria for Evaluating Elementary Mathematics Books

Key for Evaluating Material

Key A	Key B
0 = Not included	0 = Not applicable
1 = Poor	1 = Strongly disagree
2 = Fair	2 = Disagree
3 = Good	3 = Uncertain
4 = Very good	4 = Agree
5 = Outstanding	5 = Strongly agree

PUBLISHER:					
5	4	3	2	1	0

CONTENT OF PUPIL'S TEXT—Use Key A

	5	4	3	2	1	0
Concept development: uses one procedure to introduce a topic, develops it fully before moving to another topic						
Concept development: moves smoothly and slowly from concrete to the symbolic level						
Appropriateness of base 10 models for developing place value concepts (size and clarity of models)						
Problem solving: use of systematic approach with adequate practice—includes situational problems, nonroutine, and word problems.						
Explanations are clear, adequate, and appropriate for children to understand						
Terminology and reading vocabulary appropriate in regard to grade level and maturity level						
Quality and quantity of practice to develop and reinforce each concept (varied format for problems?)						
Provision to challenge better students in terms of quantity and quality						
Provision for frequent and regular evaluation of concepts and skills						
Provision for regular review and skill maintenance						
Numeration—number systems (place value, counting, number order, etc.)						
Number sense—number relationships						
Measurement						

(continued)

Geometry—spatial sense					
Fractions—decimals					
Basic Whole Number Skills					
Graphs					
Money					
Estimation—mental math					
Statistics—probability					
Calculators					
Computers					

TEACHER'S EDITION—Use Key A	5	4	3	2	1	0
Quality and quantity of testing—varied assessment techniques						
Quantity, usefulness, and appropriateness of ideas for teaching the lesson						
Quantity and quality of enrichment and expansion suggestions for each unit (related activities and projects)						
Provision for quantity and quality of reteaching and remedial suggestions and materials per unit						
Suggestions for using manipulatives for concept development						
Physical make-up of book: practical, well organized, useful						
PHYSICAL CHARACTERISTICS—PUPIL'S TEXT—Use Key B						
Illustrations and color are functional as well as motivational						
Arrangement of problems on a page in a non-cluttered manner with appropriate type						
Well-developed organization and sequence of topics and concepts						
Provides for different races, both sexes, and different life styles						
Usable and adequate table of contents, glossary, and index						

Approximately how much time did you spend evaluating these textbooks?

☐ 30 minutes or less   ☐ 1–2 hours hours
☐ 30 minutes–1 hour   ☐ 2 or more hours

SCHOOL: _____   GRADE LEVEL: _____
NAME: _____   DATE: _____

CLASSROOM PRESENTATION

Student: _____

Teacher: _____

Date: _____

Grade: _____

Presentation Number: _____

School: _____

Be sure to conference with your cooperating teacher *both* before and after your lesson. Use a pre-conference to get ideas and suggestions for methodologies and material resources appropriate for your assigned lesson. Use the post conference to get feedback and suggestions for improvement.

1. Lesson topic:

2. Lesson objectives:

3. List and briefly describe your planned activities.

4. Presentation/methodologies planned and used in order of presentations.

5. Material resources (include computer software).

6. How did *you* integrate the computer into your lesson topics?

7. a) Student assessment.

   b) Lesson assessment.

8. Comments based on feedback and personal observations.

9. Sources and references.

10. Attach your lesson plan following format of Chapter 2, exercise C1.

CLASSROOM OBSERVATION

Student: _____

Teacher: _____

Date: _____   Grade: _____

Observation Number: _____   School: _____

Observe an entire math class. Be sure to arrive before the class begins and remain after the class. If possible, discuss the lesson with the teacher before the observation and then find time after the observation for further questions and clarification.

1. What is the lesson topic?

2. What are the stated/unstated objectives?

3. Make notes on the lesson introduction, presentation, student/teacher interactions, student/student interactions, resources used, activities, assignments, computer use, calculator use, etc.

4. What did you learn from this observation?

5. Give details of your post lesson discussion with the teacher.

TEACHER INTERVIEW

Student: _____

Teacher: _____

Grade: _____

Date: _____   School: _____

1. What are your likes and dislikes about teaching math?

2. What do you see as strengths and weaknesses in mathematics teaching at your school?

3. What are approaches used in the teaching of math that work well for you and what does not work well?

4. What are the areas of weakness or difficulty for students at your grade level and what do you do to overcome these difficulties?

5. Describe the manipulatives you feel are useful for difficult topics.

6. What can be done to make math fun/enjoyable?

7. Why did you select teaching as a career?

8. In what ways have you used calculators and computers to teach mathematics?

9. Additional questions/comments.

Index